W0090658

Low-Dimensional Systems: Theory, Preparation, and Some Applications

NATO Science Series

A Series presenting the results of scientific meetings supported under the NATO Science Programme.

The Series is published by IOS Press, Amsterdam, and Kluwer Academic Publishers in conjunction with the NATO Scientific Affairs Division

Sub-Series

I. Life and Behavioural Sciences	IOS Press
II. Mathematics, Physics and Chemistry	Kluwer Academic Publishers
III. Computer and Systems Science	IOS Press
IV. Earth and Environmental Sciences	Kluwer Academic Publishers
V. Science and Technology Policy	IOS Press

The NATO Science Series continues the series of books published formerly as the NATO ASI Series.

The NATO Science Programme offers support for collaboration in civil science between scientists of countries of the Euro-Atlantic Partnership Council. The types of scientific meeting generally supported are "Advanced Study Institutes" and "Advanced Research Workshops", although other types of meeting are supported from time to time. The NATO Science Series collects together the results of these meetings. The meetings are co-organized bij scientists from NATO countries and scientists from NATO's Partner countries – countries of the CIS and Central and Eastern Europe.

Advanced Study Institutes are high-level tutorial courses offering in-depth study of latest advances in a field.
Advanced Research Workshops are expert meetings aimed at critical assessment of a field, and identification of directions for future action.

As a consequence of the restructuring of the NATO Science Programme in 1999, the NATO Science Series has been re-organised and there are currently Five Sub-series as noted above. Please consult the following web sites for information on previous volumes published in the Series, as well as details of earlier Sub-series.

http://www.nato.int/science
http://www.wkap.nl
http://www.iospress.nl
http://www.wtv-books.de/nato-pco.htm

Series II: Mathematics, Physics and Chemistry – Vol. 91

Low-Dimensional Systems: Theory, Preparation, and Some Applications

edited by

Luis M. Liz-Marzán
Universidade de Vigo,
Vigo, Spain

and

Michael Giersig
Hahn-Meitner Institut Berlin,
Germany and
Poznan Technology University,
Poznan, Poland

Kluwer Academic Publishers

Dordrecht / Boston / London

Published in cooperation with NATO Scientific Affairs Division

Proceedings of the NATO Advanced Research Workshop on
Dynamic Interactions in Quantum Dot Systems
Puszczykowo, Poland
16–19 May 2002

A C.I.P. Catalogue record for this book is available from the Library of Congress.

ISBN 1-4020-1168-7

Published by Kluwer Academic Publishers,
P.O. Box 17, 3300 AA Dordrecht, The Netherlands.

Sold and distributed in North, Central and South America
by Kluwer Academic Publishers,
101 Philip Drive, Norwell, MA 02061, U.S.A.

In all other countries, sold and distributed
by Kluwer Academic Publishers,
P.O. Box 322, 3300 AH Dordrecht, The Netherlands.

Printed on acid-free paper

Printed in the Netherlands.

TABLE OF CONTENTS

APPLICATIONS

PREFACE

This volume contains papers presented at the NATO Advanced Research Workshop (ARW) *Dynamic Interactions in Quantum Dot Systems* held at Hotel Atrium in Puszczykowo, near Poznań, Poland, May 16-19, 2002.

The term *low-dimensional systems*, which is used in the title of this volume, refers to those systems which contain at least one dimension that is intermediate between those characteristic of atoms/molecules and those of the bulk material. Depending on how many dimensions lay within this range, we generally speak of *quantum wells*, *quantum wires*, and *quantum dots*. As such an intermediate state, some properties of low-dimensional systems are very different to those of their molecular and bulk counterparts. These properties generally include optical, electronic, and magnetic properties, and all these are partially covered in this book.

The main goal of the workshop was to discuss the actual state of the art in the broad area of nanotechnology. The initial focus was on the innovative synthesis of nanomaterials and their properties such as: quantum size effects, superparamagnetism, or field emission. These topics lead us into the various field based interactions including plasmon- magnetic – spin- and exciton coupling. The newer , more sophisticated methods for characterization of nanomaterials were discussed, as well as the methods for possible industrial applications. In general, chemists and physicists, as well as experts on both theory and experiments on nanosized regime structures were brought together, to discuss the general phenomena underlying their fields of interest from different points of view.

The book is thus divided in three different sections, mainly devoted to theory, preparation, and applications. In the Theory section, a theoretical description of several systems (metal clusters, quantum dots, carbon nanotubes, amorphous carbon) and their structures, as well as the existing interactions during various dynamic processes, such as growth or change in crystallographic structures of nanostructures, are presented. In the second section (Preparation and Characterization) papers are devoted to the synthesis of new structures and clear interpretations of selected properties of metallic, semiconductor, and magnetic nanoparticles and nanostructured materials. The new materials show properties like quantum size effects because of their size and differences from bulk materials in their electronic and crystallographic structure. Some examples are the synthesis and characterization of anisotropic metal nanoparticles, creation of nanosystems within polymers or during photoetching, or more complicated structures like aligned carbon nanotubes. The growth process of two dimensionally ordered carbon nanotubes is a good example of the collaboration between chemists and physicists. One further and important class of systems includes arrays of magnetic particles two-dimensionally ordered by various methods. These systems require new characterization methods to take into account interparticle interactions. Properties like superparamagnetism, ferromagnetism and paramagnetism as well their dynamics were quantitatively well explained in theoretic interpretation. For the detection of such small systems and their properties it is necessary to use sophisticated methods, which are also discussed in this section. Not only high resolution transmission electron microscopy but especially for the study of dynamic processes, STM and AFM were successfully applied and some results are presented here. Finally, in the third and final section, possible applications of these nanosystems were presented. These range from electrochemical applications of carbon nanotubes to the production of quantum computers or their elements.

We would like to acknowledge the financial support from the NATO Science Programme, which made possible the organization of this workshop. Additional funds were received from Poznań University of Technology, Netzsch, NanoLab Boston, and Poznań Government, which are also gratefully acknowledged. We would also like to thank Ryszard Czajka, Jacob Rybczynski, Nelli Sobal, Philipp Giersig, and Viktoria Giersig, for their assistance during the workshop.

Luis M. Liz-Marzán

Michael Giersig

Poznan University of Technology
Faculty of Technical Physics
Poland
NATO ADVANCED RESEARCH WORKSHOP
Dynamic Interactions in Quantum Dot Systems
Poznań/Puszczykowo, Poland
May 16-19, 2002

LIST OF CONTRIBUTORS

M. Aeschlimann
Department of Physics,
University of Kaiserslautern,
Erwin Schroedinger Str. 46, D-67663
Kaiserslautern, Germany

O. Andreyev
Department of Physics,
University of Kaiserslautern,
Erwin Schroedinger Str. 46, D-67663
Kaiserslautern, Germany

M. Apostol
Department of Theoretical Physics,
Institute of Atomic Physics,
Magurele-Bucharest, MG-6,
P.O. Box MG-35, Romania

Yu. Barnakov
Center for Interdisciplinary Research,
Tohoku University, Aramaki-Aza Aoba,
Aoba-ku, Sendai 980-8578, Japan

J. Beesley
Department of Physics,
University of Kaiserslautern,
Erwin Schroedinger Str. 46, D-67663
Kaiserslautern, Germany

F. Beguin
CRMD, CNRS-Universite,
1B rue de la Ferollerie,
45071 Orleans,France

L. D. Buda
URA CEA-CNRS SPINTEC
CEA-Grenoble, 17 Avenue des Martyrs,
38054 Grenoble, France

B. R. Bułka
Institute of Molecular Physics, Polish
Academy of Sciences,
ul. M.Smoluchowskiego 17,
60-179 Poznań, Poland

P. Byszewski
Institute of Physics PAS, al. Lotników
32/46, 02-668 Warsaw, Poland

E. Canadell
Institut de Ciència de Materials de
Barcelona (ICMAB-CSIC)
Campus de Bellaterra,
08193 Barcelona, Spain

R. Chura
Boston College,
Department of Physics, Chestnut Hill,
MA 02467, USA

H. Cölfen
Max Planck Institute of Colloids and
Interfaces,
Department of Colloid Chemistry,
MPI Research Campus Golm, D-14424,
Potsdam, Germany

L. C. Cune
Department of Theoretical Physics,
Institute of Atomic Physics,
Magurele-Bucharest, MG-6,
P.O. Box MG-35, Romania

R. Czajka
Institute of Physics, Faculty of
Technical Physics, Poznan University
of Technology, ul. Nieszawska 13a,
60-965 Poznań, Poland

S. Delpeux
CRMD, CNRS-Universite,
1B rue de la Ferollerie,
45071 Orleans,France

I. Dmitruk
Kyiv National Taras Shevchenko
University, Kyiv, Ukraine

J. Dumas
Département de Physique des Matériaux, Université Claude Bernard Lyon 1, CNRS UMR 5586, 43 Bd. Du 11 Novembre, 69622 Villeurbanne Cedex, France

U. Ebels
URA CEA-CNRS SPINTEC CEA-Grenoble, 17 Avenue des Martyrs, 38054 Grenoble, France

M. Farle
Institut für Physik, Gerhard-Mercator-Universität Duisburg, Lotharstr. 1, 47048 Duisburg, Germany

E. Frackowiak
Institute of Chemistry and Technical Electrochemistry, Poznan University of Techology, 60-965 Poznan, ul. Poiotrowo 3,Poland

K. Fukuta
Department of Applied Physics, Faculty of Engineering, Nagoya University, Chikusa-ku, Nagoya 464-8603, Japan

M. Giersig
Hahn-Meitner Institut Berlin GmbH, Glienicker Str. 100, 14109 Berlin, Germany, and Poznan University of Technology, ul. Nieszawska 13a, 60-965, Poland

A. Graja
Institute of Molecular Physics, Polish Academy of Sciences, Smoluchowskiego 17, 60-179 Poznań, Poland

J. Gutek
Institute of Physics, Faculty of Technical Physics, Poznan University of Technology, ul. Nieszawska 13a, 60-965 Poznań, Poland

Y. Hamanaka
Department of Applied Physics, Faculty of Engineering, Nagoya University, Chikusa-ku, Nagoya 464-8603, Japan

M. Hanson
Department of Solid State Physics, Chalmers University of Technology and Göteborg University, SE-412 96 Göteborg, Sweden

G. V. Hartland
Department of Chemistry and Biochemistry, University of Notre Dame,
251 Nieuwland Science Hall, Notre Dame, IN 46556-5670, USA

E. Hernández
Institut de Ciència de Materials de Barcelona (ICMAB-CSIC) Campus de Bellaterra, 08193 Barcelona, Spain

T. Hihara
Nagoya Institute of Technology (NIT), Department of Materials Science and Engineering, Gokisho-cho, Showa-ku, Nagoya 465-0097, Japan

M. Hilgendorff
Hahn-Meitner Institut Berlin GmbH, Glienicker Str. 100, 14109 Berlin, Germany

Z. P. Huang
NanoLab, Inc., Brighton, MA 02135, USA

V. I. Ivanov-Omskii
Ioffe Physical-Technical Institute RAS St. Petersburg, 194021 Russia

J. Junquera
Departamento de Física de la Materia Condensada, Universidad Autónoma de Madrid, 28049, Madrid, Spain

K. Jurewicz
Institute of Chemistry and Technical Electrochemistry,
Poznan University of Techology,
60-965 Poznan, ul. Poiotrowo 3,Poland

A. Kasuya
Center for Interdisciplinary Research,
Tohoku University, Aramaki-Aza Aoba,
Aoba-ku, Sendai 980-8578, Japan

O. Kazakova
Department of Solid State Physics,
Chalmers University of Technology and Göteborg University,
SE-412 96 Göteborg, Sweden

K. Kempa
Boston College,
Department of Physics, Chestnut Hill,
MA 02467, USA

Z. Klusek
Department of Solid State Physics,
University of Łódź, ul. Pomorska
149/153, 90-236 Łódź, Poland

M. Kohls
Lehrstuhl für Silicatchemie, Bayerische
Julius-Maximilians-Universität
Würzburg, Röntgenring 11, D-97070
Würzburg, Germany

K. Král
Institute of Physics, Academy of
Sciences of Czeck Republic,
Na Slovance 2, 18221 Prague 8,
Czeck Republic

W. Z. Li
Boston College, Department of Physics,
Chestnut Hill, MA 02467, USA

S. Lipiński
Institute of Molecular Physics, Polish
Academy of Sciences,
ul. M.Smoluchowskiego 17,
60-179 Poznań, Poland

L. M. Liz-Marzán
Departamento de Química Física
Universidade de Vigo, 36200, Vigo,
Spain

G. McMahon
CANMET, Materials Technology
Laboratory, 568 Booth Street, Ottawa,
Ontario K1A0G1, Canada

G. Müller
Lehrstuhl für Silicatchemie, Bayerische
Julius-Maximilians-Universität
Würzburg, Röntgenring 11, D-97070
Würzburg, Germany

P. Mulvaney
School of Chemistry, The University of
Melbourne,
Victoria 3010, Australia

J. Mugnier
Laboratoire de Physico-Chimie des
Matériaux Luminescents, Université
Claude Bernard Lyon 1, CNRS UMR
5620, 43 Bd. du 11 Novembre,69622
Villeurbanne Cedex, France

A. Nakamura
Department of Applied Physics,
Nagoya University, Chikusa-ku,
Nagoya 464-8603, Japan

M. Natali
ICIS-CNR, Corso Stati Uniti 4, 35128
Padova, Italy

T. Ohms
Department of Physics,
University of Kaiserslautern,
Erwin Schroedinger Str. 46, D-67663
Kaiserslautern

P. Ordejón
Institut de Ciència de Materials de
Barcelona (ICMAB-CSIC)
Campus de Bellaterra,
08193 Barcelona, Spain

I. Pastoriza-Santos
Departamento de Química Física
Universidade de Vigo, 36200, Vigo, Spain

J. C. Plenet
Département de Physique des Matériaux, Université Claude Bernard Lyon 1, CNRS UMR 5586,
43 Bd. Du 11 Novembre,
69622 Villeurbanne Cedex, France

W. Polewska
Institute of Physics, Faculty of Technical Physics, Poznan University of Technology, ul. Nieszawska 13a,
60-965 Poznań, Poland

R. Porath
Department of Physics,
University of Kaiserslautern,
Erwin Schroedinger Str. 46, D-67663 Kaiserslautern

I. L. Prejbeanu
URA CEA-CNRS SPINTEC
CEA-Grenoble, 17 Avenue des Martyrs,
38054 Grenoble, France

Z. F. Ren
Boston College, Department of Physics,
Chestnut Hill, MA 02467, USA

J. Rybczynski
Hahn-Meitner Institut Berlin GmbH,
Glienicker Str. 100, 14109 Berlin, Germany, and
Poznan University of Technology,
ul. Nieszawska 13a, 60-965, Poland

J. Sader
Department of Mathematics and Statistics,
The University of Melbourne, Victoria 3010, Australia

M. Scharte
Department of Physics,
University of Kaiserslautern,
Erwin Schroedinger Str. 46, D-67663 Kaiserslautern

M. Sennett
U.S. Army Soldier and Biological Chemical Command, Natick Soldier Center, Materials Science Team,
Natick, MA 01760, USA

J. M. Soler
Departamento de Física de la Materia Condensada, Universidad Autónoma de Madrid, 28049, Madrid, Spain

L. Spanhel
Laboratoire Verres et Céramiques,
Université de Rennes 1,
CNRS UMR 6512, Institute de Chimie de Rennes, CS 74205,
35042 Rennes Cedex, France

M. Spasova
Institut für Physik,
Gerhard-Mercator-Universität Duisburg,
Lotharstr. 1, 47048 Duisburg, Germany

P. Stefański
Institute of Molecular Physics, Polish Academy of Sciences,
ul. M.Smoluchowskiego 17,
60-179 Poznań, Poland

D. S. Su
Department of Inorganic Chemistry
Fritz Haber Institute of the Max Planck Society, Faradayweg 4-6, D-14195 Berlin, Germany

J. Tamuliene
Institute of Theoretical Physics and Astronomy, A. Gostauto 12, 2600 Vilnius, Lithuania

A. Tamulis
Institute of Theoretical Physics and Astronomy, A. Gostauto 12, 2600 Vilnius, Lithuania

V. Tamulis
Institute of Theoretical Physics and Astronomy, A. Gostauto 12, 2600 Vilnius, Lithuania

Y. Tu
Boston College, Department of Physics, Chestnut Hill, MA 02467, USA

C. Urlacher-Leluyer
Laboratoire Verres et Céramiques, Université de Rennes 1, CNRS UMR 6512, Institute de Chimie de Rennes, CS 74205, 35042 Rennes Cedex, France

D. Z. Wang
Boston College, Department of Physics, Chestnut Hill, MA 02467, USA

J. G. Wen
Boston College, Department of Physics, Chestnut Hill, MA 02467, USA

M. Wessendorf
Department of Physics, University of Kaiserslautern, Erwin Schroedinger Str. 46, D-67663 Kaiserslautern

C. Wiemann
Department of Physics, University of Kaiserslautern, Erwin Schroedinger Str. 46, D-67663 Kaiserslautern

S. Yastrebov
A.F.Ioffe Physicotechnical Institute St.Petersburg, 194021, Russia

S.-H. Yu
Max Planck Institute of Colloids and Interfaces, Department of Colloid Chemistry, MPI Research Campus Golm, D-14424, Potsdam, Germany

P. Zdeněk
Institute of Physics, Academy of Sciences of Czeck Republic, Na Slovance 2, 18221 Prague 8, Czeck Republic

THEORY OF ATOMIC CLUSTERS

Metallic Clusters Deposited on Surfaces

L. C. CUNE AND M. APOSTOL
Department of Theoretical Physics, Institute of Atomic Physics, Magurele-Bucharest, MG-6, POBox MG-35, Romania

Abstract. The quasi-classical theory of matter aggregation is briefly reviewed and the guiding principles of formation of the atomic clusters are discussed. The interaction potential of a metallic ion with a semi-infinite solid exhibiting a free plane surface is derived and atomic clusters deposited on surfaces are constructed. Binding energies, ground-states, magic geometries, isomers, inter-atomic distances, vibration spectra and monolayers are thus obtained, and further developments are outlined.

1. Introduction

It is well-known that a large amount of work has been done over the last two decades on the physical and chemical properties of the atomic clusters. In this respect the reader is referred to the comprehensive review articles given in Ref. 1 and Ref. 2. Recently, there is an increasing interest in atomic clusters, especially in connection with the progress recorded in the nanosciences. In this context, atomic clusters deposited on surfaces enjoy a particular attention. They bring together the atomic clusters field and the surface physics and chemistry. Such clusters can be viewed as a limiting case of quantum dots of much lower size, which could be called atomic quantum dots. The main source of interest in such nanostructures originates in their unusual properties associated with a functionality on an ultra-miniatural scale. Quantum effects and finite size effects are essential in this respect. However, in spite of the efforts made toward the physical and chemical characterization of such nano-objects, little progress is still recorded in knowing their individual, quantitative properties, the amount of knowledge in the field being rather limited to indirect, macroscopic and phenomenological aspects.[3, 4] The progress in this direction will be brought very likely by

L.M. Liz-Marzán and M. Giersig (eds.),
Low-Dimensional Systems: Theory, Preparation, and Some Applications, 1–17.

an extensive use of refined scanning probe microscopy methods, and other similar techniques.

On the theoretical side, the main task is to understand and predict possible physical and chemical properties of such atomic aggregates, either isolated, or deposited, or with various other environmental constraints, indicate what features might be experimentally testable, and highlight particular functionalities that may lead to technological applications. In this respect, the first main question is the cohesion of the atomic clusters. This is an old problem in quantum chemistry and quantum mechanics, and it has been successfully solved since long for molecules consisting of a rather limited number of atoms. However, major difficulties arise for an increasing number of atoms, originating both in the increasing computational resources required and in the consistency of the conceptual approaches. A direct extension of the quantum-chemistry methods from molecules to nanostructures is difficult from a practical standpoint, because of the large number of degrees of freedom, though impressive efforts have been made with the so-called ab-initio wavefunction methods.[5] On the other hand, the very large number of degrees of freedom suggests quasi-classical approaches, and such methods, or those derived from them, are generically known as density-functional methods.[6] In both cases certain approximations are involved, or semi-empirical assumptions, whose validity is seldom assessed, and adjustments are often made to get an agreement, when possible, with the experimental data within cca $3-5\%$, an accuracy considered satisfactory. Such computations may frequently be plagued with lack of convergence and instabilities in processing the complex iterative schemes, without ad-hoc control procedures or semi-empirical control parameters. This state of the art of the computational methods for nanostructures requires a re-examination of the basic theoretical concepts and procedures as derived from first principles.

In this sense, the quasi-classical description of matter aggregation has been revisited recently, and a consistent, iterative approach has been devised on this basis for the chemical bond in nanostructures in particular, in two or three steps.[7, 8, 9, 10] This theory is based on the estimation of the order of magnitude of various contributions in a hierarchical scheme, made possible by the quasi-classical description. The approach starts with electron charges distributed in atomic orbitals of the upper valence shell which are allowed to be partly delocalized in extended chemical-bond orbitals. The delocalized charges participating in the chemical bond are input parameters, they being self-consistently determined after carrying the iterative scheme through. This step may be circumvented in some simple cases, where the effective charges can be estimated from the beginning by making use of the atomic screening theory for heavy atoms for instance. This basic aspect of the theory is discussed in Ref. 9 and Ref. 10. The first step in carrying out this computational scheme is the identification of the Hartree term as bringing the main contribution, providing the Thomas-Fermi equations are linearized via a variational parameter related to the electron density. In some simplifying cases, such a variational parameter turns out to play the role of an effective screening wavevector. This procedure, fully based on the consistent quasi-classical description, may ensure cohesion; it was probably suggested for the first time by Schwinger.[11, 12] At this level of computations, the main role is played by the self-consistent potential, and, making use of it, one may estimate the structures of the atomic aggregates, *i.e.* the atom positions and the geometric forms of the nanostructures, compute the binding energy, vibration spectra, stability with respect to the number of atoms, *i.e.* the magic clusters forms and numbers, both for the ground-state and the isomers (*i.e.* clusters whose atoms occupy slightly different positions with respect to each other, and which differ by a small amount of energy in comparison with the ground-state energy). This is already a lot of information which might be tested experimentally. The second step is to compute the so-called quantum corrections by solving Schrodinger's equation for the electron energy levels and wavefunctions, with the self-consistent potential derived in the previous step. The exchange contribution to the Hartree-Fock equations is included in this step. It may be remarked that the exchange term plays in this scheme of computation the role of a quantum correction. Indeed, this is based on its well-known properties of "rigidity" and "non-locality", emphasized probably for the first time by Slater.[13] The quantum corrections lead to the electronic single-particle properties which account for various clusters spectroscopies. Such quantum corrections include the effect of the abrupt variations of the electronic wavefunctions near the ionic cores (including the self-consistent determination of the fractional charges participating in the bond), and their contribution is estimated[7, 9] to cca 17%. At this level, one may also employ the exchange integrals for computing spin-dependent properties, like, for instance, the magnetism of the nanostructures. Similarly, we may estimate now the response of the nanostructures to various external perturbations, including electric polarizability, diamagnetic susceptibility, transport properties, etc. The next iterative step would bring a second-order correction of the order of $0.17 \times 17\% \cong 3\%$, which is estimated as being comparable to the lifetime effects of the quasiparticles. As such effects originate in the Hartree-Fock type of single-particle decomposition of the wavefunction, there is no point in going further on with the accuracy of the computations, unless genuine many-particle wavefunctions are used. However, this is a task which may be left aside for the time being. It is worth noting that this level of accuracy predicted by the present theory coincides with the level of accuracy accepted as being satisfactory on semi-empirical grounds by the current computational methods. It is also worth noting here that this computational scheme described briefly above is valid for large values of the cluster size N (N being the number of atoms in the cluster), as it is based on a quasi-classical description. This is precisely the range of the nanostructures, involving a number of atoms from $N \sim 10 - 20$ up to very large numbers. On the other hand, it differs from the methods currently used for bulk solids, as the latter employ essentially the translational symmetry of the crystals.

The above program of computation is being carried out at this moment to a rather limited extent, namely up to the first step in the scheme described above. This gives us results concerning cohesion, structure and related information about nanostructures within an accuracy of cca 17%, as said above. The usefulness of pursuing such a theoretical approach resides in that it is consistently derived from first principles, provides a convergent simple iterative scheme with only two, or three, steps, it is free of any semi-empirical or ad-hoc assumptions or adjustments, and produces meaningful results already in the first step with rather limited computational resources; these results may be used as input data for getting more refined results in the next step. In addition, the computations are restricted at this moment to homo-atomic nanostructures consisting of some simple metallic ions with a model point-like charge distribution. The latter is a simplifying model assumption which does not affect qualitatively the results. The study of more realistic atomic-like orbitals is underway.

The present theory has been tested on computing the binding energy, quantum corrections included, of heavy atoms, which have been used as one of the benchmarks of the present approach. Similarly, the bulk metals have been tackled in the continuum approximation, reproducing in a highly satisfactory manner the cohesion energy, sound velocity, electron-phonon interaction, plasmons and the whole spectrum of results concerning the normal liquid theory for electronic quasi-particles. The emerging overall picture is that of a model metal consisting of screened ionic cores weakly interacting through a two-body potential, very similar to a Wigner metal.[14, 15, 16] Ionization potentials for atomic clusters, as well as electric polarizability and diamagnetic susceptibility have also been computed in a satisfactory agreement with the experimental data. All this information can be found in Ref. 9.

It has also been shown that several results obtained within this theory are in good agreement both with experimental data, where available, and with theoretical results derived by means of different approaches. For instance, we get stable, icosahedral structures in agreement with both theoretical and experimental results for isolated clusters like (iron) Fe_{13}, or (palladium) Pd_{13}, or (barium) Ba_{45}, including inter-atomic distances, binding energies and vibration spectra. A similar agreement was obtained for the iron core of the more complex iron-hydrocarbon cluster $Fe_{13}(C_2H_2)_6$. These results are reported in Ref. 7 and Ref. 8, and are discussed briefly in the next section. It has also been shown that the present theory leads, within certain approximations, to a deformed-harmonic oscillator potential of the Clemenger-Nilson type, which is extensively employed in assessing magic numbers of atomic clusters (see, for instance, Ref. 9). In addition, we derive here the work function for some metals in good agreement with previous computations and experimental data (see Section 3). All this gives support to our computational scheme.

In the present paper we report upon new results concerning metallic clusters deposited on surfaces. We establish the effective potential of a free surface in the continuum approximation, in perfect agreement with the experimental data concerning the work function, and employ it for constructing metallic clusters deposited on such surfaces. We report upon structural data of such clusters, their geometrical forms, cohesion energy, stability, isomers, magic numbers, and indicate also the way such clusters may diffuse into the bulk. The calculations concerning the electronic structure and the effect of the surface upon the electronic structure of isolated clusters are underway. The results presented here are reliable as they are derived from first principles, are approximate within an accuracy which we know, as discussed above, can be refined quantitatively within the next step of the theory, and offer structures that may be tested experimentally, most directly by scanning probe microscopy.

2. Self-consistent potential

Solid atomic aggregates occur through delocalization of the electrons in the upper valence shells. The chemical bond originates in a superposition of atomic-like orbitals and extended orbitals that vary slowly in space. Such a picture is amenable to a quasi-classical description of the Hartree-Fock equations. An atomic aggregate looks like an ensemble of electrons spinning around and the ionic cores left behind.[7] The ionic cores have effective charges z_i^*, where i is the label of the ion. These charges are distributed radially-symmetric, as for s-orbitals, or directionally, as for p, d, f-orbitals; several electrons in d- and f-orbitals may approximately be viewed as radially-symmetric. In addition, such atomic-orbital charges with spherical symmetry may also be approximated for the beginning by point-like distributions. Most of the metallic ions fall in this class of s-, or approximately spherical d, f-orbitals. The effective charges may be estimated for atoms sufficiently heavy by making use of the atomic screening theory. For instance, effective charges are $z^* = 0.44$ for Na (sodium) and $z^* = 0.57$ for Fe (iron).[7, 8, 9, 10] The point-like charge distribution has a limited validity for s-orbitals, or for those d- and f-orbitals with several electrons, which may be approximately viewed as radially symmetric. In addition, the model assumption of a δ-type distribution in the ionic core allows simple calculations, and does not affect qualitatively the results. The approximations involved are comparable with those derived from the general theory as discussed in Introduction.[7, 8, 9, 10]

Under the circumstances given above, within the quasi-classical description, the electrons in an atomic aggregate move in a self-consistent Hartree potential

$$\varphi(\mathbf{r}) = \sum_{i=1}^{N} \frac{z_i^*}{|\mathbf{r} - \mathbf{R}_i|} e^{-q|\mathbf{r}-\mathbf{R}_i|} \ , \tag{1}$$

where N is the number of ions, $\mathbf{R}_i$ denote their positions and q is a screening wavevector to be determined variationally. The derivation of this potential is fully discussed in Refs. 7-10. The well-know atomic units are used here, namely the Bohr radius $a_H = \hbar^2/me^2 = 0.53\text{Å}$ and twice the rydberg $e^2/a_H = 27.2\text{eV}$. According to the general theory,[7, 8, 9, 10] the self-consistency implies the electron density n being dependent on the potential φ by $n = (q^2/4\pi)\varphi$, so that we obtain the potential energy

$$E_{pot} = -\frac{3}{4} q \sum_{i=1}^{N} z_i^{*2} + \frac{1}{2} \sum_{i \neq j=1}^{N} \Phi(R_{ij}) \tag{2}$$

of the interacting electrons and ions (electron-electron and ion-ion Coulomb repulsions included), where

$$\Phi(R_{ij}) = -\frac{1}{2} q z_i^* z_j^* (1 - \frac{2}{qR_{ij}}) e^{-qR_{ij}} \tag{3}$$

are effective interaction (pseudo-) potentials between ions separated by distance $R_{ij} = |\mathbf{R}_i - \mathbf{R}_j|$. The derivation of the potential energy given by (2) can be found in Refs. 7-10. It is worth emphasizing that the effective potentials (3) are spherically symmetric, as expected from a point-like charge distribution in the ionic cores, and are pairwise potentials. In general, many-body contributions to effective ionic potentials, like three- or four-body terms, are difficult to be rigorously justified from first principles, on one hand; on the other hand, in those cases where their derivation is proved, their contribution to the cohesion turns out to be small. In the present theory, the effect of such "correlations effects" is actually taken into account implicitly by the self-consistency of the theory. Indeed, the screening wavevector q, which is determined variationally, depends on the ions positions, more precisely on the inter-ionic distances R_{ij}. This is a weak dependence arising from the interacting terms in the potential energy (2), so that we may write the screening wavevector as $q = q_0 + \delta q(R_{ij})$, where q_0 includes only the main contribution to the potential energy as expressed by the first term in (2) (this is the ionic self-energy). On the other hand, we may expand the potential (3) in powers of δq, which leads to many-body contributions to the pairwise potential (3), written this time with the screening wavevector q_0. As one can see, such a formal series expansion is a perturbational scheme, where the higher-body terms bring a small contribution, and the main two-body term (first-order term) is in fact less accurate than the closed formula (3). Another source of many-body contribution resides in the effective charges z_i^*, and their environment dependence. Such dependence is included in the self-consistent determination of the effective charges, as discussed in Introduction, and the effect of such dependence is of the order of the quantum corrections, as the latter account for the short-scale variations. However, in some cases, like Na and Fe, chosen here for illustrating the theory, the effective charges are satisfactorily estimated from the atomic screening theory, and, consequently such effects are already included. The potential energy given by (2) is minimized with respect to $\mathbf{R}_i$ (actually $q\mathbf{R}_i$) in order to find the ionic equilibrium positions; this way, we determine the geometric forms of the atomic aggregates, both for their ground-states and isomers. Thereafter, the kinetic energy $E_{kin} = (27\pi^2/640)q^4 \sum_i z_i^*$ is added, and the quasi-classical energy $E_q = E_{kin} + E_{pot}$ is minimized with respect to the screening wavevector q; finally the exchange energy $E_{ex} = -(9/32)q^2 \sum_i z_i^*$ is included to obtain the binding energy $E = E_q + E_{ex}$. As one can see the exchange interaction is taken into account at this level of computations.

This theoretical approach has been applied to homo-atomic metallic clusters, where geometric magic numbers have been obtained, together with binding energies, inter-atomic distances and vibration spectra (up to $N \sim 160$).[7] Leaving aside the small contribution of the interacting part in the potential energy at equilibrium, the screening wavevector reads approximately $q \simeq 0.77 z^{*^{1/3}}$ in this case, and the average inter-atomic distance may be estimated as $a = \overline{R_{ij}} \sim 2.73/q$; all the same, the binding energy is given by $E = -N(0.43 z^{*7/3} + 0.17 z^{*5/3})$. Similarly, the theory has been used to estimate other, more complex structures, as, for instance, the equilibrium Fe-core structure of the iron-hydrocarbon $Fe_{13}(C_2H_2)_6$-cluster.[8] We emphasize that icosahedral structures of Fe clusters, as Fe_{13} for instance, are currently reported in the literature, both experimental and theoretical, and our quantitative results agree satisfactorily with these data, where available.[1, 2, 17, 18] For instance, we obtain $\simeq -5.3$eV a cohesion energy per atom for Fe_{13}, which agrees well with −5.2eV reported in Ref. 19 by using density-functional methods, and a similar agreement holds also for inter-ionic distances (of cca 2Å; see also Refs. 20-22). The iron-hydrocarbon cluster $Fe_{13}(C_2H_2)_6$ has been synthesized experimentally,[23] and the structure derived theoretically in Ref. 8 agrees well with the experimental data, including atomic positions in the Fe_{13}-core, inter-atomic distances, core contribution to the binding energy, stability and vibration spectra. Such an agreement is also reported in Ref. 24, with regard to the vibration spectrum as computed by means of the density-functional methods. We may also note here that structures, magic numbers and binding energies have recently been reported for Pd (palladium) clusters up to $N = 20$ in close agreement with the values obtained by us ($z^*_{Pd} = 0.40$, cohesion energy per atom $\simeq -2.5$eV for $N = 20$), by using a theoretical model of an embedded-atom potential.[25] Including more realistic atomic-like charge distribution (instead of the point-like distribution) we expect to get greater inter-atomic distances, but the cohesion energy and other relevant results will not change drastically.

The above theoretical description is to be developed along two directions at least. First, the directional character of the atomic-like orbitals (as well as their radial dependence) must be included in order to obtain, for instance, *p*- or *sp*-orbitals atomic aggregates (as well as directional *d*- and *f*-orbitals aggregates). Secondly, the quantum corrections must be included in the quasi-classical treatment, in order to get a more accurate knowledge of the electronic single-particle properties, like energy levels (or bands), ionization potential, chemical affinity, optical properties, polarizability, magnetic properties, etc. An error of cca 17% is estimated without quantum corrections, while including them may lead to an accuracy of up to cca 3%, at most, as discussed in Introduction.[7, 8, 9, 10] Various hetero-atomic aggregates could then be studied with more confidence.

Until then, the present theory can be employed to get a description of metallic surfaces or interfaces, or metallic clusters deposited on such surfaces, or atomic aggregates with various others geometric constraints.

3. Metallic Surface

The summation over ions in (1) can be restricted to half a space, as for a semi-infinite solid with a free plane surface perpendicular to, say, the x-direction at $x = 0$. Such a surface is shown in Fig. 1. In the continuum approximation we obtain the self-consistent potential

$$\begin{aligned} \varphi(x) &= \tfrac{4\pi z^*}{q^2a^3}\left(1 - \tfrac{1}{2}e^{qx}\right)\ ,\ x < 0\ , \\ \varphi(x) &= \tfrac{2\pi z^*}{q^2a^3}e^{-qx}\ ,\ x > 0\ , \end{aligned} \tag{4}$$

where z^* is the average effective charge and a denotes the average interatomic distance.

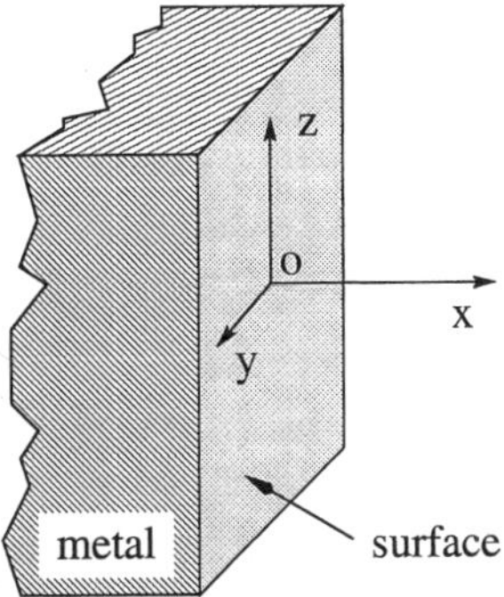

Figure 1. A free plane surface of a semi-infinite solid

When compared to the bulk contribution $\varphi = 4\pi z^*/q^2a^3$ one can see that a dipolar potential $\delta\varphi$ occurs at thc surface, which gives rise to a spill over of the electrons and a surface charge double layer. The electron density $n = (q^2/4\pi)\varphi$ is shown in Fig. 2, and the total charge density is plotted in Fig. 3 *vs* x.

The work function of the solid as computed from (4) is φ, as expected. The interaction energy $-(1/2)\int dx \cdot \delta\varphi\delta n$ associated with the electronic double layer (per unit area) is $-\pi z^{*2}/2q^3a^6$, and it acts like an additional uncertainty in the quasi-particle energy giving rise to boundary (finite-size) lifetime; it also leads to a weak relaxation of the ionic positions at the surface.

It is worth noting that the surface potential (4) and the corresponding surface charge distribution has been suggested long time ago on semi-empirical grounds[26] (see also Ref. 27), and used for computing work function of metals. The screening wavevector given in Ref. 26 and Ref. 27 for Na is 1.27Å corresponding to a work function 2.9eV as compared with the experimental 2.4eV. We obtain a screening wavevector $q = 1.17$Å from our theory ($q = 0.77z^{*1/3}$), leading to a 2.7eV work function, which is closer to the experimental value. A similar agreement holds also for other metals.

On the other hand, the potential energy can be estimated from (2) and (3) for a semi-infinite solid; in the continuum approximation we obtain

$$E_{pot} = -\frac{3}{4}qz^{*2}N + \frac{\pi z^{*2}}{2q^3a^6}A \ , \tag{5}$$

where A is the area of the cross-section; therefore, the potential energy (5) includes a surface contribution $(\pi z^{*2}/2q^3a^6)A$, beside the bulk contribution given by the first term (the interacting part vanishes in the bulk continuum limit); the surface tension of the solid is $\sigma = (\pi z^{*2}/2q^3a^6)$, and it agrees with the double layer energy given above.

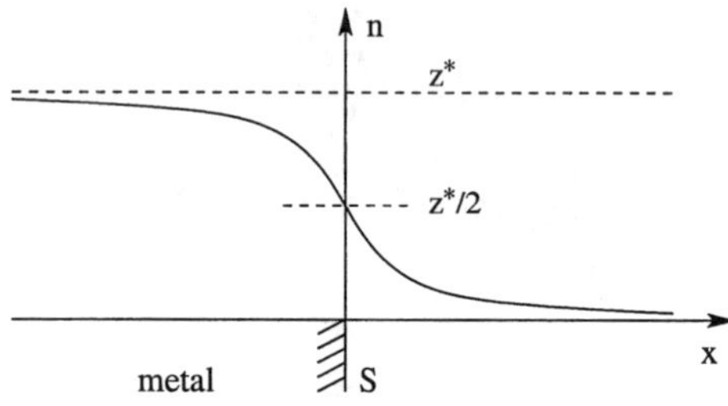

Figure 2. Electron density at the surface

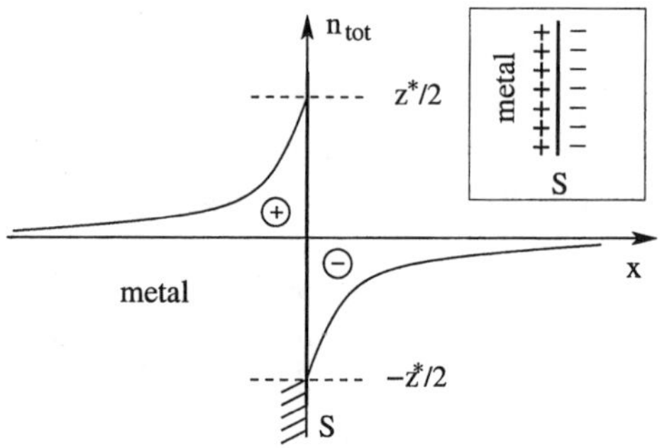

Figure 3. Charge distribution at the surface by-layer

Similarly, we can estimate the interaction potential between a semi-infinite solid and an ion with an effective charge z_0^* placed at distance x from the surface; indeed, making use of (2) and (3), we obtain

$$E_{pot} = E_s - \frac{3}{4}qz_0^{*2} - \frac{\pi z^* z_0^*}{qa^3}xe^{-q|x|} \ , \tag{6}$$

where E_s is the potential energy of the solid as given by (5); the second term in (6) is the self-energy of the added atom and the third term represents the interaction potential of the atom with the semi-infinite solid; it is shown in Fig. 4.

This interaction potential exhibits a potential barrier just beneath the surface, and has an attractive part above; the latter is responsible of adsorbing additional atoms on the surface, and of stabilizing deposited atomic clusters.

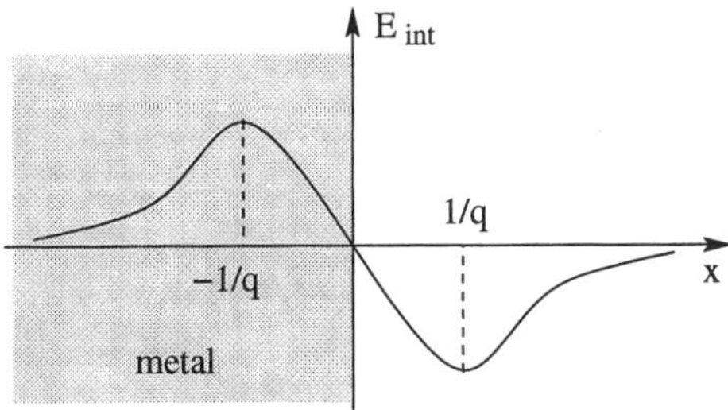

Figure 4. Interaction potential between an atom and a semi-infinite solid

Now it is easy to write down the potential energy of an ensemble of N atoms of effective charges z_i^* deposited on the surface; it reads

$$E_{pot} = E_s - \frac{3}{4}q\sum_{i=1}^{N} z_i^{*2} + \frac{1}{2}\sum_{i\neq j=1}^{N} \Phi(R_{ij}) - \frac{\pi z^*}{qa^3}\sum_{i=1}^{N} z_i^* X_i e^{-q|X_i|} \ , \qquad (7)$$

where the potentials $\Phi(R_{ij})$ are given by (3) and X_i is the x-coordinate of $\mathbf{R}_i$. It is worth noting that the screening wavevector q is the one corresponding to the solid, as the latter prevails upon the deposited cluster in the thermodynamic limit. In this respect the deposited clusters differ from the isolated clusters, which have their own screening wavevector as it results from the minimization of their quasi-classical energy. The binding energy of a deposited cluster is given by $E = E_q + E_{ex}$, where the quasi-classical energy is $E_q = (27\pi^2/640)q^4 \sum_i z_i^* + E_{pot} - E_s$, and the exchange energy is given by $E_{ex} = -(9/32)q^2 \sum_i z_i^*$; the potential energy given by (7) is minimized with respect to the ionic positions $\mathbf{R}_i$. It is worth noting that an interaction energy

$$E_{int} = -\frac{\pi z^*}{qa^3}\sum_{i=1}^{N} z_i^* X_i e^{-q|X_i|} \qquad (8)$$

can be defined from (7), between the deposited cluster and the solid, which may serve as a measure of the energy needed to separate the cluster off the surface (the difference in the cluster energy must be added, arising from its own screening wavevector corresponding to the cluster relaxation). One can also notice that the interaction energy (8) for the halves of a solid compensates exactly the surface energies of the two faces, as given by (5). If two distinct solids are put in contact there is a diffusion of one into another across the interface, according to the tunneling through the interaction potentials given by (8). Finally, we note that the continuum approximation

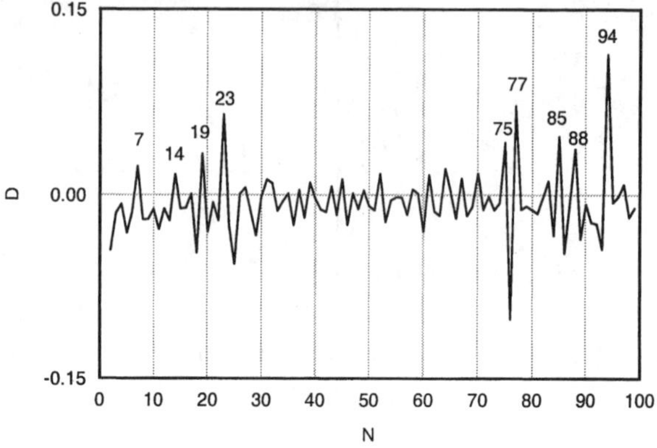

Figure 5. Ground-state abundance spectrum and magic clusters deposited on surfaces

is not necessary, and we can treat the cluster deposition by preserving the discrete summations over fixed ionic positions in solid; we have adopted the continuum approximation here for the sake of the simplicity; the errors introduced on this occasion refer to the few atomic layers in the vicinity of the surface, and of course to the matching problem of the lattice constants.

4. Clusters Deposited on Surface

The main problem of depositing atomic clusters on a surface is the minimization of the potential energy (7) with respect to the ionic positions $\mathbf{R}_i$ (in fact with respect to $q\mathbf{R}_i$).

Initially, we give positions $\mathbf{R}_i$ randomly distributed in space and let the ions move step by step along the forces until a local equilibrium is reached (corresponding to forces less than 10^{-4}eV/Å); this equilibrium is checked by computing the corresponding vibration spectra. For each number N of atoms the procedure is repeated for a few hundreds times, in order to get the ground-state and the isomers; the latter are clusters higher in energy with slightly different ionic positions. This procedure has been applied to Fe-clusters ($z^* = 0.57$) deposited on Na-surface ($z^* = 0.44$) up to $N = 100$. The original ionic positions are randomly distributed in space both below and above the surface of the solid; we find that equilibrium positions are reached mostly above the surface, as for deposited clusters. The binding energies $E(N)$ have been computed for the ground-state of these clusters as indicated before, and abundance spectra $D = \ln(I_N^2/I_{N+1}I_{N-1}) = E(N+1) + E(N-1) - 2E(N)$ have been obtained, where I_N is Boltzmann's

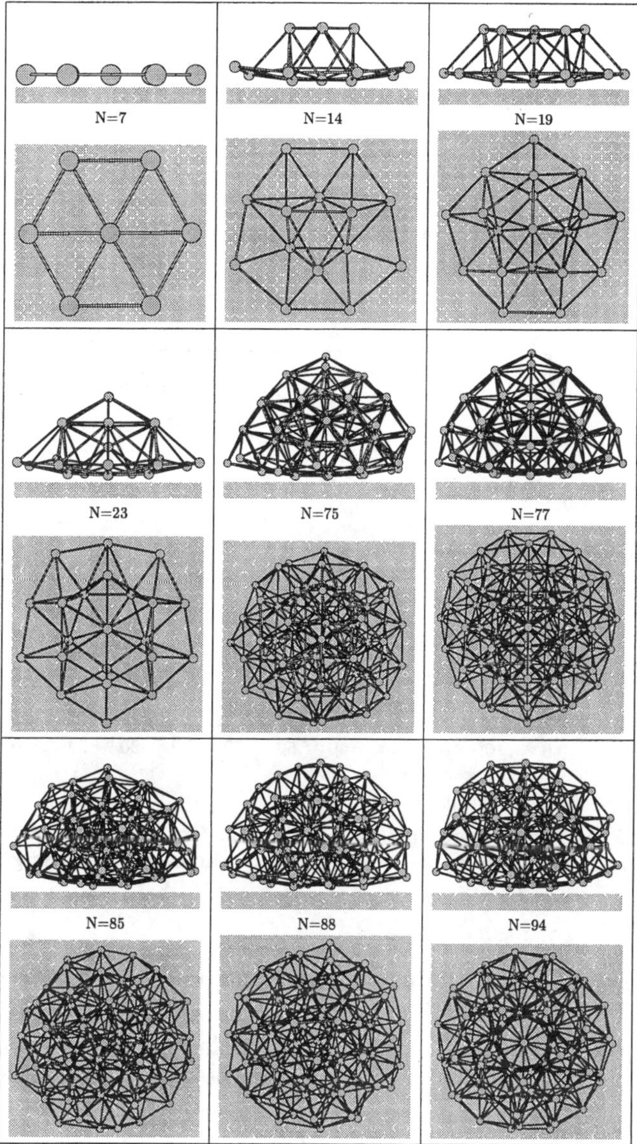

Figure 6. Magic clusters deposited on a surface

statistical weight.

Such an abundance spectrum D is shown in Fig. 5; these spectra depend weakly on the nature of the clusters and of the substrate. Magic clusters deposited on surface are to be noted in Fig. 5, as, for instance, those corresponding to $N = 7, 14, 19, 23, 75, 77, 85, 88, 94...$; they acquire highly symmetric forms, as shown in Fig. 6. The rather structureless island between $N = 23$ and $N = 75$ is intriguing in Fig. 5. As a general rule, for small values of N the atoms are adsorbed on the surface as a monolayer, forming up more-or-less regular polygons. On increasing the number of atoms, they distribute themselves both horizontally and vertically, giving rise to multilayer structures, with various, intricate geometries, and sometimes beautiful symmetries, as those corresponding for instance to $N = 23, 77, 94$.

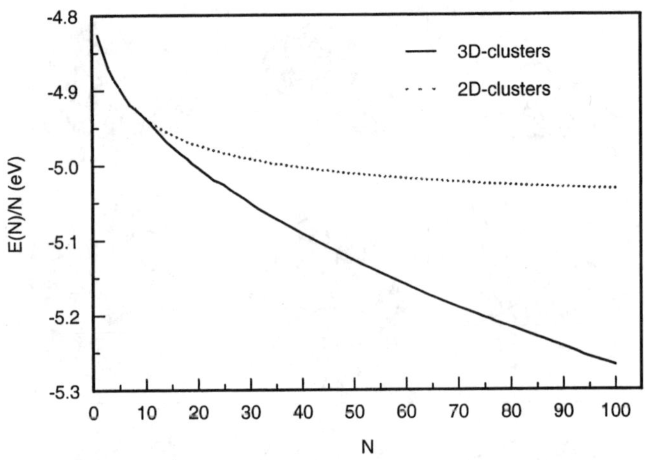

Figure 7. Ground-state energy per atom for Fe-clusters (3D, solid line) deposited on Na-surface *vs* number of atoms, as compared with monolayer cluster energy (2D, dashed line)

It is worth noting that their binding energies are higher in comparison with their monolayer (two-dimensional) versions (which are isomers), *i.e.* growing up vertically helps stabilizing the clusters; such a comparison is shown in Fig. 7. In general, there is a competition between the two directions of growth, horizontal and vertical, and it is difficult to predict which would prevail for a given number of atoms.

Bound states can also be obtained for clusters deposited on surfaces with parts pervading beneath the surface, as shown in Fig. 8. Indeed, the first two pictures in Fig. 8 shown a 50-atoms cluster diffusing into solid, while the last picture in Fig. 8 exhibits a 100-atoms cluster developing an interface with the solid. These states are isomeric, and, in some cases, atoms may escape into the solid where they acquire free positions, *i.e.* they are no more bound to the cluster. Similar formations can be obtained for a large

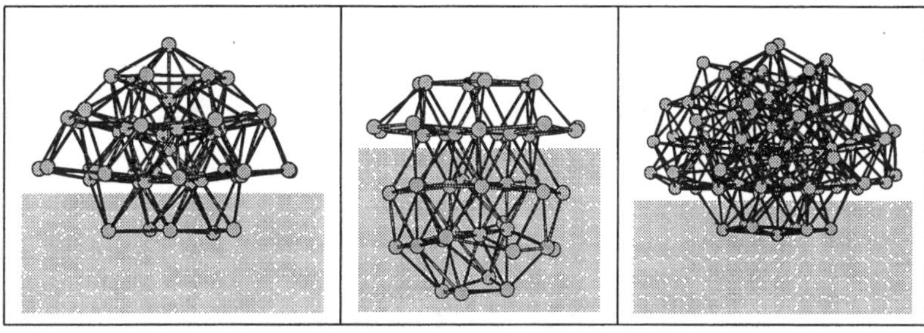

Figure 8. A 50-atoms cluster diffusing into solid (first two pictures), a 100-atoms one developing an interface with the solid (last picture)

variety of situations, including both geometric constraints, like a deposition surface, and dynamic constraints, like applying external forces.[3, 28]

As regards the comparison of the present structural results with experimental data concerning homo-metallic clusters deposited on surfaces, or incipiently diffusing into metals just beneath the surface, we are not aware of distinct, clear-cut experimental results yet, amenable to such a comparison. Direct observations have been reported to our knowledge for Si (silicon) and Ga (gallium) clusters deposited on Si-surfaces by electron scanning microscopy, indicating the existence of stable, abundant structures and corresponding magic numbers. It is relevant in this context the following excerpt from Ref. 28:

"Although speculations about the existence of magic clusters on surfaces were raised in a paper in 1992, which was based on a study of the Pt/Pt(111) surface using He scattering,[29] no SCM *(surface magic clusters, our note)* were found in the later scanning tunneling microscopy (STM) study of this surface.[30] The first demonstration of such clusters exhibiting enhanced stability and abundance had to wait for several years until SCM were directly observed on the $\sqrt{3} \times \sqrt{3}R30°$-reconstructed Ga/Si(111) surface.[31, 32] Soon after this work, Si islands with magic numbers of unit cells on the Si(111) 7×7 surface were reported.[33] This was then followed by the observation of a type of Si magic cluster on Si(111).[34]"

The study of Si, or Ga clusters, and others alike, requires the inclusion of the directional character of the atomic-like orbitals. Such an investigation is underway.

5. Concluding Remarks

In conclusion, we may say that interaction potentials can be identified in the quasi-classical description of atomic aggregation, between atoms and semi-infinite solids, which allow to analyze the deposition of the atomic clusters on surfaces. At the present level of computations the geometric forms of deposited metallic clusters are obtained, as well as binding energies, inter-atomic distances and vibration spectra. Magic clusters are identified, deposited on surfaces, exhibiting, in general, high symmetries, both horizontally and vertically. Increasing the number of atoms they may intrude beneath the surface, giving thus the possibility of building up interfaces, and contacts, between two solids. Further investigations are pursued into extending the theory to directional chemical bonds and electronic single-particle properties, and to increase the degree of accuracy of the results.

Acknowledgments

The authors gratefully acknowledge the critical reading of the manuscript by the NATO ARW reviewers and editors, who made very useful suggestions and remarks. This work has been partially supported by the Swiss National Science Foundation SCOPES Programme, Grant #7BUPJ062407.00/1 - FCST, and by Romanian Government CERES Research Programme #65/ 2001.

References

1. de Heer, W. A. (1993) The physics of simple metal clusters: experimental aspects and simple models, *Revs. Mod. Phys.* **65**, 611-676.
2. Brack, M. (1993) The physics of simple metal clusters: self-consistent jellium model and semiclassical approaches, *Revs. Mod. Phys.* **65**, 677-732.
3. Meiwes-Broer, K.-H. (ed) (2000) *Metal Clusters at Surfaces: Structure, Quantum Properties, Physical Chemistry (Cluster Physics)*, Springer.
4. Binns, C. (2001) Nanoclusters deposited on surfaces, *Surf. Sc. Rep.* **44**, 1-49.
5. Pople, J. A. (1999) Nobel Lecture: Quantum chemical models, *Revs. Mod. Phys.* **71**, 1267-1274.
6. Kohn, W. (1999) Nobel Lecture: Electronic structure of matter-wave functions and density functionals, *Revs. Mod. Phys.* **71**,1253-1266.
7. Cune, L. C. and Apostol, M. (2000) Ground-state energy and geometric magic numbers for homo-atomic metallic clusters, *Phys. Lett.* **A273**, 117-124.
8. Cune, L. C. and Apostol, M. (2001) Iron-hydrocarbon cluster $Fe_{13}\,(C_2H_2)_6$, *Chem. Phys. Lett.* **344**, 287-291.
9. Cune, L. C. and Apostol, M. (2000) *Metallic Binding*, apoma, Magurele-Bucharest.
10. Cune, L. C. and Apostol, M. (2002) Atomic Clusters: Chemical Bond in Condesed Matter, in Graja, A., Bulka, B. R. and Kajzar, F. (eds) *Molecular Low-Dimensional and Nanostructured Materials for Advanced Applications*, , Kluwer Academic Publishers, Dordrecht, pp 221-231.
11. Schwinger, J. (1980) Thomas-Fermi model: The leading correction, *Phys. Rev.* **A22**, 1827-1832.
12. Schwinger, J. (1981) Thomas-Fermi model: The second correction, *Phys. Rev.* **A24**, 2353-2361.
13. Slater, J. C. (1979) *The Calculations of Molecular Orbitals*, Wiley, NY.
14. Wigner, E. and Seitz, F. (1933) On the Constitution of Metallic Sodium, *Phys. Rev.* **43**, 804-810.
15. Wigner, E. and Seitz, F. (1934) On the Constitution of Metallic Sodium. II, *Phys. Rev.* **46**, 509-524.

16. Wigner, E. (1934) On the Interaction of Electrons in Metals, *Phys. Rev.* **46**, 1002-1011.
17. Doye, J. P. K. and Wales, D. J. (1997) Structural consequences of the range of the interatomic potential: a menagerie of clusters, *J. Chem. Soc., Faraday Trans.* **93**, 4233-4244.
18. Rayane, D., Melinon, P., Tribollet, B., Chabaud, B., Hoareau, A. and Broyer, M. (1989) Binding energy and electronic properties in antimony clusters: Comparison with bismuth clusters, *J. Chem. Phys.* **91**, 3100-3110.
19. Dunlap, B. I. (1990) Symmetry and cluster magnetism, *Phys. Rev.* **A41**, 5691-5694.
20. Castro, M. and Salahub, D. R. (1993) Theoretical study of the structure and binding of iron clusters: Fe_n ($n \leq 5$), *Phys. Rev.* **B47**, 10955-10958.
21. Christensen, O. B. and Cohen, M. L. (1993) Ground-state properties of small iron clusters, *Phys. Rev.* **B47**, 13643-13647.
22. Wang, Q., Sun, Q., Sakurai, M., Yu, J. Z., Gu, B. L., Sumiyama, K. and Kawazoe, Y. (1999) Geometry and electronic structure of magic iron oxide clusters, *Phys. Rev.* **B59**, 12672-12677.
23. Huisken, F., Kohn, B., Alexandrescu, R. and Morjan, I. (2000) Reactions of iron clusters with oxygen and ethylene: Observation of particularly stable species, *J. Chem. Phys.* **113**, 6579-6584.
24. Knickelbein, M. B., Koretsky, G. M., Jackson, K. A., Pederson, M. R. and Haznal, Z. (1998) Hydrogenated and deuterated iron clusters: Infrared spectra and density functional calculations, *J. Chem. Phys.* **109**, 10692-10700.
25. Karabacak, M., Ozcelik, S. and Guvench, Z. B. (2002) Structures and energetics of Pd_n ($n = 2 - 20$) clusters using an embedded-atom model potential, *Surf. Science* **C507-510**, 636-642.
26. Smith, J. R. (1969) Self-Consistent Many-Electron Theory of Electron Work Functions and Surface Potential Characteristics for Selected Metals, *Phys. Rev.* **181**, 522-529.
27. Jones, W. and March, N. H. (1973) *Theoretical Solid-State Physics*, Wiley-Interscience, London, vol. II, p. 1062 and *ff.*
28. Wang, Y. L. and Lai, M. Y. (2001) Formation of surface magic clusters: a pathway to monodispersed nanostructures on surfaces, *J. Phys.: Condens. Matter* **13**, R589-R618.
29. Rosenfeld, G., Becker, A. F., Poelsema, B., Verheij, L. K. and Comsa, G. (1992) Magic clusters in two dimensions?, *Phys. Rev. Lett.* **69**, 917-920.
30. Michely, T., Hohage, M. , Esch, S. and Comsa, G. (1996) The effect of surface reconstruction on the growth mode in homoepitaxy, *Surf. Sci. Lett.* **349**, L89-L94.
31. Lai, M. Y. and Wang, Y. L. (1998) Direct Observation of Two Dimensional Magic Clusters, *Phys. Rev. Lett.* **81**, 164-167.
32. Lai, M. Y. and Y. L. Wang, Y. L. (1999) Gallium-induced nanostructures on Si(111): From magic clusters to incommensurate structures, *Phys. Rev.* **B60**, 1764-1770.
33. Voigtlander, B., Kastner, M. and Smilauer, P. (1998) Magic Islands in Si/Si(111) Homoepitaxy, *Phys. Rev. Lett.* **81**, 858-861.
34. Hwang, I.-S., Ho, M.-S. and Tsong, T. T. (1999) Dynamic Behavior of Si Magic Clusters on Si(111) Surfaces, *Phys. Rev. Lett.* **83**, 120-123.

SUB-WETTING LAYER CONTINUUM STATES IN QUANTUM DOT SAMPLES

K. KRÁL AND P. ZDENĚK
Institute of Physics, Academy of Sciences of Czech Republic
Na Slovance 2, 18221 Prague 8, Czech Republic

Abstract. In the polar semiconductor samples of the self-organized quantum dots, grown by the Stranski-Krastanow growth method, the lowest energy extended states of the electronic excitations are assumed to be the wetting-layer states. The coupling between these extended states and the electronic states localized in the individual quantum dots, may influence the optical spectra of such samples in the sub-wetting layer region of energy. This effect is studied assuming the Fröhlich's coupling between the electrons and polar optical phonons. The contribution of this interaction to the appearence of the sub-wetting layer continuum in the optical spectra and to the level broadening of the localized states, pointed out in some experiments, is estimated.

1. Introduction

The semiconductor nanoparticles are promissing from the point of view of both basic science and applications [1]. One of their significant features is that the charge carriers can be localized within these nanoparticles. The quantum dots are thus regarded as a realization of artificial atoms with the nearly delta function-like spectral density of the electronic states. The quantum dot aggregates are prepared very often by the Stranski-Krastanow growth technique [1]. In the self-assembled quantum dot (SAQD) samples, prepared by this method, the quantum dots are grown on the top of the wetting layer (WL), with which the quantum dots interact. The optical absorption spectra of nanoparticles dispersed in a polymer film, which also seem to display an absorption continuum background increasing with the energy of the absorbed photon [2, 3], are not discussed here.

L.M. Liz-Marzán and M. Giersig (eds.),
Low-Dimensional Systems: Theory, Preparation, and Some Applications, 19–35.

In the type I quantum dots [1] in the undoped SAQD samples, the lowest energy excitations are the states with the electrons excited over the semiconductor band gap to the conduction band electronic quantum dot localized states, leaving the holes in the valence-band states of the quantum dot. In the SAQD samples it is expected that the lowest energy electronic excitations, which are not localized in the quantum dots, are the excitations to the wetting layer states [4, 5]. The scheme of the electronic states available to the electronic excitations in the SAQD samples is drawn in the Fig. 1.

The electronic system can also be excited to the extended states belonging to the substrate and to the electronic states in the cladding layers [6]. With a certain simplification, the electronic states localized in the quantum dots can be expected to have a delta function like density of states. This may be seen in contrast with experiments [4, 6, 7, 8, 9, 10, 11, 12, 13, 14] showing that the spectral densities may be different from delta functions and that the width of them may depend on the intensity of laser light excitating the SAQD sample. Besides this observation, the experiments on luminescence and optical absorption suggest that in the sub-wetting layer energy region, where we expect only bound electronic states, a continuum background is observed.

The continuous background, which seems characteristic of the SAQD samples, occurs often in the measurements on samples with a rather high density of the quantum dots and a large density of electrons excited to the WL states [14]. The presence of high densities of charge carriers may lead us to consider the electrostatic coupling among the charge carriers [15, 16, 17]. We shall not consider this mechanism in the present work. Also, we shall neglect any direct coupling between quantum dots, like the electron tunneling [18], electrostatic coupling between the dots, or any other. Recently, the Fröhlich's coupling [19] between the electrons and dispersionless polar optical phonons has been applied to the interpretation of the energy relaxation of electrons in quantum dots [20, 21] and to other properties [20, 22, 23]. The theory of the energy relaxation in the two level system with the dispersionless phonons has recently been supported by an exact calculation [24]. The purpose of this work is to show the possible contribution of the Fröhlich's coupling to the origin of the sub-WL continuous background spectral density and to the electronic level broadening in the SAQD samples.

2. The theoretical model of the SAQD system

Trying to study the optical spectra in the rather complicated samples, as the SAQD are, we are led to make a number of simplifications, which make

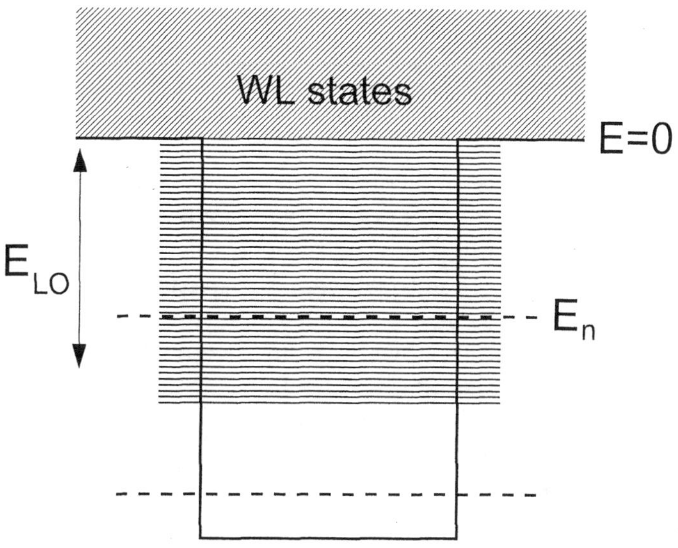

Figure 1. Schematic picture of the energetic structure of the single-electron states in the self-assembled quantum dot (SAQD) sample. The full line is the potential profile of the electrons in the quantum dot. The dashed horizontal lines denote the states bound in the quantum dots. The oblique hatched area denotes the extended electronic states in the wetting-layer. The horizontal hatching marks the sub-wetting layer area of electronic energies. E_n is the energy of a selected bound state in the sub-wetting layer energy region, E_{LO} is optical phonon energy.

the problem tractable, but which hopefully do not influence principally the main conclusions of the work. Dealing with the electron-phonon Fröhlich's interaction we shall neglect the influence of the interfaces in the SAQD heterostructure on the structure of the optical phonons. From the reasons given e. g. in the ref. [25] we shall simply assume that the charge carriers interact with the dispersionless longitudinal phonons of the bulk crystal of GaAs. Neglecting the effect of the electrostatic interaction among electrons and holes, we neglect also the exciton effect. We shall assume the limit of very heavy holes, in which their influence is reduced only to the static electrostatic field contributing to the effective potential in which the electrons move in the quantum dot potential in the conduction band.

In order that the details of the present model Hamiltonian are better understood, let us remind the principal features of the structure of the SAQD sample. With a certain simplification, in the Stranski-Krastanow method the surface of a substrate is first covered by a thin layer of the material, called wetting-layer, from which the quantum dots are to be grown. After depositing this layer the sample is left to obey a spontaneous process. In this process the wetting layer changes its structure, namely, due to the difference in the lattice constant between the substrate material and that of the wetting layer, the wetting layer starts to become thinner and the material of the wetting layer forms small nanocrystals on the top of the wetting-layer. These small nanostructures may have sometimes a shape of a pyramid. Because of the shape and the size these small structures are called quantum dots. The process stops at some size of the nanocrystals and at some remaining thickness of the wetting-layer. The final state is a result of an enegretic equilibrium in the whole system of the interatomic forces of the system. In order to prevent the mutual transfer of the matterial between the wetting layer and the quantum dots after the growth is ended, all the structure is covered from the top by a capping layer of a material, which can be similar to the substrate material. In this structure the electrons appear to be localized in the quantum dots. In order to overcome this localization they either have to gain an energy to transfer to the capping layer, or they need to get an energy sufficient for to transfer to the lowest energy states in the very thin weting layer. Because of the small thickness of the wetting layer, the lowest energies in this structure are rather large. Usually however, the lowest energetic barrier for the electron to leave the quantum dot is that to the wetting layer. The density of the quantum dots, with which they are distributed on the surface of WL, depends on the technological conditions of the Stranski-Krastanow process.

The wetting-layer is a quasi-two dimensional quantum well nanostructure. In the SAQD sample the electrons of the quantum dots are expected to be coupled to electronic states of the wetting layer. The complete orthonormal set of the electronic states in the quantum dots and in the wetting layer depend on the details of the geometry of the SAQD sample. Realizing that besides the WL extended states there are also other extended states in the real sample, namely those associated with the substrate and capping layers [4], which may be energetically not very far from the wetting-layer states, and in order to simplify the theory, we shall assume that the wetting layer is a three-dimensional subsystem with the electronic states extended throughout the whole SAQD sample. The impact of this dimensionality assumption can be verified by recalculating the present results with a two-dimensional wetting layer assumption. This will be postponed to a further work. Our model, using the three-dimensional wetting-layer continuum, may therefore correspond better to the vertically coupled three-dimensional stacks of the quantum dots [26], or, to the Volmer-Weber island growth technology without the wetting layer (see e. g. [27]).

We shall approximate the electronic bound states in the individual dots by the electronic eigenstates in the three dimensional infinitely deep quantum wells. Although we shall assume that the density of the quantum dots in the sample is rather high, namely 8×10^{21} m^{-3}, which by the mean inter-dot distance corresponds to the usual two-dimensional density of quantum dots in the measured samples [13], we shall assume that the single electron states in the individual quantum dots are mutually orthogonal. In this work, confining ourselves to the basic estimates, we shall need to consider only one bound state per a single dot. In real samples, the individual quantum dots have only a finite potential depth. In our model we shall therefore assume that the single-electron spectrum in a quantum dot is shifted by a certain energy, so that the energy of the bound state under consideration is positioned in the sub-wetting layer region of electronic energy and we shall neglect the other bound states in the dots.

The electronic motion in the WL states can be approximated by plane waves, with the electronic energy $E_{\mathbf{k}} = \hbar^2 \mid \mathbf{k} \mid^2 /(2m)$, m being the electron conduction band energy in GaAs. The plane waves together with the bound states would not make an orthogonal set of states. The problem of nonorthogonality of the basis can be approximately avoided upon restricting the magnitude k of the electron wave vectors $\mathbf{k}$ in the WL states to the values $k \ll k_m$. In this work we shall assume that the electronic wavelength corresponding to k_m is $1.5 * d$, d being the lateral size of the cubic quantum dot. We assume $d = 20\,\mathrm{nm}$. In GaAs this assumption limits the WL electronic energy from above by about 24 meV.

The restriction put on the extent of the WL space of states will be utilized also in another context. We know little about the state of electronic statistical distribution in the course of the experiment, so that we use a simple assumption of the population of the WL states. Namely, we shall assume that the quantum mechanical electronic WL states are populated by electrons with a homogeneous density $\overline{N_{\mathbf{k}}}$, independent of the electron wave vector. This homogeneity assumption simplifies the theoretical treatment in a certain way. The overall density of of the WL electrons is then obviously determined by $\overline{N_{\mathbf{k}}}$ and k_m.

In the above specified single-electron basis the full Hamiltonian consists of free electron Hamiltonian in the bound and WL states, H_e, free phonon Hamiltonian, H_{ph}, and the electron-phonon coupling H_1 between them:

$$
\begin{aligned}
H &= H_e + H_{ph}, \\
H_e &= \sum_{i,n} E_n c_{i,n}^{+} c_{i,n} + \sum_{\mathbf{k}} E_{\mathbf{k}} c_{\mathbf{k}}^{+} c_{\mathbf{k}}, \\
H_{ph} &= \sum_{\mathbf{q}} E_{LO} b_{\mathbf{q}}^{+} b_{\mathbf{q}}, \\
H_1 &= \sum_{\lambda,\mu,\mathbf{q}} A_q \Phi(\lambda, \mu, \mathbf{q})(b_{\mathbf{q}} - b_{-\mathbf{q}}^{+}) c_{\lambda}^{+} c_{\mu}, \qquad (1)
\end{aligned}
$$

Here λ and μ are the indexes of single electron states, either (i, n), or $\mathbf{k}$, where i is the quantum dot number and n is the electron orbital number.

c and b are, respectively, electron and phonon annihilation operators, E_{LO} is phonon energy. The coupling constant is [19]

$$A_q = -ieq^{-1}[E_{LO}(\kappa_\infty^{-1} - \kappa_0^{-1}]^{1/2}(2\varepsilon_0 V)^{-1/2}. \quad (2)$$

Here κ_∞ and κ_0 are, respectively, the high frequency and static dielectric constants, ε_0 is the permittivity of the free space, V is the volume of the sample and e is the electronic charge. The form-factor in (1)

$$\Phi(\lambda, \mu, \mathbf{q}) = \int d^3\mathbf{r}\psi_\mu^*(\mathbf{r})e^{i\mathbf{qr}}\psi_\lambda(\mathbf{r}). \quad (3)$$

modifies the Fröhlich's coupling to the case of quantum dot. Because the Fröhlich's coupling does not change the electron spin, we treat the electronic subsystems with the given spin separately. The functions ψ are the single electron orbitals in the SAQD sample. The terms of H_1 corresponding the Fröhlich's coupling between two bound states localized in different quantum dots will be considered as zero.

3. Spectral density

We shall calculate the electronic spectral density for the Hamiltonian specified above. This basic quantity will be used to compare the theoretical results with some experimental data in photoluminescence, photoluminescence excitation and optical absorption experiments. The electronic spectral density $\sigma_{i,n}(E)$, E being the energy variable, is related to the retarded Green's function $G_{i,n}(E)$ with help of the formula $\rho_{i,n}(E) = -\frac{1}{\pi}ImG_{i,n}(E)$. Similarly, spectral densities and the corresponding Green's functions $G_{\mathbf{k}}(E)$ are introduced for the wetting layer electronic states.

The electronic Green's function can be determined by the corresponding electronic self-energy, $M_{i,n}(E)$. Only the diagonal terms, in the electronic orbital quantum number n, of the Green's function and self-energies, will be considered. The reader is referred to the earlier works on the electron energy relaxation in individual quantum dots [21, 28, 29, 30]. The bound states electronic self-energy in the self-consistent Born approximation reads:

$$\begin{aligned} M_{i,n}(E) &= \sum_m \alpha_{n,m}\left\{\frac{1 - N_{i,m} + \nu_{LO}}{E - E_m - E_{LO} - M_{i,m}(E - E_{LO})}\right. \\ &+ \left.\frac{N_{i,m} + \nu_{LO}}{E - E_m + E_{LO} - M_{i,m}(E + E_{LO})}\right\} \\ &+ \sum_{\mathbf{k}} \alpha_{n,\mathbf{k}}\left\{\frac{1 - N_{\mathbf{k}} + \nu_{LO}}{E - E_{\mathbf{k}} - E_{LO} - M_{\mathbf{k}}(E - E_{LO})}\right. \\ &+ \left.\frac{N_{\mathbf{k}} + \nu_{LO}}{E - E_{\mathbf{k}} + E_{LO} - M_{\mathbf{k}}(E + E_{LO})}\right\}, \end{aligned} \quad (4)$$

while for the wetting layer electrons we have:

$$\begin{aligned} M_{\mathbf{k}}(E) &= \sum_{m} \alpha_{\mathbf{k},\mathbf{k}'} \left\{ \frac{1 - N_{\mathbf{k}'} + \nu_{LO}}{E - E_{\mathbf{k}'} - E_{LO} - M_{\mathbf{k}'}(E - E_{LO})} \right. \\ &+ \left. \frac{N_{\mathbf{k}'} + \nu_{LO}}{E - E_{\mathbf{k}'} + E_{LO} - M_{\mathbf{k}'}(E + E_{LO})} \right\} \\ &+ \sum_{i,s} \alpha_{\mathbf{k},i,s} \left\{ \frac{1 - N_{i,s} + \nu_{LO}}{E - E_{s} - E_{LO} - M_{i,s}(E - E_{LO})} \right. \\ &+ \left. \frac{N_{i,s} + \nu_{LO}}{E - E_{s} + E_{LO} - M_{i,s}(E + E_{LO})} \right\}. \end{aligned} \tag{5}$$

Here i is the number of quantum dot, while n and s are the numbers of electron bound state orbitals in the given dot. In these equations for the electronic self-energy, the constants α are generally given by

$$\alpha_{\lambda,\mu} = \sum_{\mathbf{q}} \mid A_q \mid^2 \mid \Phi(\mu, \lambda, \mathbf{q}) \mid^2, \tag{6}$$

where the summation extends over all optical phonon wave vectors $\mathbf{q}$. The above equations for the electronic self-energy are self-consistent. Generally, they correspond to including an infinite number of terms of the self-energy expansion in the powers of H_1. The use of the self-consistency, corresponding to the inclusion of the multiple-phonon scattering, proved important in the case of the electron-energy relaxation among the discrete energy levels of an electron in quantum dots [20, 21] and would be important when considering multiple bound states in a single dot, coupled via H_1. Those terms in the equations (4,5), corresponding to the scattering between one bound state and one extended state, and those, corresponding to the scattering of the electron between two extended states, will be included in the bare Born approximation only.

Let us now specify the constants α in the three specific cases of the choice of the electronic states. In the case when both λ and μ denote the states localized in the same quantum dot, $\lambda = (i, n)$, $\mu = (i, m)$, we find that the corresponding constant $\alpha_{n,m} = \alpha_{m,n}$ in the equation (6) does not depend on the quantum dot site index i. This is because the form-factor depends on the site index i in the following way:

$$\Phi((i, m), (i, n), \mathbf{q}) = e^{i\mathbf{q}\mathbf{r}} \Psi^{(irr)}(m, n, \mathbf{q}), \tag{7}$$

where the irreducible part of the form-factor is defined as

$$\Psi^{(irr)}(m, n, \mathbf{q}) = \int d^3\mathbf{r} \psi^*_{(0,n)}(\mathbf{r}) e^{i\mathbf{q}\mathbf{r}} \psi_{(0,m)}(\mathbf{r}). \tag{8}$$

Here the site index 0 means the position of the quantum dot at the origin of coordinates. We therefore drop out the site index i from the constant α completely for the coupling between the bound states, and we have:

$$\alpha_{nm} = \sum_{\mathbf{q}} | A_q |^2 | \Psi^{(irr)}(m, n, \mathbf{q}) |^2 . \tag{9}$$

The constant $\alpha_{n,\mathbf{k}} = \alpha_{\mathbf{k},n}$, which characterizes the mutual coupling of the bound and extended states, is determined with help of the form-factor

$$\Phi((i,n), \mathbf{k}, \mathbf{q}) = V^{-1/2} \int d^3\mathbf{r} e^{i\mathbf{k}\mathbf{r}} e^{i\mathbf{q}\mathbf{r}} \psi_m(\mathbf{r}), \tag{10}$$

V is the volume of the sample. When the electron wave vector $\mathbf{k}$ is put zero, the latter integral is the Fourier transform of the electronic wave function. Because this wave function is localized in the cube of the lateral dimension d, the Fourier transform will be a function of $\mathbf{q}$, the width of which in the $\mathbf{q}$-space will be about π/d. Having confined our space of available extended states to the plane waves the $\mathbf{k}$-vectors of which fulfill the condition $| \mathbf{k} | \ll \pi/d$, we can expect that in the limit of $\mathbf{k}$-vector going to zero, the magnitude of $\mathbf{k}$ in the form-factor $\Phi((i,n), \mathbf{k}, \mathbf{q})$ will not have any important influence on the value of the corresponding constant α. In practice we will assume that $| \mathbf{k} | \leq k_m$, $2\pi/k_m = 1.5 \times d$. We get:

$$\alpha_{n,\mathbf{k}} \approx \alpha_{n,\mathbf{k}=0} = \sum \sum_{\mathbf{q}} | A_q |^2 | \Phi^{(irr)}(\mathbf{k}=0, n, \mathbf{q}) |^2, \tag{11}$$

where

$$\Phi^{(irr)}(\mathbf{k}=0, n, \mathbf{q}) = V^{-1/2} \int d^3\mathbf{r} e^{i\mathbf{q}\mathbf{r}} \psi_{0,n}(\mathbf{r}). \tag{12}$$

It is then straightforward to get $\alpha_{n,\mathbf{k}} = B_n/V$, where the constant B_n is

$$B_n = \frac{e^2 E_{LO}(\kappa_\infty^{-1} - \kappa_0^{-1})}{16\pi^3 \varepsilon_0} | \int d^3\mathbf{q} e^{i\mathbf{q}\mathbf{r}} \psi_{0n}(\mathbf{r}) |^2 . \tag{13}$$

In the equation (5), the term which introduces the coupling between the extended states, is (we just rewrite the original Fröhlich's coupling)

$$\alpha_{\mathbf{k},\mathbf{k}'} \equiv \sum_{\mathbf{q}} | A_q |^2 | \Phi(\mathbf{k}, \mathbf{k}', \mathbf{q}) |^2, \tag{14}$$

with $\Phi(\mathbf{k}, \mathbf{k}', \mathbf{q}) = \delta_{\mathbf{k},\mathbf{k}',+\mathbf{q}}$. Then $\alpha_{\mathbf{k}',\mathbf{k}} = C_{\mathbf{k}'-\mathbf{k}}/V$, and

$$C_{\mathbf{k}'-\mathbf{k}} = \frac{e^2 E_{LO}(\kappa_\infty^{-1} - \kappa_0^{-1})}{2\varepsilon_0 | \mathbf{k}' - \mathbf{k} |^2} = \frac{p}{| \mathbf{k}' - \mathbf{k} |^2}. \tag{15}$$

4. The influence of the wetting layer states

We shall now consider the influence of the wetting layer states on the spectral density in the sub-wetting layer region of electronic energy. We have set the bottom of the wetting layer states at the energy $E = 0$. The electronic states localized in quantum dots are then at $E < 0$.

Inspecting the equations (4) and (5) we see that only some of the terms at the right hand side give singularities of the self-energy in the sub-wetting layer region, in the lowest order of iteration of these equations. These terms are only the following ones: the term corresponding to the last fraction in equation (4) and similarly the term containing the second fraction in equation (5). These terms bring phonon satellites into the electronic spectral density at the sub-WL region. Also, these terms give a contribution, which is proportional to the electronic population of the wetting-layer states, which property appears to be connected with the presence of the discrete level broadening in the sub-WL region [14].

The experiments, with which we wish to compare our results, are usually performed at 10 K. Therefore, we take $\nu_{LO} = 0$ for the population of the phonons. So that, unless the temperature of the lattice optical phonons is increased substantially, the above self-energy terms are nonzero only when there is a nonzero population of the wetting layer electron states. In the present numerical estimates we shall assume a rather large population in the wetting-layer states. The mechanism of achieving such a WL population is not discussed here in any detail. The large population of the WL states corresponds to the numerical estimates in the experimental papers [14].

4.1. A SINGLE BOUND STATE

Let us see the influence of the wetting layer states on the spectral density of a single bound electronic state. Let us assume the simple case, when there is only a single bound state in every quantum dot, which we denote as the state $n = 2$, the energy of which will be denoted as E_2 and $E_2 = -E_{LO} + \eta$. The parameter η determining the position of the bound state energy level is chosen to be 10 meV. In this work we use the material parameters of Gallium Arsenide. Namely, the electron relative effective mass is 0.067, the static dielectric constant is κ_0 =12.91 (SI units) and the high frequency dielectric constant is κ_∞ =10.91. The bulk crystal phonon energy in GaAs is 36.2 meV. The wave function of this state will be assumed to be that of an electron in one of the states of the triply degenerate first excited energy level of an infinitely deep cubic quantum dot with the lateral size d. Namely, this wave function is

$$\psi_{02}(\mathbf{r}) = (2/d)^{3/2} cos(\pi x/d) cos(\pi y/d) sin(2\pi z/d) \qquad (16)$$

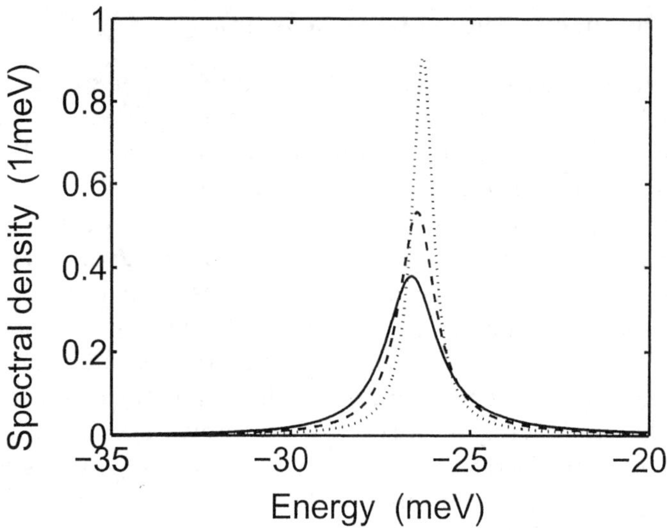

Figure 2. Spectral density $\sigma_n(E)$ of the bound state of a single quantum dot interacting with the wetting-layer states at the WL polulation: $\overline{N_{\mathbf{k}}}$ =0.25 (dotted line), 0.5 (dashed) and 0.75 (full).

(reminding that the index 0 means the wave function placed at the origin of coordinates). The natural line width of the spectral density of the bound state is taken here to be $\Delta = 0.05\,\mathrm{meV}$. The same natural line-width is assumed for the states in the wetting-layer. This means that the spectral density of the unperturbed bound state would be a Lorenzian, having the halfwidth of 0.05 meV, being placed at the energy $-E_{LO} + 10\,\mathrm{meV}$. The electronic self-energy term, expressed in the Born approximation to the coupling between the bound state ψ_{02} and the WL states, and being proportional to the WL population at low temperatures, is simply

$$M_n(E) = \sum_{\mathbf{k}} \alpha_{n,\mathbf{k}} \frac{N_{\mathbf{k}}}{E - E_{\mathbf{k}} + E_{LO} + i\Delta}. \tag{17}$$

We omit the index of the quantum dot here. We assume here that the population of the WL states is constant for all $\overline{N_{\mathbf{k}}}$. Under the above made assumption, according to which the wetting-layer states $\mathbf{k}$-vector magnitude is restricted to be not larger than k_m, this means that the average density of the electrons (of both spins) is about $1.5\times10^{23}\,\mathrm{m}^{-3}$ for $\overline{N_{\mathbf{k}}} = 0.5$. We assume that the electrons occupy only the WL states.

Reminding the above introduced approximation, according to which the wavevector dependence of the constant $\alpha_{n,\mathbf{k}}$ is neglected, the integral in the equation (17) can be integrated analytically [31]. The result of the numerical evaluation of the electronic spectral density $\sigma_n(E)$ of the electronic state $n = 2$, bound in a quantum dot, is presented in Fig. 2 for three values of the population of the wetting layer states. The spectral peak of the bound state is broadened, from the unperturbed value of the full width at half maximum being 0.1 meV, to the value of about 2 meV. This spectral width is of the same order of magnitude as it may be found in the experimental papers (see e. g. [14]. The Figure 2 shows the trend of the discrete state spectral density to broaden with the increase of the wetting layer states population. This behaviour of the bound state line-width appears to be in accord with the experiment [14]. The purpose of the present study is to pay attention to the main features of the effects under consideration. From this reason we use the material parameters of GaAs, as a typical material among the polar semiconductors. The quantum dot samples studied in experiment are often the InGaAs quantum dots, and the corresponding weting layer, both grown on the substrate of GaAs. Although the parameters characterizing InGaAs are rather similar to those of GaAs, we nevertheless do not compare our results quantitatively with the experiment. In the case of the bound state energy level broadenig, the reader is referred to the paper [14], namely to the Figure 3 in that paper, for the experimental data on the level broadening as it depends indirectly on the WL electron density. Let us emphasize that in the experimental works, with which we compare our results, the optical spectra are obtained in such a way, that the they are measured in the limit of a single dot measurement and therefore the inhomogeneous broadening of the optical spectra is avoided. In this sense, we can therefore compare the experimental data with our "homogeneously" broadened theoretical shape.

Besides the broadening of the discrete electronic level, the coupling of the bound state to the WL states tends to give a weak continuous background. Quantitatively, this background appears rather weak at the presently used values of the input parameters of the material, quantum dot size and shape, and at the present assumption about the populated states in the wetting-layer continuum. We may conclude that, the interaction of the bound state in the dot with the wetting layer states, under rather high population per quantum mechanical state, leads mainly to the bound state level broadening in the sub-wetting layer region.

4.2. PHONON SATELLITE OF THE WETTING LAYER STATES

We do not pay attention to the changes of the spectral density in the WL region, which come from the coupling of the WL states to the bound states. In the absence of the quantum dots in the sample, the electron-phonon interaction between the wetting-layer states leads, as it is well known [32], to the formation of the polaron state of an electron, in which the electron

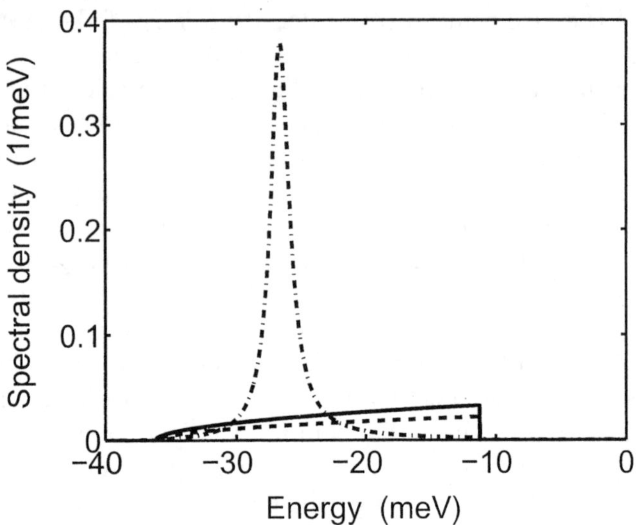

Figure 3. The full and the dash-dot lines give, respectively, the spectral densities of the phonon satellite of the wetting-layer states $\sigma_{tot}(E)$, related to a single quantum dot, and the spectral density $\sigma_n(E)$ of the discrete state of a single dot, both computed for the wetting-layer population $\overline{N_{\mathbf{k}}} = 0.75$. The dashed line is the spectral density of the phonon sattelite of the WL states $\sigma_{tot}(E)$ computed for the WL states population $\overline{N_{\mathbf{k}}} = 0.5$.

is wrapped by the cloud of the optical phonons. The depth of the effective electronic polaron potential hole in the crystal of GaAs may be about 2.5 meV [19]. In this case the spread of the polaron cloud may be estimated as several tens of nanometers. When the density of the quantum dots in the sample is large enough, so that the inter-dot separation is comparable to the size of the polaron, the existence of the polaron states may be not well established. In an overall agreement with the current experiments [4, 14] we shall assume the density of the quantum dots in the sample to be $8\times10^{21}\,\mathrm{m}^{-3}$. Realizing that the polaronic spectral density peak would be only about several meV below the low-energy edge of the wetting-layer states in GaAs, and having in mind the intention to study a mechanism leading to the appearance of a broad continuous spectral density in the broad range of energy from about $E = -E_{LO}$ until the wetting-layer edge $E = 0$, we shall leave the question of the polaron effect of the carriers in the WL states open and give our attention to that term in the equation (5), which is at $T = 0$ proportional to the electronic population of the wetting layer $N_{\mathbf{k}}$.

That term of the self-energy of the wetting-layer state $\mathbf{k}$, which is proportional to the wetting-layer population, will be in the lowest order approximation:

$$M_{\mathbf{k}}(E) = \sum_{\mathbf{k}'} \alpha_{\mathbf{k},\mathbf{k}'} \frac{N_{\mathbf{k}'}}{E - E_{\mathbf{k}'} + E_{LO} + i\Delta}. \tag{18}$$

Substituting for $\alpha_{\mathbf{k},\mathbf{k}'}$ we get

$$M_{\mathbf{k}}(E) = \frac{1}{(2\pi)^3} \iiint d^3\mathbf{k}' \frac{p}{|\mathbf{k}' - \mathbf{k}|^2} \frac{N_{\mathbf{k}'}}{E + E_{LO} - E_{\mathbf{k}'} + i\Delta}. \tag{19}$$

The integration is extended over the range of $\mathbf{k} < k_m$. Changing the integration variable from $\mathbf{k}'$ to $\mathbf{w} = \mathbf{k}' - \mathbf{k}$, we get

$$M_{\mathbf{k}}(E) = \frac{1}{(2\pi)^3} \iiint d^3\mathbf{w} \frac{p}{|\mathbf{w}|^2} \frac{N_{\mathbf{w}+\mathbf{k}}}{E + E_{LO} - E_{\mathbf{w}+\mathbf{k}} + i\Delta}. \tag{20}$$

When going to the spherical coordinates, the factor $|\mathbf{w}|^2$ in the denominator cancels with the factor coming from the Jacobian of the transformation. In the rest of the integrand we shall make the approximation of "forward scattering", which will consist in neglecting the scattering vector $\mathbf{w}$ in $N_{\mathbf{w}+\mathbf{k}}$ and in $E_{\mathbf{w}+\mathbf{k}}$. The integration limits obviously must obey simultaneously the complicated condition of $|\mathbf{k}| < k_m$ and $|\mathbf{w} + \mathbf{k}| < k_m$. We shall substitute this condition by the condition $|\mathbf{w}| < k_m$ which is obviously exact in the limit of $\mathbf{k} = 0$. The preliminary detailed numerical evaluation of the integral shows, that the simplified evaluation of the integral in (20) is plausible.

The above introduced approximations allow for obtaining the following simple approximate analytical form of the self-energy, namely,

$$M_{\mathbf{k}}(E) = \frac{\gamma}{E + E_{LO} - E_{\mathbf{k}} + i\Delta}, \tag{21}$$

in which $\gamma = pk_m\overline{N_{\mathbf{k}}}/(2\pi^2)$. The distribution of the electrons in all the wetting-layer states $\mathbf{k}$ is constant and equal to $\overline{N_{\mathbf{k}}}$.

The Green's function of the state $\mathbf{k}$ in the wetting layer then is:

$$G_{\mathbf{k}}(E) = \frac{1}{E - E_{\mathbf{k}} - \frac{\gamma}{E + E_{LO} - E_{\mathbf{k}} + i\Delta} + i\Delta}. \tag{22}$$

Looking for the approximate single pole behaviour of this Green's function in the region of energy $E \approx E_{LO}$, we find the following approximate expression of the wetting-layer electron Green's function:

$$G_{\mathbf{k}}(E) = \frac{\gamma/E_{LO}^2}{E + E_{LO} - E_{\mathbf{k}} + i\Delta}. \tag{23}$$

Summing up all the satellite spectral densities of the individual WL states $\mathbf{k}$ within the range of $\mathbf{k} < k_m$ gives the total spectral density $\sigma_{tot}(E)$ of all the extended states. The total spectral density of the phonon satellite of the WL states, related to a single quantum dot, is displayed in the Fig. 3 for the values 0.5 and 0.75 of the population $\overline{N_{\mathbf{k}}}$. In the same graph the single dot bound state spectral density for $\overline{N_{\mathbf{k}}} = 0.75$ is displayed for the purpose of comparing the spectral density of the single dot bound state spectral line with the spectral density of the phonon satellite per one quantum dot.

The total spectral density of the WL satellite, $\sigma_{tot}(E)$, is proportional to the first power of the WL population $\overline{N_{\mathbf{k}}}$, while the integral of the discrete state $\sigma_n(E)$ is constant and equal one. Therefore, increasing the WL population leads to a relative weakening of the discrete state spectral density with respect to the $\sigma_{tot}(E)$. A similar trend is observed in the experimental paper [14], in which the discrete peaks in the sub-wetting layer region of the spectra appear to weaken with respect to the continuous background signal, when the WL population increases.

In the present model of the SAQD sample, the set of the WL states included in the theory is identical with the set of states which are populated with nonzero $N_{\mathbf{k}}$. Only these states contribute then to the formation of the phonon satellite of the WL states. This may lead us to expect, that the experimental spectra of the sub-WL optical response might be influenced also by the distribution function of the population of the WL states.

In the present approach the WL satellite continuum is found in the interval of energy $(-E_{LO}, 0)$, while in the experiment the continuous background may be observed even at the lower energy side of this energy interval. We expect that our model may give the extended range of the continuum upon going to higher orders in the iterative solution of the equations (4,5).

The numerical estimate of the dimensionless quantity γ/E_{LO}^2 gives the value of about 2.8×10^{-2}. The spectral density of the wetting layer satellite should thus be about 36 times smaller than that of the WL spectral density itself. This number is not easy to compare with the available experimental data, see e. g. [33], because, e. g., we have to be aware of the fact that the optical spectra may be not simply proportional to the spectral densities and also, because our model, assuming the three dimensional wetting layer, may not allow for such a direct comparison of these spectral densities.

5. Conclusions

The present numerical estimate shows that the Fröhlich's mechanism of the electron-phonon coupling in the self-assembled quantum dot samples can contribute considerably to the spectral features observed in the PLE experiments, providing that the population of the electrons in the wetting-layer states is sufficiently large. The optical signal in the sub-wetting layer region of the excitation energy can be expected to have a twofold origin. First, the energy levels of the bound states are broadened by the coupling to the populated WL states. Second, there is a continuous background in the sub-WL region, which is ascribed to the low-energy phonon sattelite of the wetting-layer states and which intensity increases linearly with the WL population.

The present conclusions, saying that the electron-LO-phonon coupling can provide the explanation of the continuous features in the optical spectra, do not mean an exclusion of other contributions, like the direct carrier-carrier coupling. These effects remain to be considered in detail, together with the influence of the polaron states and the dimensionality of the WL states.

Acknowledgements

The work was supported by the grants IAA1010113, OCP5.20, RN19982003014 and by the project AVOZ1-010-914.

References

1. Yoffe, A. D. (2001) Semiconductor quantum dots and related systems: electronic, optical, luminescence and related properties of low dimensional systems, *Adv. Phys.* **50**, 1-208.
2. Mittleman, D. M., Shoenlein, R. W., Shiang, J. J., Colvin, V. L., Alivisatos A. P., and Shank, C. V. (1994) Quantum size dependence of femtosecond electronic dephasing and vibrational dynamics in CdSe nanocrystals, *Phys. Rev. B* **49**, 14435-14447.
3. Banin, U. Cerullo, G., Guzelian, A. A., Bardeen, C. J., Alivisatos, A. P., Shank, C. V., (1997) Quantum confinement and ultrafast dephasing dynamics in InP nanocrystals, *Phys. Rev. B* **55**, 7059-7067.
4. Toda, Y., Moriwaki, O., Nishioka, M., and Arakawa, Y. (1999) Efficient carrier relaxation Mechanism in InGaAs/GaAs self-assembled quantum dots based on the existence of continuum states, *Phys. Rev. Lett.* **82**, 4114-4117.
5. Hinooda S., Loualiche, S., Lambert, B., Bertru, N., Paillard, M., Marie, X., and Amand, T. (2001) Wetting layer carrier dynamics in InAs/InP quantum dots, *Appl. Phys. Lett.* **78**, 3052-3054.
6. Schmidt, K. H., Medeiros-Ribeiro, G., Oestreich, M., Petroff, P. M., and Döhler, G. H. (1996) Carrier relaxation and electronic structure in InAs self-assembled quantum dots, *Phys. Rev. B* **54**, 11346-11353.
7. Steer, M. J., Mowbray, D. J., Tribe, W. R., Skolnick, M. S., Sturge, M. D., Hopkinson, M., Cullis, A. G., Whitehouse, C. R., and Murray, R. (1996) Electronic energy levels and energy relaxation mechanisms in self-organized InAs/GaAs quantum dots, *Phys. Rev. B* **54**, 17738-17744.
8. Hessman D., Castrillo, P., Pistol, M.-E., Pryor, C., and Samuelson, L. (1996) Excited states of individual quantum dots studied by photoluminescence spectroscopy, *Appl. Phys. Lett.* **69**, 749-751.
9. Heitz, R., Veit, M., Letentsov, N. N., Hoffmann, A., Bimberg, D., Ustinov, V. M., Kop'ev, P. S., and Alferov, Zh. I. (1997) Energy relaxation by multiphonon processes

in InAs/GaAs quantum dots, *Phys. Rev. B* **56**, 10435-10445.
10. Ledentsov, N. N., Ustinov, V. M., Shchukin, V. A., Kop'ev, P. S., and Alferov, Zh. I., (1998) Quantum dot heterostructures: fabrication, properies, laser (Review), *Semiconductors* **32**, 343-365.
11. Sauvage S., Boucaud, P., Gérard J.-M., and Thierry-Mieg, V. (1998) Resonant excitation of intraband absorption in InAs/GaAs self-assembled quantum dots, *J. Appl. Phys.* **84**, 4356-4362.
12. Lemaître, A., Ashmore, A. D., Finley, J. J., Mowbray, D. J., Skolnick, M. S., Hopkinson, M., Krauss, T. F. (2001) Enhanced phonon-assisted absorption in single InAs/GaAs quantum dots, *Phys. Rev. B* **63**, R161309-R161312.
13. Finley, J. J., Ashmore, A. D., Lemaître, A., Mowbray, D. J., Skolnick, M. S., Itskevich, I. E., Maksym, P. A., Hopkinson, M., Krauss, T. F. (2001) Charged and neutral exciton complexes in individual self-assembled In(Ga)As quantum dots, *Phys. Rev. B* **63**, 073307-073310.
14. Nakaema, M. K. K., Brasil, M. J. S. S., Iikawa, F., Ribeiro, E., Heinzel, T., Ensslin, K., Medeiros-Ribeiro, G., Petroff, P. M., Brum, J. A. (2002) Microphotoluminescence of self-assembled quantum dots in the presence of an electron gas, *Physica E* **12**, 872-875.
15. Gelmont, B. L. (1978) Three-band Kane model and Auger recombination, *Zh. Eksp. Teor. Fiz.* **75**, 536-544.
16. Kempa, K., Bakshi, P., Engelbrecht, P., Zhou, Y. (2000) Intersubband electron transitions due to electron-electron interactions in quantum well structures, *Phys. Rev. B* **61**, 11083-11087.
17. Morris, D., Perret, N., and Fafard, S. (1999) Carrier energy relaxation by means of Auger processes in InAs/GaAs self-assembled quantum dots *Appl. Phys. Lett.* **75**, 3593-3595.
18. Král, K., Khás, Z., Zdeněk, P., Čerňanský, M., and Lin, C. Y. (2001) Electron-energy relaxation in polar semiconductor double quantum dots, *Int. J. Mod. Phys.* **27**, 3503-3512.
19. Callaway, Joseph (1974) *Quantum theory of the solid state*, Academic Press, New York.
20. Král, K., and Khás, Z. (2001) Femtosecond to picosecond electron-energy relaxation and Fröhlich coupling in quantum dots, *arXiv:cond-mat*/0103061.
21. Tsuchiya, H., Miyoshi, T. (1998) Nonequilibrium Green's function approach to high-temperature quantum transport in nanostructure devices, *J. Appl. Phys.* **83**, 2574-2585.
22. Král, K., and Khás, Z. (2000) Homogeneous linewidth of optical transitions and electronic energy relaxation in quantum dots, *Optical Properties of Semiconductor Nanostructures*, pp. 405-420. Proceedings of the NATO SCIENCE PARTNERSHIP SUB-SERIES: 3: "High Technology", vol. 81. Kluwer Academic Publishers, Dordrecht 2000.
23. Král, K., Khás, Z., and Lin, C. Y. (2001) Optical line-shape and electronic energy relaxation in quantum dots *Phys. Status Solidi B* **224**, 453-456.
24. Menšík, M. (1995) Nonradiative recombination of a localized exciton, *J. Phys. Condens. Matter* **7**, 7349-7366.
25. Rücker, H., Molinari, E., and Lugli, P. (1991) Electron-phonon interaction in quasi-two-dimensional systems, *Phys. Rev. B* **44**, 3463-3466.
26. Ledentsov N. N., Shchukin, V. A., Grundmann, M., Kirstaedter, N., Böhrer, J., Schmidt, O., Bimberg, D., Ustinov, V. M., Egorov, A. Yu., Zhukov, A. E., Kop'ev, P. S., Zaitsev, S. V., Gordeev, N. Yu., Alferov, Zh. I., Borovkov, A. I., Kosogov, A. O., Ruvimov, S. S., Werner, P., Gösele, U., and Heydenreich, J. (1996) Direct formation of vertically coupled quantum dots in Stranski-Krastanow growth, *Phys. Rev. B* **54**, 8743-8750.
27. Matsumura, N., Kimura, Y., Endo, H., Saraie, J. (2000) Self-assembling CdTe quantum dots on ZnSe by alternate supplying and molecular beam epitaxial method, *J.*

Crystal Growth **214**, 694-697.
28. Král, K., Khás, Z. (1997) Hot-electron relaxation rate in quantum dots, *Phys. Stat. Sol. B* **204**, R3-R4.
29. Král, K., and Khás, Z. (1998) Electron self-energy in quantum dots, *Phys. Rev. B* **57**, R2061-R2064.
30. Král, and Khás, Z. (1998) Absence of phonon bottleneck and fast electronic relaxation in quantum dots, *Phys. Stat. Sol. B* **208**, R5-R6.
31. Dwight, H. B. (1961) *Tables of integrals and other mathematical data*, 4th ed., The Macmillan Company, New York.
32. Mahan, G. D. (1990) *Many-Particle Physics*, 2nd ed., Plenum Press, New York.
33. Heitz, R., Veit, M., Ledentsov, N. N., Hoffmann, A., Bimberg, D., Ustinov, V. M., Kop'ev, P. S., Alferov, Zh. I. (1997) Energy relaxation by multiphonon processes in InAs/GaAs quantum dots, *Phys. Rev. B* **56**, 10435-10445.

PLASMONS IN CARBON NANOTUBES

K. KEMPA AND R. CHURA
Boston College,
Department of Physics, Chestnut Hill, MA 02467, USA

Abstract. Carbon nanotubes posses a very rich spectrum of plasmon modes. For single wall, metallic carbon nanotubes, in addition to the high frequency plasma modes in the UV range, and the depolarization shifted van Hove plasma resonances, there are also low frequency, quasi-1D plasmons in the far-infrared frequency range, which interact with phonons. We study these plasmon modes in a unified random phase approximation approach, including effects of phonon coupling. We discuss our calculations in the context of recent experiments.

Carbon Nanotubes (CNT's) constitute a new form of carbon with unique physical properties [1]. CNT's can be metals, with conductivity comparable to that of Cu and the mechanical strength exceeding that of diamond, excellent thermal conductors, semiconductors with variable gap (0 - 1 eV), and superconductors. They can be grown in various forms, such as periodic arrays of aligned CNT's [2]. There are also numerous potential, as well as, existing applications of CNT's, ranging from novel high strength composites, to field emission devices and displays, to nano-transistors, to chemical catalysts, to bio-sensors.

In spite of the intensive theoretical and experimental work, understanding of the basic physics of CNT's is still far from complete. One area still under intensive investigation is the electromagnetic response. While there has been a remarkable progress in understanding of the Raman response of individual CNT's, which is primarily due to numerous phonon excitations [3], there is an evidence for a subtle interaction of those with the electronic degrees of freedom in the far infrared (IR) range [4-5]. Electron energy loss spectroscopy (EELS) revealed a rich spectrum of collective electronic excitations (plasmons) in CNT's in the IR/visible and UV frequency ranges [6].

In the simplest form, a CNT is a tube, with diameter ranging from 0.4 nm to 20 nm, made of a single graphene sheet [1]. This is a single wall nanotube (SWNT). It can be viewed as a result of "rolling" a plane graphene sheet in a direction which allows for a continuous, seamless self-joint. Obviously, there are infinitely many of such "rolling" directions, each described by the chiral vector **C**, which defines the circumference of a CNT (**C** = n**a** +m**b**, where n,m are integers, and **a**,**b** are the unit cell base vectors of the graphene sheet). SWNT's can be semiconductors or metals, depending on **C**. The reason for this is well understood, and basically due to the circumferential quantization of the electronic motion combined with the details of the band structure of the graphene sheet [1]. It can be shown that, CNT's for which n-m = multiple of 3, are metallic. Metallic SWNT's are excellent metals. The typical Fermi energy EF, measured from the bottom of the main,

L.M. Liz-Marzán and M. Giersig (eds.),
Low-Dimensional Systems: Theory, Preparation, and Some Applications, 37–43.

essentially parabolic band, is of the order of 10 eV. The effective Bohr radius is a* ≈ 0.1 nm, and the Fermi wave vector kF ≈ 10 nm^{-1}. The Wigner radius is rs ≈ 0.5, which is much closer to the "perfect metal" (rs <<1) than Al. This implies a very strong collective (plasmon) behavior of electrons in SWNT's.

A simple understanding of the topological structure of the plasmon spectrum of a SWNT can be achieved by noticing that the density of states (DOS) of a SWNT is strongly peaked at the van Hove singularities [1]. This is reflection of the quasi-1D nature of SWNT's, and allows to approximate the DOS as a series of δ-function-like peaks. This immediately implies, via a simple RPA argument, that the dielectric function can be roughly approximated by a sum of simple harmonic oscillator terms:

$$\varepsilon(\omega) = 1 - \sum_{i=0}^{N} \frac{\alpha_i}{\omega^2 - \omega_i^2} \tag{1}$$

where α_i is the i-th oscillator strength, the frequency of the i-th van Hove resonance is $\omega_i = (E_k - E_l)/\hbar$. Ek ($E_l$) is the k-th (l-th) van Hove energy, and ω is the frequency. A sketch of the $\varepsilon(\omega)$, as given by (1), is shown in Fig. 1. Plasmon modes occur if $\varepsilon(\omega) = 0$, and this condition is satisfied at various frequencies.

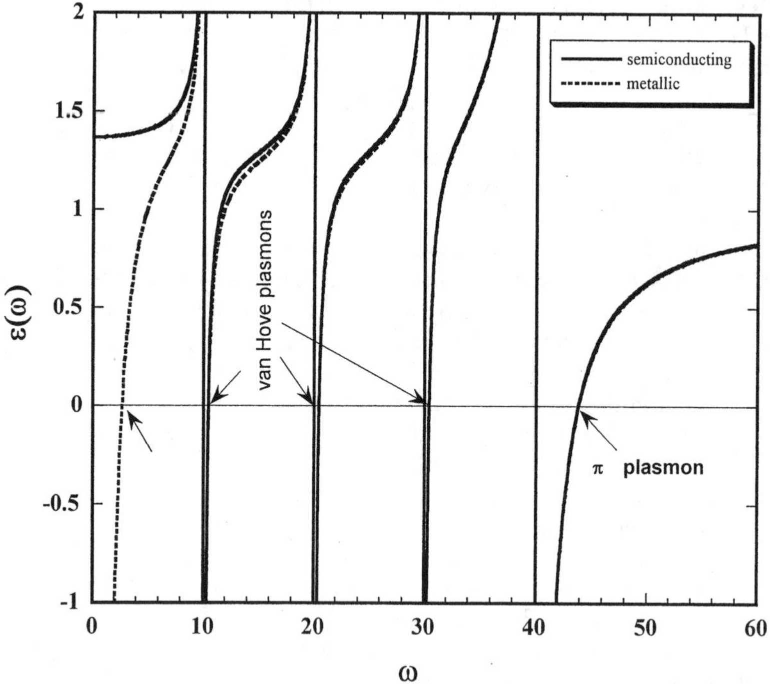

Figure 1. Plot of the simple model of the dielectric function of SWNT versus frequency. Solid line is for semiconducting and the dotted for the metallic tubes. The vertical lines indicate the singular points associated with the van Hove singularities of DOS, i.e. the van Hove resonances. Plasmons occur for ε(ω) = 0, i.e. at the crossing points of the curve with the horizontal thin line. Note the occurrence of the quasi-1D plasmon mode for metallic tubes.

At high frequencies, i.e. for $\omega^2 >> \omega_i^2$, it yields $\omega \approx \omega_p$, where the effective plasma frequency is $\omega_p = (\Sigma \alpha_i)^{1/2}$. This is the so called π-plasmon (ω_i span the π band only), and its frequency is in the 6 eV range [6]. It has a 3D character, i.e. disperses quadratically, like a conventional 3D plasmon. It also occurs in other nanocarbons and graphite. There is also an even higher 3D plasmon mode (about 23 eV), the π-σ plasmon, resulting from additional contributions from the σ band [6].

At intermediate frequencies, as Fig. 1 shows, there is also a plasmon mode between each pair of van Hove resonances ω_i. These weakly dispersive modes [6] are similar to the inter-subband plasmons in quantum wells [7], and occur in the range of frequencies between 0.5 to 3 eV.

Finally, at very low frequencies, in the case of the *metallic* SWNT's, there is yet another plasmon mode (see Fig. 1). This is a strongly dispersive quasi-1D plasmon mode [8]. It represents propagating charge density waves along the nanotube. These waves are circumferentially quantized, leading to an infinite, but discrete, set of spiral modes, propagating along the tube. The only non-spiral mode which is propagating directly along the tube axis (without perpendicular component) is gapless, i.e. it has a quasi-acoustic dispersion. To show this dispersive behavior one has to consider the full RPA response [9] of the electron gas in a bundle of SWNT's, each modeled as a cylinder of radius d made of a 2D electron gas. The axis of the cylinder is along the z-direction.

The problem essentially reduces to the calculation of the screened Coulomb interaction via the corresponding Dyson equation, which immediately yields the following expression for the relevant dielectric function :

$$\varepsilon(\Omega,Q) = 1 + v(Q)\Pi 0(\Omega,Q) \tag{2}$$

where Ω and Q are respectively the frequency and the z-component of the wave vector of a plasmon in Fermi units (i.e. normalized so that $\Omega = h\omega/E_F$, $Q=q/k_F$). Taking T=0 (excellent approximation even room temperature), and $E(k_z + Q) - E(k_z) = 2k_zQ + Q^2$, the single electron susceptibility is given by

$$\Pi_0(\Omega,Q) = \frac{1}{\alpha Q}\ln\left|\frac{(Q-2+\Omega/Q)(Q-2-\Omega/Q)}{(Q+2+\Omega/Q)(Q+2-\Omega/Q)}\right| \tag{3}$$

where $\alpha = 2\pi a^* k_F$. The bare Coulomb interaction is given by

$$v(Q) = \sum_p \sum_l 2K_m(QA)I_m(Qd)\exp(i\mathbf{KB}) \tag{4}$$

The plasmon wave vector is defined as (**K**,Q), i.e. it has components along the tube axis (Q) and perpendicular to it (**K**). The bundle lattice vector is given by **B**=(pax,lay), with p and l integer and ax, ay the components of the bundle lattice vector. Here A=d (the tube diameter) for p=l=0, and A=|**B**| otherwise. The p and l summations run over all nanotubes in the bundle. Km and Im are the modified Bessel functions. The result for a single tube follows from p=l=0. For non-spiral, gapless plasmon mode we take m=0.

Plasmon existence condition, ε(Ω,Q) = 0, leads immediately (through Eqs. 2 and 3) to a simple analytical formula for the plasmon mode dispersion

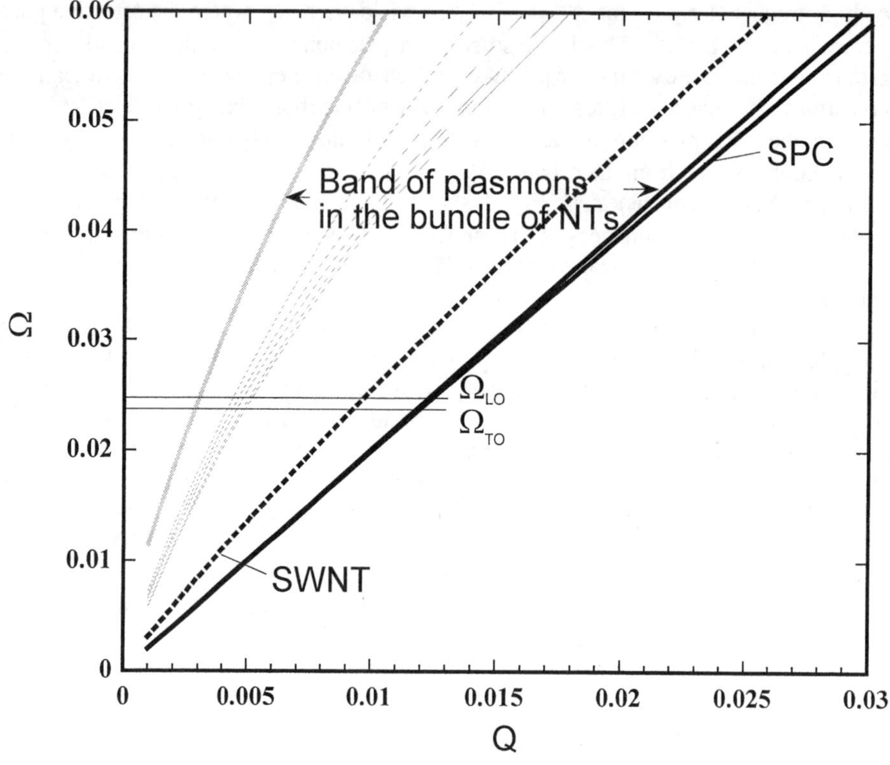

Figure 2. Dispersion of the quasi-1D gapless (m=0) plasmon mode for a (9,0) SWNT (dotted line) and for a bundle consisting of 25 such nanotubes (thin solid lines). Ω and Q are respectively the frequency and the z-component of the wave vector of a plasmon in Fermi units (i.e. normalized so that Ω =hω/EF, Q=q/kF). SPC is the single particle continuum.

$$\Omega = 2Q\sqrt{Q\coth\left(\frac{\alpha Q}{2\mathrm{v}(Q)}\right)+\frac{Q^2}{4}+1} \tag{5}$$

Fig. 2 shows the calculated plasmon dispersion, from Eq. 5, for the gapless plasmon mode (m = 0) of a single nanotube (dotted line) in the small frequency range ($\Omega << 1$). In this range, excitations of the electron gas can interact with phonon modes. Also shown is the single particle continuum (SPC). It is clear that the plasmon is quasi-acoustic, and that it follows the SPC. With decreasing diameter of the tube, the plasmon mode moves away from SPC. Fig. 2 shows also that in the bundle of nanotubes, the band of gapless plasmons forms, covering a large area of the energy-momentum space. Only a few plasmons from this band are shown in the figure. We find also, that the width of the plasmon band increases with the number of tubes in the bundle. Formation of the band of plasmons is a known phenomenon in semiconductor superlattices [10], and is essentially similar to the formation of the electronic band from a single level of an atom in the crystal lattice.

Also shown in Fig. 2, are positions of the two phonon modes of the Raman active G band. Interaction of these modes with the electron excitations leads to appearance of the so called Breit-Wigner-Fano (BWF) line [4]. We can easily study this effect by incorporating the phonon response into the dielectric function in Eq. 2. It was shown recently [11], that a curvature-induced charge polarization occurs in carbon nanotubes. This immediately allows us to employ the well known (Lyddane-Sachs-Teller) form of dielectric function contribution, valid for polar crystals. The new dielectric function for a nanotube bundle is

$$\varepsilon(\Omega,Q) = (\omega^2 - \omega_{LO}^2)/ (\omega^2 - \omega_{TO}^2) + v(Q)\Pi_0(\Omega,Q) \qquad (6)$$

where ω_{LO} and ω_{TO} are the frequencies of the longitudinal optical and transverse optical phonon modes of the G band, respectively [4]. Eqs. 3 and 4 remain valid, but in this case, no analytic formula can be derived for Ω, so that the spectral function, $F(\Omega,Q) = \mathrm{Im}[1/\varepsilon(\Omega,Q)]$, must be obtained numerically.

It is obvious, that phonon modes will interact strongly with excitations of the electron gas only in the vicinity of the mode crossing, since only there, both, energy and momentum are conserved, and an efficient interaction can take place.

Fig. 2 shows that this occurs for $Q \approx 0.012$ in the case of phonon-single electron interaction, for $Q \approx 0.0093$ for phonon-plasmon of the single tube interaction, and for only $Q \approx 0.003$ for phonon-plasmon of the bundle interaction.

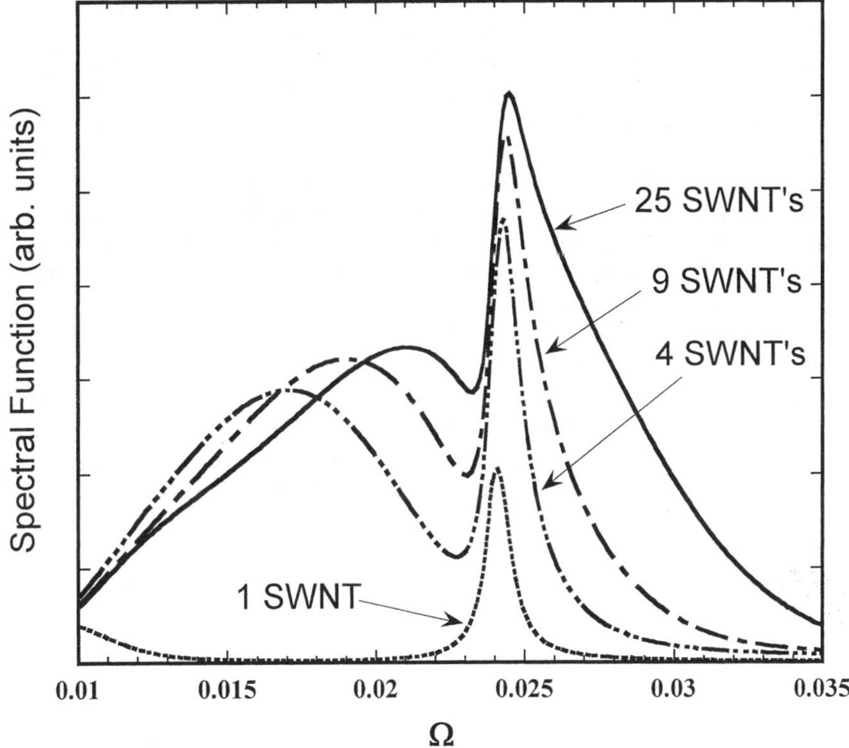

Figure 3. Spectral response function $S(\Omega,Q) = \mathrm{Im}[1/\varepsilon(\Omega,Q)]$ versus Ω for a single SWNT, and bundles of 4, 9 and 25 such nanotubes.

This fact alone explains the recent experimental observations of the BWF line. In the Raman experiments the amount of the available photon momentum is very small, $Q_0 \approx 10^{-4}$, which is obviously insufficient to excite the system in the strong interaction domain of the energy-momentum space. The remaining extra momentum must be provided by some symmetry breaking, such as finite tube length, intra nanotube defects, roughness of the substrate, etc. If a source of the additional momentum is available, it provides a spectrum of momenta, described by G(Q). One might expect that for randomly distributed defects, G(Q) is a monotonically decreasing (away from Q_0) function of Q, and this will result in a very weak coupling of photons to plasmons (and SPC) for single nanotube, but increasingly strong coupling for nanotube bundles with increasing number of tubes. This is fully confirmed (see Fig. 3) by our calculation of the momentum averaged spectral function

$$\bar{F}(\Omega) = \int dQ\, F(\Omega, Q)\, G(Q) \tag{7}$$

where G(Q) was assumed to be a gaussian centered at Q_0. For the single SWNT there is only one peak corresponding to the direct excitation of the LO phonon mode of the G band. There is only a slight shoulder below this pronounced peak. This is the BWF line, and is due to the combined phonon-plasmon and phonon-SPC interactions of this single tube. On the other hand, the relative strength of the BWF line is strongly increasing with the number of tubes in a bundle, as photons can more readily excite the coupled phonon-bundle plasmon excitation. This is in full agreement with recent experimental results [5]. We have also confirmed that a similar evolution of the spectrum, from a single peak to the peak and a broad BWF band, happens also for the decreasing distance between nanotubes in the bundle, again in full accord with the experiment [5]. Note, that the random variations of chirality of the tubes within a bundle, make the bundle look like a random lattice of metallic nanotubes. Such a lattice will posses a band of plasmons, but its width will be different for different bundles, and in general narrower, due to a larger on average distance between the nanotubes. This will lead to a random behavior of the BWF line, from bundle to bundle, indeed observed experimentally [5]. For a single nanotube, the strength of the BWF line increases with the decreasing nanotube diameter. This is consistent with recent experimental results of the MIT group [4].

In conclusion, we have studied plasmon modes of SWNT. In a unified RPA approach, we classify all the possible plasma modes for metallic and semiconducting SWNT's and their bundles. We show that the quasi-acoustic, gapless plasmon mode in a SWNT and its band in the bundles, can interact with phonons, leading to the BWF line. Formation of the broad BWF band is sensitive not only to the electron-phonon interaction, but also to the spectrum of available wave vectors. This, in turn, can vary from nanotube to nanotube (or bundle to bundle), due to variations in length, defect concentrations, etc. Our calculations are in agreement with recent experimental results.

References

1. Saito, R., Dresselhaus, G., and Dresselhaus, M. S. (1999) *Physical Properties of Carbon Nanotubes*, Imperial College Press, London.
2. Ren, Z. F., Huang, Z. P., Xu, J. W., Wang, J. H., Bush, P., Siegal, M. P., and Provencio, P. N., (1998) Synthesis of large arrays of well-aligned carbon nanotubes on glass, *Science* **282**, 1105 – 1107.

3. Dresselhaus, M. S. and Eklund, P. C. (2000) Phonons in carbon nanotubes, *Adv. Phys.* **49**, 705-814.
4. Brown, S. D. M., Jorio, A., Corio, P., Dresselhaus, M. S., Dresselhaus, G., Saito, R., and Kneipp, K. (2001) Origin of the Breit-Wigner-Fano lineshape of the tangential G-band feature of metallic carbon nanotubes, *Phys. Rev. B* **63**, 155414/1-155414/8.
5. Jiang, C., Kempa, K., Zhao, J., Schlecht, U., Kolb, U., Basché, T., Burghard, M. and Mews, A., Strong enhancement of the Breit-Wigner-Fano Raman line in carbon nanotube bundles caused by plasmon band formation, *to be published.*
6. Pichler, T., Knupfer, M., Golden, M. S., Fink, J., Rinzler, A., and Smalley, R.E. (1998) Localized and delocalized electronic states in single wall carbon nanotubes, *Phys. Rev. Lett.* **80**, 4729-4732.
7. Kempa, K., Broido, D., Beckwith, C., and Cen, J. (1989) d-function Approach to the Electromagnetic Response of Semiconductor Heterostructures, *Phys. Rev. B* **40**, 8385-8392.
8. Longe, P. and Bose, S. M. (1993) Collective Excitations in metallic graphene tubules, *Phys. Rev. B* **48**, 18239-18243.
9. Kempa, K., Liebsch, A., and Schaich, W. L. (1988) Comparison of Calculations of Dynamical Screening at Jellium Surfaces, *Phys. Rev. B* **38**, 12645-12648.
10. Kempa, K. (1987) Influence of a constant current on Raman spectra of high mobility superlattices, *Appl. Phys. Lett.* **50**, 1185-1187; Bakshi, P., Cen, J., and Kempa, K. (1988) Amplification of surface modes in type II semiconductor superlattices, *J. Appl. Phys.* **64**, 2243-2245.
11. Dumitrica, T., Landis, C. M., and Yakobson, B. I. (2002) Curvature-Induced Polarization in Carbon Nano shells, abstract submitted to NT02, Boston College, July 6-11. (http://dielc.kaist.ac.kr/nt02/abstracts/P178.shtml)

MOLECULAR DYNAMICS SIMULATIONS OF NANOTUBE GROWTH

Eduardo Hernández, Pablo Ordejón and Enric Canadell
Institut de Ciència de Materials de Barcelona (ICMAB-CSIC)
Campus de Bellaterra, 08193 Barcelona, Spain
ehe@icmab.es

Javier Junquera and José M. Soler
Departamento de Física de la Materia Condensada
Universidad Autónoma de Madrid, 28049 Madrid, Spain

Abstract We present some recent results of simulations of carbon single-walled nanotube growth. Our simulations are based on Density Functional Theory electronic structure calculations, and they allow us to gain important understanding on the physico-chemical processes driving nanotube growth, and the conditions under which these take place.

Keywords: Computer simulation, carbon nanotubes, molecular dynamics.

1. Introduction

Carbon has always been regarded as an important chemical element, the essential component of organic chemistry and bio-chemistry, and having two technologically important allotropes such as graphite and diamond, but by the 1980's it was thought that carbon had essentially given up all its secrets. Therefore the discoveries of the Fullerenes in 1985 [1] and that of carbon nanotubes in 1991 [2] took the scientific community by surprise. Today, more than a decade later, these discoveries are regarded as the starting point of a scientific revolution in physics, chemistry and materials science [3]. Fullerenes, and especially nanotubes, have become the paradigm of the multidisciplinary field of nanotechnology, one of the most vibrant and dynamical fields of current cutting-edge research. Nanotubes, in particular, are the focus of attention, due to their many interesting properties and their large potential for technological applications. Let us review briefly why nanotubes are re-

L.M. Liz-Marzán and M. Giersig (eds.),
Low-Dimensional Systems: Theory, Preparation, and Some Applications, 45–56.

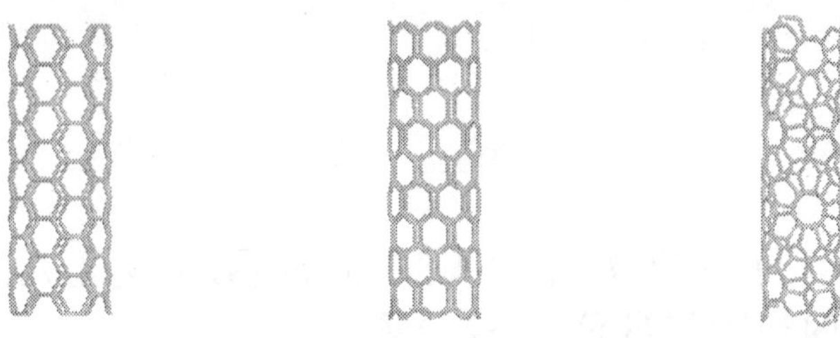

Figure 1. Examples of single-walled carbon nanotubes with different structures. The structure on the left is a (5,5) arm-chair nanotube, which is metallic; the central nanotube is an (8,0) zig-zag nanotube, which is semi-conducting. Both the (5,5) and (8,0) are achiral. The structure on the right is a chiral nanotube, denoted (6,3) nanotube, and is metallic.

garded with so much interest: firstly, they have a quasi one-dimensional structure: their diameter ranges from a few Å to up to 2 nm in the case of single-walled nanotubes (SWNT's), or a few tens of nm in the case of multi-walled nanotubes (MWNT's), yet their length is usually several orders of magnitude larger. They have many structural possibilities [see Fig. (1)], and this in turn results in different electronic conduction properties; indeed SWNT's may be metallic or semi-conducting. They have a high thermal conductivity, and extraordinary mechanical properties: the Young's modulus of carbon nanotubes has been reported to be approximately 1.2 TPa, as determined both experimentally and from theoretical calculations, thus being the highest of any known material (the Young's modulus of diamond is approximately 600 GPa).

All the above listed properties result in a large potential for applications. Indeed nanotubes have been used to fabricate field emission devices, because they can emit electrons from localised electronic states at their tips when placed in an electric field. They have also been used as tips of scanning probe microscopy instruments (atomic force and scanning tunnelling microscopes) [4], and also as constituents of nanoelectronic devices [5]. They have also found application as gas sensors, in lubrication, composite reinforcement, etc. Many of these applications have been demonstrated at the level of laboratory prototypes, but at present they cannot be exploited commercially, mostly because as yet nanotubes are too costly to produce, and they cannot be synthesised in sufficiently large quantities. Furthermore, present day synthesis techniques do not have a sufficient degree of structural selectivity, the kind of selectivity that would allow one to obtain only metallic (or only semiconducting) tubes, or tubes of specific structural characteristics, which may be required for certain applications. Clearly, if things are to be improved, it is necessary to have a detailed understanding of the growth process of nanotubes, as only then it will be possible to improve the yield and selectivity of their production.

It is known that SWNT's are obtained in the presence of a transition metal (usually a mixture thereof, involving Ni, Fe, Co, etc), and observing nanotube samples with high-resolution transmission electron microscopy (HRTEM) it appears that nanotubes extrude from clusters containing large amounts of these metals. It is currently assumed [6] that nanotubes form as the result of a precipitation process, when the mixture of carbon and transition metals in the form of nano-sized droplets at high temperature formed during the initial stages of synthesis (either by laser ablation or arc discharge) becomes unstable toward phase separation as the temperature drops. Indeed, carbon is known to have a very low miscibility in most transition metals except at high temperatures. The aim of this work is to gain some understanding of the process of nanotube formation, on the basis of Density Functional Theory (DFT) first-principles electronic structure calculations. We have studied liquid alloys formed by Ni and C in different conditions of composition and temperature, both as bulk liquids using periodic boundary conditions, and in a cluster geometry. These calculations allow us to obtain a detailed description of the physical properties of these systems (structure, diffusivity, etc) and to corroborate that indeed, under certain conditions of temperature and composition, C:Ni alloys are unstable toward phase separation, resulting in precipitation of C, which could have important consequences in our efforts to understand nanotube formation and growth.

2. Methodology

The calculations described below have been carried out using the SIESTA [7] code (Spanish Initiative for Electronic Simulation with Thousands of Atoms). This program permits the realisation of Density Functional Theory (DFT) [8] calculations using a basis set of localised atomic orbitals centred on the atoms, combined with the pseudopotential approximation. The basis set functions are strictly localised, going to zero at a predefined range, which allows for a rapid and efficient construction of the matrix representation of the Kohn-Sham Hamiltonian. The exchange-correlation energy has been accounted for using the Local Density Approximation (LDA) with the parametrisation of Perdew and Zunger [9] adjusted to the results of Ceperley and Alder [10]. SIESTA has the possibility of performing simulations in a fully self-consistent

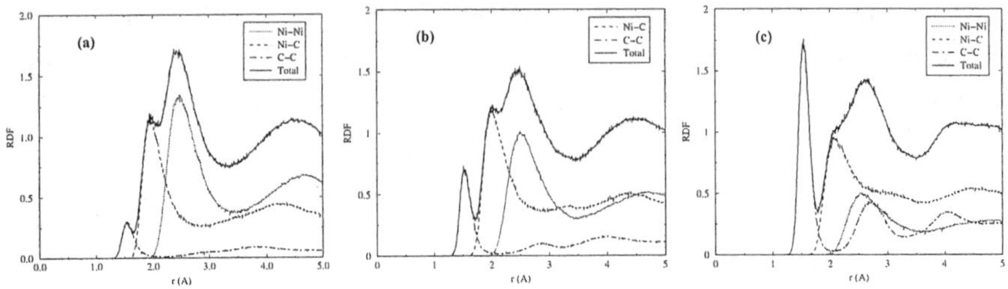

Figure 2. Radial distribution functions of the different Ni:C liquid alloys considered in this work. (a) $x_C = 0.25$, (b) $x_C = 0.33$, and (c) $x_C = 0.50$.

fashion using the Kohn-Sham [8] functional, or using an approximate functional due to Harris [11], which is not self-consistent. We have used both these facilities, as described below.

Molecular Dynamics (MD) simulations have been undertaken using the Verlet [12] algorithm with a time-step of 2 fs, which is sufficiently short to ensure adequate conservation of the total energy during the dynamics. Simulations were conducted in the microcanonical ensemble, considering both a bulk liquid structure in periodic boundary conditions and a cluster geometry. Our calculations with periodic boundary conditions were carried out using the non-self-consistent Harris functional; the use of this functional is not recommended in a cluster geometry due to the rapid changes in the electronic density at the cluster surface. Therefore, for the cluster geometries we employed the more reliable but more costly Kohn-Sham functional.

3. Results

3.1 Bulk liquid simulations

In order to investigate the stability and properties of Ni:C liquid alloys, we have conducted simulations using the Harris functional for a set of systems consisting of a total of 108 atoms in periodic boundary conditions. The carbon fractions considered were $x_C = 0.25$ (27 carbon atoms), $x_C = 0.33$ (36 carbon atoms) and $x_C = 0.50$ (54 carbon atoms). Each liquid alloy was simulated at temperatures of 1500, 2000 and 2500 K. At all these temperatures and compositions the systems are clearly in the liquid phase, displaying a liquid-like diffusive behaviour (see below).

From these simulations we can obtain a wealth of information on the structure and dynamics of these alloys. Let us first comment on their structural properties. Fig. (2) shows the radial distribution functions (RDF) obtained at a target temperature of 2000 K for each of

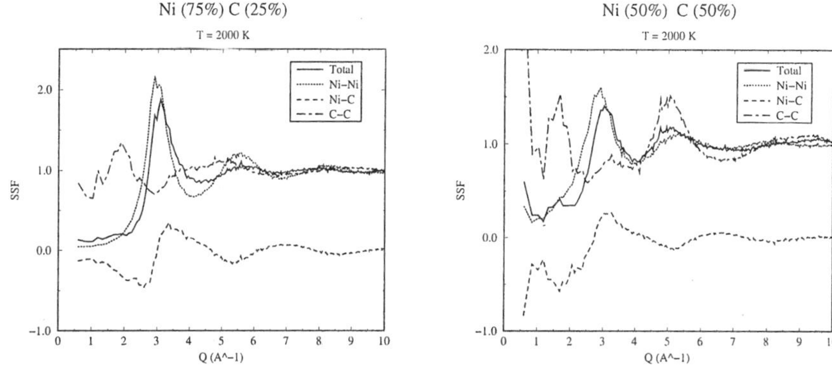

Figure 3. Static structure factors for the liquid alloys; on the left the static structure factors obtained at $x_C = 0.25$ and T = 2000 K are shown; on the right those obtained at $x_C = 0.50$ and the same temperature are displayed. Notice the large divergences observed in this case for small values of the wave vector q.

the different carbon fractions considered. In each plot four curves are drawn, showing each of the contributions to the total RDF (Ni-Ni, Ni-C and C-C). It can be seen that each type of pair contributes a nearest-neighbour peak, with the C-C nearest neighbour peak occurring at the shortest distance, approximately 1.54 Å. The C-Ni nearest-neighbour peak occurs at around 2 Å, and the Ni-Ni at 2.5 Å. The position of the Ni-Ni nearest neighbour peak is very similar to that found in pure liquid Ni at similar temperatures both from experiment [13] and simulations using empirical potentials [14]. The position of the C-C and Ni-Ni peaks is largely unaffected by x_C; that of C-Ni shifts slightly from 1.95 Å at $x_C = 0.25$ to 2.06 Å at $x_C = 0.5$. The effect of varying x_C is more noticeable in the relative intensities of the different peaks, as is to be expected. The C-C peak is very small at $x_C = 0.25$, and beyond this first peak there is very little structure in the C-C RDF. Essentially each carbon atom is surrounded by Ni atoms, and very few C atoms have another C atom as nearest neighbour. The intensity of the C-C nearest neighbour peak increases as x_C is raised, as expected. Simultaneously, a second and third nearest neighbour peaks appear in the C-C RDF at *c.a.* 2.9 and 4.0 Å respectively. The effect of varying the temperature is to widen or narrow the peaks in the RDF according as to whether the temperature is increased or decreased, but the overall structure of the RDF's is unaffected.

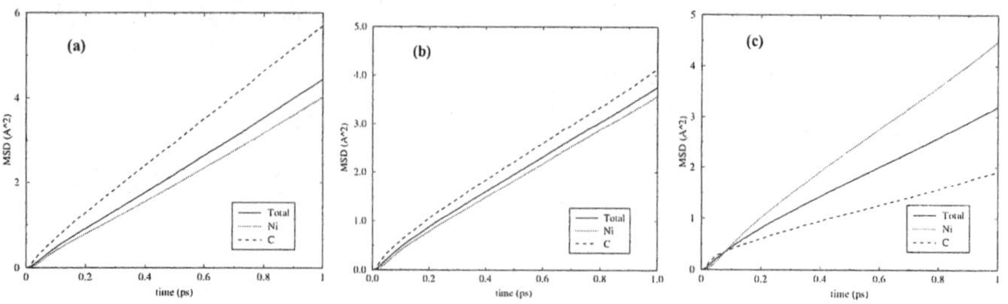

Figure 4. Mean-squared displacements obtained for the the liquid alloys considered in this work. (a) $x_C = 0.25$, (b) $x_C = 0.33$, and (c) $x_C = 0.50$.

Another interesting structural parameter to monitor is the static structure factor (SSF), defined as:

$$S_{\alpha\beta}(q) = < \rho_\alpha^\star(\mathbf{q}) \cdot \rho_\beta(\mathbf{q}) >, \tag{1}$$

where the brackets indicate a thermal average; $\rho_\alpha(\mathbf{q})$ is the Fourier component at wave vector $\mathbf{q}$ of the number density for species α, *i.e.*:

$$\rho_\alpha(\mathbf{q}) = \frac{1}{N_\alpha^{1/2}} \sum_{i=1}^{N_\alpha} e^{-i\mathbf{q}\cdot\mathbf{r}_i^{(\alpha)}}, \tag{2}$$

where the sum goes over all atoms of species α. The SSF's contain the same information as the RDF's, but in reciprocal space. However, the SSF can give us additional information on the stability of the system with respect to phase separation. This is because the SSF's provide information on fluctuations in the number densities of the different species. If the system has a tendency toward segregation into its different constituents, in the limit of low $\mathbf{q}$ (large wave lengths) the SSF will diverge, indicating the presence of large density fluctuations for each species. This intuitive argument can be formulated more precisely [15, 16], but it will be sufficient here to consider it at this intuitive level. In Fig. (3) the SSF's for Ni:C mixtures at 2000 K can be seen for the three different values of x_C considered. Note how for the $x_C = 0.25$ and $x_C = 0.33$ cases there appears to be no unusual behaviour in the limit $q \to 0$; however, as we increase x_C to 0.5 the values of the C-C and Ni-C SSF's take very large (absolute) values, which we take as a clear indication that under these conditions the system is unstable and will tend toward phase separation. Naturally, such phase separation cannot be monitored using periodic boundary conditions, as we do in these simulations. Nevertheless, we re-address the issue of phase separation in a cluster geometry below.

		T = 1500 K	T = 2000 K	T = 2500 K
	Total	3.62	7.36	10.4
$x_C = 0.25$	Ni	3.4	6.74	8.80
	C	4.26	9.23	15.3
	Total	4.45	5.97	11.8
$x_C = 0.33$	Ni	4.42	5.78	10.5
	C	4.52	6.34	14.5
	Total	2.15	4.87	9.92
$x_C = 0.50$	Ni	3.36	7.07	12.9
	C	0.93	2.65	6.92

Table 1. Diffusion coefficients for the Ni:C liquid alloys as a function of temperature and composition. The values are given in units of $10^5 \times \text{cm}^2/\text{s}$.

Let us now address the issue of the atomic dynamics in these liquid alloys. As stated above, under the different conditions of temperature and chemical composition considered here, these alloys display a distinctly liquid-like behaviour. This is clear from the mean-squared displacements, which have the linear time-dependent behaviour expected in the liquid phase, as can be appreciated in Fig. (4). From the slope of those curves it is easy to determine the diffusion coefficient of each chemical species involved in the alloy using Einstein's relation:

$$6D_\alpha t = < |\mathbf{r}_i^{(\alpha)}(t) - \mathbf{r}_i^{(\alpha)}(0)|^2 >, \qquad (3)$$

where D_α is the diffusion coefficient of species α. The calculated diffusion coefficients obtained from our simulations are given in Table (1). The first observation that can be extracted from the data given there is that the magnitude of the diffusion coefficients is typical of liquids, *c.a.* $\approx 10^{-5}$ cm^2/s, corroborating our earlier statement that indeed at these temperatures and compositions these alloys are indeed liquid. Another interesting observation is that the diffusivity of the different species is a function of the composition of the alloy. This in itself is not surprising; however, it is somewhat unexpected that Ni actually diffuses faster than C when $x_C = 0.5$. Intuitively, C, being the lighter chemical species, should diffuse faster, and indeed this is what we find at $x_C = 0.25$, regardless of the temperature of the simulation. But as the carbon fraction is increased, its diffusivity decreases. At $x_C = 0.33$ both Ni and C have very similar diffusivities, and at $x_C = 0.5$ Ni is actually the faster diffusing species. These changes in diffusivity are also apparent in Fig. (4). Obviously, the slowing down of the diffusion of C as x_C is increased must be somehow correlated to the structural changes occurring in the liquid. We know from Fig. (2) that as x_C is increased the C-C nearest neighbour peak becomes more and more prominent, indicating that somehow C atoms tend to aggregate if x_C is sufficiently high. In Fig. (5) we can see a typical structure resulting from our simulations at $x_C = 0.5$. This image shows a clear tendency of C atoms to cluster together, forming structures containing chains and rings; both a hexagon and a pentagon can be observed in this structure. Such large objects clearly will diffuse very slowly, due to their large and ramified shape; they may also act as traps for other carbon atoms. As a result of this, the overall diffusivity of C atoms will decrease sharply, as indeed we have observed in the MSD curves [see Fig. (4) and Table (1)]. We believe that the formation of these large structures of carbon in the liquid alloy at large x_C is responsible for the destabilisation of the liquid, and that such objects will tend to segregate from the Ni atoms. To address this point we studied the dynamics of precipitation by performing simulations of Ni:C alloys in a cluster geometry, simulations which are described in the following subsection.

3.2 Cluster simulations

Our bulk simulations described above have provided a detailed description of the structure and dynamics of liquid Ni:C alloys as a function of temperature and composition. However, bulk simulations such as those are not appropriate for discussing phase separation, even if they may hint at a possible phase instability (see our discussion on the behaviour of the SSF above). In order to try to investigate the dynamics of phase separation in a Ni:C alloy the only possibility is to study the system in a cluster geometry, and this is what we have done. We have simulated two clusters of a total of 108 atoms at a temperature of 1200 K. The first cluster contained 33 C atoms ($x_C = 0.31$), while the second one contained 57 ($x_C = 0.53$). While these clusters are rather small compared to the clusters observed in nanotube samples, they constitute an upper limit of the system sizes that can currently be addressed with first-principles molecular dynamics. We found it necessary to study these systems using the self-consistent Kohn-Sham functional. The Harris functional, which we also tried in this case, led to results inconsistent with those of the more accurate Kohn-Sham functional, possibly due to the large variations of the electron density in the proximity of the cluster surface.

In the cluster with low carbon content ($x_C = 0.31$) we observe no clear tendency of the carbon atoms to aggregate or migrate toward the surface; rather, they tend to remain in the interior of the cluster, forming C pairs or dimers. In the second cluster, with a dominance of carbon ($x_C =$

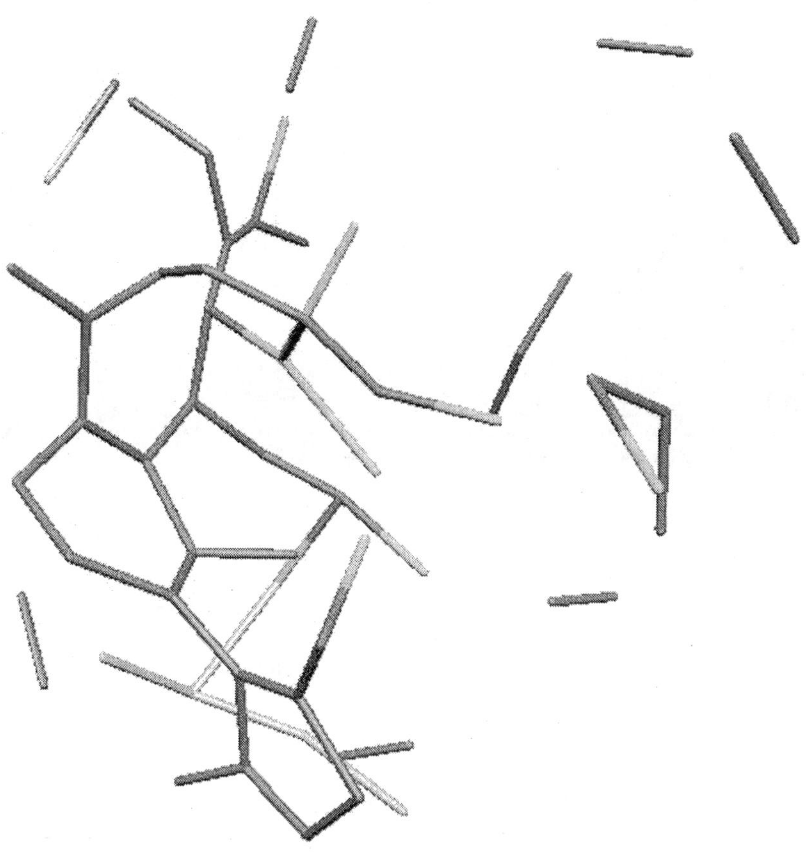

Figure 5. Typical structure of a carbon aggregate found in the liquid alloy at $x_C = 0.50$. To clarify the image other atoms not directly bonded to this structure are not displayed. Carbon atoms are indicated by a darker shade.

0.53), the tendency is different, however. In this case we see a clear trend of C atoms to agglomerate and migrate toward the surface, where they form ring-containing structures very similar in appearance to that found in the bulk alloy at $x_C = 0.5$ [see Fig. (6)]. We believe such structures could serve as seeds for the growth of carbon nanotubes, providing the tip or cap, and that nanotubes could grow from the continuous precipitation of carbon at the cluster-nanotube interface. Similar results have been recently obtained by Gavillet *et al.* [17].

It is worth considering the role played by the Ni atoms in this process. Detailed analysis of the atom trajectories in the $x_C = 0.53$ cluster

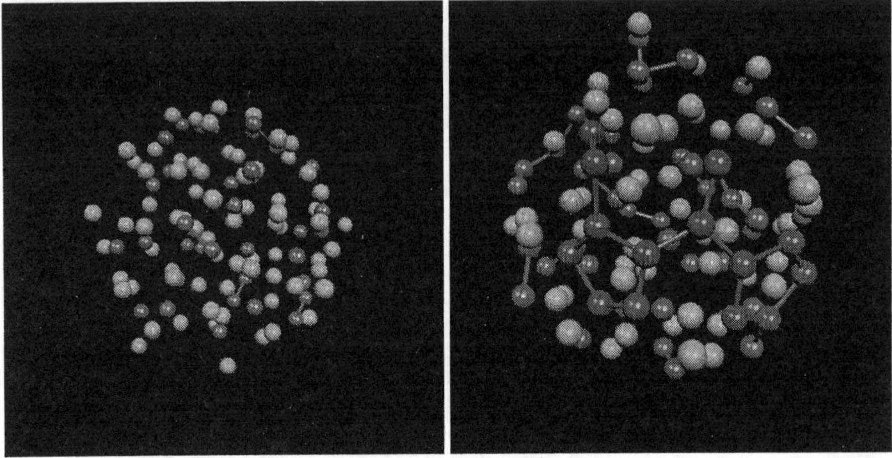

Figure 6. Cluster simulations. Blue coloured atoms represent Ni atoms, C atoms are shown in green. Left: final structure obtained with a low carbon fraction ($x_C = 0.31$). Right: final structure obtained with a high carbon fraction ($x_C = 0.53$). Note that in this case there is a marked tendency for carbon atoms to precipitate on the surface of the cluster and form ring-containing structures (highlighted in red) resembling a fullerene.

have revealed that frequently Ni atoms at the cluster surface stabilise incomplete carbon rings. We have observed the formation of several ring structures, pentagons formed by four C and a Ni atom stabilising the ring. We have also observed that wen a new carbon atom approaches such a structure the Ni atom moves away, being substituted by the approaching C atom, thus facilitating the formation of a pure C pentagonal ring. It is envisaged that the same process could lead to the formation of hexagons, though in the limited time scale of our simulations we did not observe such a process.

Due to computational limitations, our simulations cannot cover a time scale relevant to nanotube growth, which is estimated to take place in the ms range, and therefore we cannot directly monitor nanotube growth. Also the size of system that we can address is significantly smaller than that required for a direct simulation of nanotube growth. Nevertheless, we believe that these simulations set the stage for nanotube growth, and clarify considerably the conditions present during this process, and the phenomena involved.

4. Conclusions

Our simulations have provided us with a detailed characterisation of the physical properties of liquid Ni:C alloys under different conditions of temperature and composition. In all the cases considered (1500 K$\leq$ T $\leq$2500 K; 0.25$\leq$ x_C $\leq$0.5) these alloys behave as liquids, and we have analysed their structural and dynamical properties as these conditions are varied, finding that the diffusivity of carbon is strongly dependent on x_C. This is because if x_C is high (0.5) carbon atoms tend to aggregate forming large structures containing rings and chains, which diffuse slowly through the melt. Under these conditions the diffusivity of Ni is actually larger than that of C, in spite of it being the heavier element. Also, by analysis of the static structure factor, we can conclude that in these conditions the system would tend to phase-separate, although we cannot observe this separation directly due to the imposition of periodic boundary conditions in our bulk liquid simulations. Nevertheless, we do observe a clear trend toward precipitation of carbon in our cluster calculations when x_C is large, exactly as predicted by the bulk simulations. Although the time scale of our studies is too short to allow the direct observation of nanotube growth, we do observe the formation of fullerene-like structures which could serve as seeds for the growth. Also, these simulations have revealed that the transition metal plays the role of stabilising open ring structures, facilitating the incorporation of new carbon atoms into them, thereby aiding the growth process.

Acknowledgements

We thank the EU for providing financial support for this work through the IST project SATURN (project IST-1999-10593), and the Spanish Ministry of Science and Technology (projects BFM2000-1312-C02 and BFM2002-03278).

References

[1] Kroto H. W., Heath J. R., O'Brien S. C., Curl R. F. and Smalley R. E. (1985) C_{60}: Buckminsterfullerene, *Nature*, **318**, 162-3.

[2] Iijima S. (1991) Helical microtubules of graphitic carbon, *Nature*, **354**, 56-8; Iijima S. and Ichihashi T. (1993) Single-shell carbon nanotubes of 1-nm diameter, *Nature*, **363**, 603-5.

[3] For further information on Nanotubes and Fullerenes, see e.g. Ajayan P. M. and Ebbesen T. W. (1997) Nanometre-size tubes of carbon, *Rep. Prog. Phys.*, **60**, 1025-62; Dresselhaus M. S., Dresselhaus G. and Eklund P. C. (1996) *Science of Fullerenes and Carbon Nanotubes* (Academic Press, New York); Ebbesen T. W. (Ed.) (1997) *Carbon Nanotubes, Preparation and Properties* (CRC Press, Boca Raton); Terrones M., Hsu W. K., Kroto H. W. and Walton D. R. M. (1999),

Nanotubes: a revolution in materials science and electronics, *Topics in Current Chemistry*, **199**, 189-234.

[4] Dai H. J., Hafner J. H., Rinzler A. G., Colbert D. T. and Smalley R. E. (1996), Nanotubes as nanoprobes in scanning probe microscopy, *Nature*, **384**, 147-150.

[5] Dekker C. (1999) Carbon nanotubes as molecular quantum wires, *Physics Today*, **5**, 22-8.

[6] Kanzow H. and Ding A. (1999), Formation mechanism of single-wall carbon nanotubes on liquid-metal particles, *Phys. Rev. B*, **60**, 11180-6.

[7] Sánchez-Portal D., Ordejón P., Artacho E. and Soler J. M. (1997) Density-functional method for very large systems with LCAO basis sets, *Int. J. Quantum Chem.*, **65**, 453-61; Artacho E., Sánchez-Portal D., Ordejón P., García A and Soler J. M. (1999) Linear-scaling ab-initio calculations for large and complex systems, *Phys. Stat. Sol. (b)*, **215**, 809-17; Soler J. M., Artacho E., Gale J. D., García A., Junquera J., Ordejón P. and Sánchez-Portal D. (2002) The Siesta method for ab initio order-N materials simulation, *J. Phys. Condens. Matter*, **14**, 2745-79.

[8] Hohenberg P. and Kohn W. (1964) Inhomegeneous electron gas, *Phys. Rev.*, **136**, B864-71; Kohn W. and Sham L. J. (1965) Self-consistent equations including exchange and correlation effects, *Phys. Rev.*, **140**, A1133-8.

[9] Perdew J. P. and Zunger A. (1981) Self-interaction correction to density-functional approximations for many-electron systems, *Phys. Rev. B*, **23**, 5048-79.

[10] Ceperley D. M. and Alder B. J. (1980) Ground state of the electron gas by a stochastic method, *Phys. Rev. Lett.*, **45**, 566 9.

[11] Harris J. (1985) Simplified method for calculating the energy of weakly interacting fragments, *Phys. Rev. B*, **31**, 1770-9.

[12] Allen M. P. and Tildesley D. J. (1987) *Computer Simulation of Liquids*, (Clarendon Press, Oxford).

[13] Johnson M. W., March N. H., McCoy B., Mitra S. K., Page D. I. and Perrin R. C. (1976) Structure and effective pair interaction in liquid nickel, *Philos. Mag.*, **33**, 203-6.

[14] Alemany M. M. G., Rey C. and Gallego L. J. (1998) Computer simulation study of the dynamic properties of liquid Ni using the embedded-atom model, *Phys. Rev. B*, **58**, 685-93.

[15] Alfé D., Price G. D. and Gillan M. J. (1999) Oxygen in the Earth's core: a first-principles study, *Phys. Earth Planet. Interiors*, **110**, 191-210.

[16] Bhatia A. B. and Thornton D. E. (1970) Structural aspects of the electrical resistivity of binary alloys, *Phys. Rev. B*, **2**, 3004-12.

[17] Gavillet J., Loiseau A., Journet C., Willaime F., Ducastelle F. and Charlier J. C. (2001) Root-growth mechanism for single-wall carbon nanotubes, *Phys. Rev. Lett.*, **87**, 275504(1-4).

DENSITY OF STATES OF AMORPHOUS CARBON

S.YASTREBOV

A.F.Ioffe Physicotechnical Institute
St.Petersburg, 194021, Russia

1. Introduction

According to the modern conception, amorphous carbon (a-C), as it is sometimes called, diamond-like, consists of graphite-like phase embedded into diamond-like phase. The first papers where hypothesis was formed that nanosize fragments of graphite phase with sizes of some angstroms are responsible for formation of the energy gap of the fundamental absorption edge of the material is paper [1]. Basic assumption of paper [1] is separation of contribution of σ (sp^3 phase) and π states to the optical properties of amorphous carbon. In other words, while energy gap for sp^3 phase was considered greater than 5.5 eV (value for volume diamond), the energy transitions between occupied and unoccupied π states were considered smaller than this value. Aromatic hydrocarbons were used as a model of graphite phase and optical transitions between highest occupied and lower unoccupied π states were used as the model of the energy gap E_g. In this case, as was shown using tight binding modeling of electronic structure, value of E_g depends on the number of benzene rings, from which fragment of nanosize graphite consists of, as inverse function of the radical of their number. However, the approach was not presented in this paper, which should be applied for quantitative description of experimental dependences of optical absorption on the photon energy. Moreover, the shape of dependence of the imaginary part of dielectric function of the material on energy differs drastically both from similar dependence for diamond, graphite and aromatic hydrocarbons.

The main goal of present paper is to elaborate on the approach to the quantitative description of imaginary part of the dielectric function of amorphous carbon. Currently, some approaches exist to the quantitative description of the experimental data.

1. Approach proposed by Tauc, who suggested the radical dependence of density of occupied and unoccupied states for amorphous semiconductors [2].
2. Approach based on the assumption of excitation of surface plasmon within nanosize fragment of graphite with sizes of tens of nanometers [2].
3. Approach based upon the assumption of Gaussian shape of density of occupied and unoccupied states without the gap between states [3].

Mentioned approaches demonstrate some defects. Inconsistency does exist between 1 and 3 assumptions and experimental data obtained by different researchers for the fundamental absorption edge of amorphous carbon (a-C) [2]. Approach 2 is inconsistent with data obtained by such direct method of analysis of structure of amorphous carbon as transmission electron microscopy [4].

L.M. Liz-Marzán and M. Giersig (eds.),
Low-Dimensional Systems: Theory, Preparation, and Some Applications, 57–64.

In the present study, a model is developed of the energy spectrum of the material, arising from the structure of–C, as a system of quantum wells - nanosize fragments of graphite (with typical sizes of tens of angstroms), having Gaussian distribution of size quantization levels. Among other possible reasons, the reason for Gaussian distribution of size quantization levels is Gaussian size-distribution function of graphite fragments. To introduce the energy gap between occupied and unoccupied states one has to expect that size of the fluctuation is limited.

Optical transitions between two levels – highest occupied with π electrons and lower unoccupied level were considered for theoretical calculation of energy dependence of the imaginary part of the dielectric function of a single graphite fragment. An averaging procedure was performed, as was done in the paper [5], to calculate the contribution of all fragments of graphite containing in the sample.

There are following differences between the present paper and paper [5]:

1. Explicit form of density of states functions, size of fluctuation,
2. Limitation of the fluctuation size
3. Comparison of the model with original experimental data
4. Introduction of square of plasmon energy as a characteristics for number of π electrons per unit of volume

Difference between paper [3] and the present paper is in improvement of the model and in introduction of energy gap between occupied and unoccupied states.

The fit of the theoretical dependence to experimental values in 1.5-6 eV range was performed for two types of a-C samples – hydrogenated (a-C:H) and non-hydrogenated (a-C). A good agreement between theory and experiment was found. Fitting also allowed a reconstruction of the parameters of energy dependence of the density of electron states and dependence of plasmon energies on the annealing temperature. The behavior of parameters of Gaussians and the plasmon frequency with annealing temperature shows, for materials of both types, that value of the average number of electrons taking part in optical transitions between occupied and unoccupied states has trend to increase with temperature.

2. Experimental

Amorphous carbon layers were deposited by dc magnetron sputtering of a graphite target onto fused silica substrates. A 1 : 4 mixture of hydrogen and argon was used as the working gas to obtain *a*-C:H films, and pure argon, in the case of *a* -C. The substrate temperature during growth was 200 C, the gas pressure in the working chamber constituted 8–9 mTorr, and the magnetron power was maintained at 0.36 kW. The time of *a*-C and *a* -C:H film growth was chosen to be 40 and 30 min, respectively, which gave thicknesses of layer 770 and 740 Å. The transmission of the films deposited onto fused silica in the visible spectral range (200–850 nm) was measured on a Hitachi U-3410 double-beam spectrophotometer. The light spot on a sample was 0.5x0.5 cm^2 in size. Ellipsometric measurements were carried out with an LEF-3M ellipsometer at photon energy of 1.96 eV in reflection arrangement. The grown films were subjected to successive isochronous annealings in a vacuum (at residual pressure of 1 mTorr) in the temperature range 260–475° C. The spectral and ellipsometric measurements were done in air immediately after every annealing. According to ellipsometric data, the parameters of annealed films remained unchanged during a month of exposure to air. The obtained ellipsometric data (film thicknesses, refractive index of the material, and its extinction coefficient at He-Ne laser wavelength) were used to reconstruct the spectral dependence of the extinction coefficient. The dependence of the extinction coefficient on wavelength was obtained using the Kramers–Kronig relation. From the spectral dependences of the refractive index and extinction coefficient, the spectral dependence of the imaginary part of the dielectric function shown in Figure1 (a) was determined. It can be seen from Figure 1 that the imaginary part of the dielectric constant of an *a* -C:H film is a monotonic function of energy at low annealing temperatures, but, at a certain temperature, a spectral feature appears, which becomes more pronounced and shifts to lower energies with increasing annealing temperature. A similar type of behavior is observed for the dependence for *a* -C (Figure 1 (b)).

3. Dielectric function of amorphous carbon

Material is considered as a system of quantum wells - nanosize fragments of graphite, having Gaussian distribution of size quantization levels. The size of fluctuations is limited by the maximal and minimal allowed size of fragments of graphite. Maximal allowed size forms the energy gap of the material E_g.

Let us introduce the ideal system of two levels (Figure 2.). Let us resolve optical transition between highest occupied with electrons (E_G) and lowest unoccupied (E_{G*}) levels. The exact expression for the dielectric function corresponding to the optical transition from occupied to unoccupied state is:

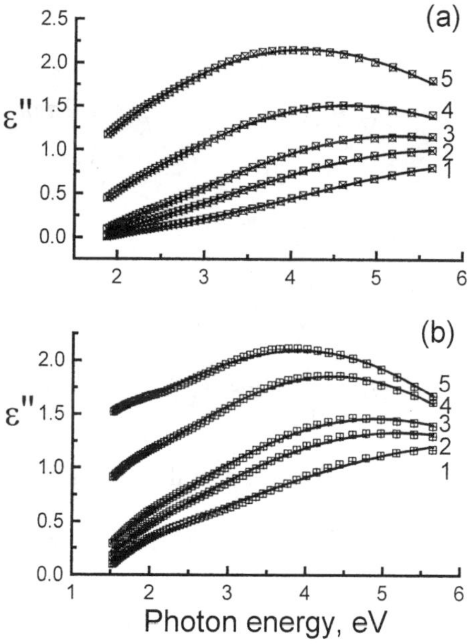

Figure 1. (a) Spectral dependences of the imaginary part of the dielectric function of (1) unannealed *a* -C:H film and films annealed at (*2*) 310, (*3*) 360, (*4*) 415, and (*5*) 475°C. Points stand for experimental data; curves, fitting with expression (5). (b) Spectral dependences of the imaginary part of the dielectric function of (*1*) unannealed *a*-C film and films annealed at (*2*) 310, (*3*) 360, (*4*) 415, and (*5*) 475°C. Points stand for experimental data; curves, fitting with expression (6).

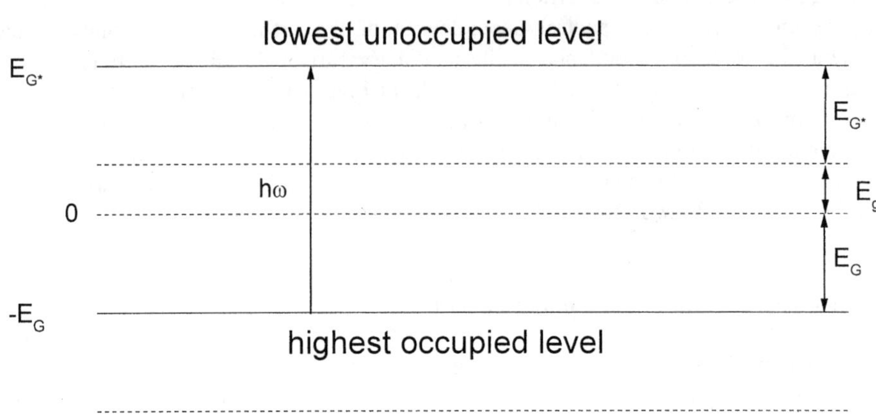

Figure 2. Diagram of optical transition for two-level system.

$$\varepsilon''_{(ou)}(\hbar\omega) = \frac{K}{(\hbar\omega)^2 a^3} |P_{ou}|^2 \delta(E_{G*} + E_g + |E_G| - \hbar\omega) =$$
$$= \frac{K}{(\hbar\omega)^2 a^3} |P_{ou}|^2 \int_0^{\hbar\omega} \delta(E - E_{G*} - E_g)\delta(E + |E_G| - \hbar\omega) dE \tag{1}$$

$$|P_{ou}|^2 = |< u^{(n)} |\nabla|_o^{(l)} >|^2 \qquad K = \left(\frac{2\pi e \hbar^2}{m} \right)^2$$

Let us introduce DOS functions:

$$N_u(E) = \frac{N}{V} \delta(E - E_{G*} - E_g)$$
$$N_o(E) = \frac{N}{V} \delta(E + |E_G|) \tag{2}$$

where N is the number of electrons taking part in the optical transitions in the sample of volume V.

It is easy to see that equivalent for (1) can be written in integral form in following way:

$$\varepsilon''_{(ou)}(\hbar\omega) = \frac{KV^2}{(\hbar\omega)^2 N^2 a^3} |P_{ou}|^2 \int_0^{\hbar\omega} N_u(E) N_o(E - \hbar\omega) dE \tag{3}$$

Equation (3) describes the microscopic dielectric function for a two-level system. In order to consider the optical properties of macroscopic sample we need to take into account all number of electrons N taking part in optical transitions. Here we suggest the scale of fluctuation of the parameter "a" is small. For the special case of statistically independent fluctuation of positions of size quantization levels both for occupied states (conduction band) and unoccupied states (valence band) with respect to the Fermi levels in the set of different wells of the sample [5], the spatial averaging gives the following formula for the overall dielectric function of the sample $\varepsilon'' = N\varepsilon''_{(ou)}$

$$\varepsilon''(\hbar\omega) = \frac{KV}{(\hbar\omega)^2}\left|P_{ou}\right|^2 \int_0^{\hbar\omega} N_u(E) N_o(E-\hbar\omega)\,dE \tag{4}$$

To derive explicit theoretical expression for the imaginary part of the dielectric function of a-C inhomogeneous broadening is considered for DOS of occupied and unoccupied energy bands: In this case (2) turns into the following form:

$$N_u(E) = \frac{N}{V}\frac{D^{1/2}}{\sqrt{2\pi}s_u}\exp\left[-\frac{1}{2}\left(\frac{E - E_{G*} - E_g}{s_u}\right)^2\right], \quad E > E_g$$

$$N_o(E) = \frac{N}{V}\frac{D^{1/2}}{\sqrt{2\pi}s_u}\exp\left[-\frac{1}{2}\left(\frac{E + \left|E_G\right|}{s_o}\right)^2\right], \quad E < 0 \tag{5}$$

where constant D can be obtained from normalization conditions. The parameters of the model are the following: $s_u = s_o = s$ - dispersion in the Gauss law; the most probable transition corresponds to the energies $|E_G| = E_{G*}$ are energies corresponding to the maximum of the Gauss distribution. Below is shown, the first approximation is sufficient to fit the experimental data with reasonable accuracy. By substituting (5) into (4) and integrating:

$$\varepsilon''(\hbar\omega) = \frac{A}{(\hbar\omega)^2}\exp\left(-\frac{(2E_{G*m} - \hbar\omega)^2}{4s^2}\right) erf\left(\frac{\hbar\omega - E}{2s}\right) \tag{6}$$

where

$$A = \frac{DKV}{2s\sqrt{\pi}}\left|P_{ou}\right|^2$$

Figure 3 shows comparison of the experimental data to different models. It is seen that best agreement was achieved for the fit Equation (6) to experimental data.

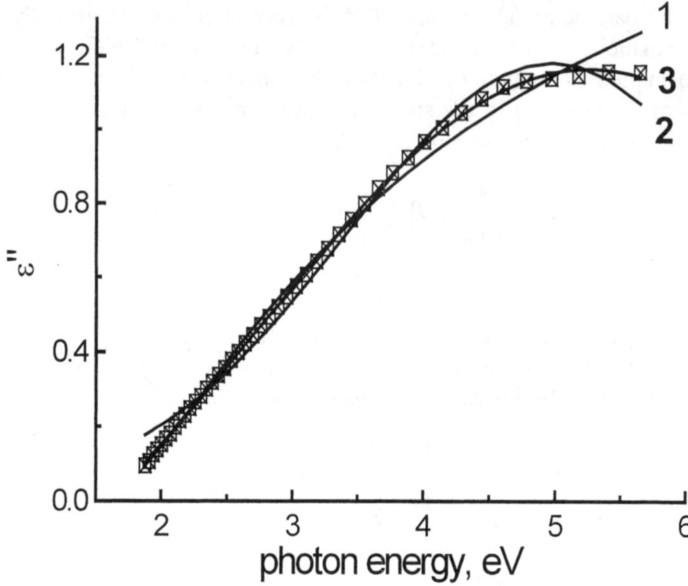

Figure 3. Squares stand for experiment (curve 4,Figure 1(a)), 1- fit to the Tauc formula, 2- fit to two-Gaussian model of DOS without gap [3], 3- fit to two-Gaussian model of DOS with the gap (formula (6)).

4. Thermal annealing impact

Let us apply the developed procedure to study the effect of thermal annealing on the fundamental absorption edge in *a*-C and *a*-C:H. We characterize the effect of annealing by the values of E_g and E_G and the concentration of electrons involved in optical transitions (an increase in this concentration corresponds to graphitization of the material in annealing). For this purpose, we use the expression for the energy corresponding to frequency of plasma oscillations of electrons, $\hbar\omega_p$, related to the imaginary part of the dielectric function of the material by (6)

$$(\hbar\omega_p)^2 = \frac{2}{\pi}\int_0^\infty \hbar\omega\varepsilon''(\omega)\, d\hbar\omega \tag{7}$$

The values, calculated using (7), are also presented in Figure 4. It follows from Figure 4 that, in the case of *a*-C, the plasmon frequency depends on the annealing temperature nonmonotonically, in agreement with the data of [2]. It was shown in [3] that for *a*-C films the temperature dependence of the film thickness has the form of a curve peaked at around $T = 360^\circ$ C, whereas the thickness of *a*-C:H films steadily decreases with temperature. This nonmonotonic behavior was attributed to relaxation of elastic strains in thermal treatment of *a*-C films. Presumably, the nonmonotonic variation of the film thickness in the course of annealing leads to nonmonotonic changes in its density and, consequently, in plasma frequency. With the temperature increasing further ($T > 360^\circ$C), the plasma frequency grows steadily with temperature.

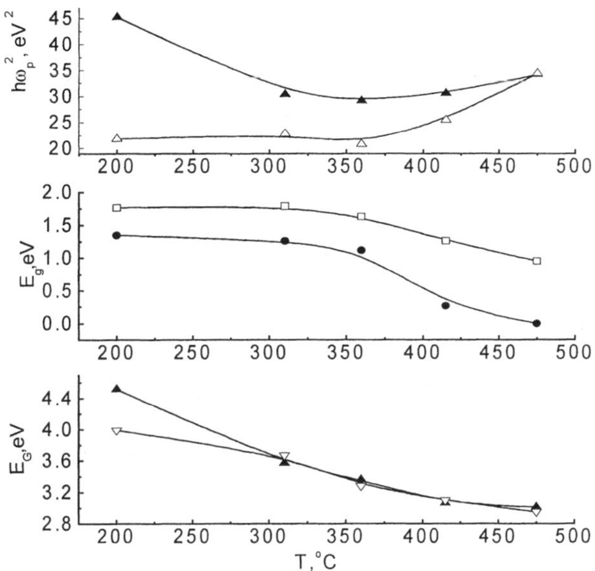

Figure 4 From top to bottom presented dependences on the annealing temperature: square of plasmon energy, energy gap between occupied and unoccupied states, maximum of density of states. Open up triangles, squares and down triangles stand for a-C:H.

At the same time, in the case of *a*-C:H films, the plasma frequency remains virtually constant in the temperature range 200–360 °C and then grows steadily, also in agreement with [3]. In both cases the energy gap E_g of *a*-C and *a*-C:H decreases steadily. Changes in the parameter E_g reflect the behavior of the maximal allowed fluctuation of the size quantization levels in annealing, and, therefore, a conclusion can be made that the size of the maximal fluctuation grows with temperature. The simultaneous increase in the plasma frequency indicates an increase in the total number of electrons per unit volume at temperatures exceeding 360°C. It also follows from the presented data that, in the optimal-size fluctuations, the distance between the centers of bands of energy levels is within the range 6–8 eV for samples of both types. The comparatively high transition energies suggest that the energy terms of optimal fragments, formed in part by σ electron states, contribute to the absorption. In the course of annealing, the distance between the centers of the bands composed of levels decreases somewhat. It should also be noted that, since all structural changes in a film in the course of its annealing are accompanied by loss of hydrogen, the role of the latter in limiting the size of the maximal allowed fluctuation both in the formation of the material and in its thermal degradation during annealing should not be underestimated.

5. Conclusions

The model was proposed of amorphous carbon structure, which consists of a set of quantum wells, - nanosize fragments of graphite. The size quantization levels of the wells distribute in accordance with Gaussian law. One of the reasons of such type of distribution is the size distribution of widths of quantum wells. The analytical formula was derived of dependence of imaginary part of the dielectric function on the photon energy. The numerical fit of the dependence to experimental values of amorphous carbon of two types, hydrogenated and nonhydrogenated, demonstrated a better agreement between theory and experiment than Tauc formula [2] and formula presented in paper [3]. The spectral dependence of the imaginary part of the dielectric function evidences the existence of a graphite fragment of a maximal size, which depends on the hydrogen concentration in a film. Number of π electrons has trend to grow with the annealing temperature.

6. Acknowledgements

Author would like to thank Prof.Ivanov-Omskii and colleagues from the IOFFE Institute and Prof.A.Tagliaferro for fruitful discussions.

Author acknowledge a grant and support from the DAAD/ZIP project 'International Quality Networks' within the subject 'Technology of New Materials' at the University of Applied Sciences Wildau, grant N 214/IQN. This study was also supported in part by the Russian Foundation for Basic Research (project no. 00-02-17004).

7. References

1. Robertson, J., and O'Reilly, E. P (1987) Electronic and atomic structure of amorphous carbon, *Phys. Rev. B* **35**, 2946-2957.
2. Ivanov-Omski, V. I., Tolmachev, A. V., and Yastrebov, S. G. (2001) Optical Properties of Amorphous Carbon Films Deposited by Magnetron Sputtering of Graphite, *Semiconductors* **35**, 220-225.
3. Dasgupta, D., Demichelis, F., Pirri, C. F., and Tagliaferro, (1991) A. Pi bands and gap states from optical absorption and electron-spin-resonance studies on amorphous carbon and amorphous hydrogenated carbon films, *Phys. Rev. B* **43**, 2131-2135.
4. Ivanov-Omskii, V.I., Lodygin, A.B., Sitnikova, A.A., and Yastrebov, S.G. (1997) Size distribution analysis of diamond and graphite-like DLC, *J. Chemical Vapor Deposition* **5**, 198-206
5. O'Leary, S. K. (1999) On the relationship between the distribution of electronic states and the optical absorption spectrum in amorphous semiconductors, *Solid State Commun.*, **109**, 589–594

ANISOTROPIC SILVER NANOPARTICLES: SYNTHESIS AND OPTICAL PROPERTIES

ISABEL PASTORIZA-SANTOS,[1] YASUSHI HAMANAKA,[2] KAZUHIRO FUKUTA,[2] ARAO NAKAMURA,[2] AND LUIS M. LIZ-MARZÁN[1]
[1] *Departamento de Química Física*
Universidade de Vigo, 36200, Vigo, Spain
[2] *Department of Applied Physics, Faculty of Engineering,*
Nagoya University, Chikusa-ku, Nagoya 464-8603, Japan

1. Introduction

The intense study devoted to metal nanoparticles is motivated by their extremely interesting optical properties. Such properties have already been described by Faraday some 150 years ago [1], and still give raise to hundreds of scientific articles every year. The origin of such a special optical behaviour can be found in the interaction between incoming light and the free conduction electrons [2]. When the wavelength of light couples with the oscillation frequency of the conduction electrons, a so-called plasmon resonance arises, which is manifested as an intense absorption band. The most spectacular cases are found when such plasmon resonance band exists in the visible wavelength range, since then the dispersions of metal nanoparticles display brilliant colours. The precise wavelength of the plasmon resonance depends on several parameters, among which particle size and shape, surface charge, and the nature of the environment are probably the most important ones [3].

In this article we describe a novel procedure for the synthesis of anisotropic silver nanoparticles, based on the use of N,N-dimethylformamide (DMF) as a solvent and as a reducing agent, in the presence of the polymer poly(vinylpyrrolidone) (PVP), which was recently reported in a preliminary account.[4] The formation of spherical silver nanoparticles in DMF has been previously demonstrated [5], and the presence of different additives was shown to induce the formation of core-shell structures, such as Ag@SiO_2 [6] or Ag@TiO_2 [7]. A detailed study was also recently carried out on the use of PVP to stabilize silver (and gold) nanoparticles formed in DMF [8]. However, in that study only low concentrations of both silver salt and polymer were employed, and therefore a spherical shape was consistently observed for the obtained nanoparticles. We however found that a noticeable increase of those concentrations leads to a dramatic change in the shape of the particles, and in turn to the optical properties of the dispersions. The interest on anisotropic metal nanoparticles is based on the different resonance possibilities that arise for the conduction electrons. It has been found for instance that metal nanorods display two distinct dipole resonances due to transversal and longitudinal oscillation [9-11]. A similar behaviour has been recently found for gold triangles [12], while for silver triangles, the transversal dipole is not very strong, but quadrupole resonances are easily observed [13].

The article is structured as follows: we first describe the experimental procedure for the formation of silver in DMF. Then, the different morphologies obtained for different concentrations will be discussed, as well as a possible mechanism for the formation of

L.M. Liz-Marzán and M. Giersig (eds.),
Low-Dimensional Systems: Theory, Preparation, and Some Applications, 65–75.

nanorods or nanoprisms from spherical particles. A study of the optical behaviour upon further addition of a stronger reducing agent will be presented, and finally we shall show how the assembly of these nanoparticles as thin films leads to anomalous spectral changes.

2. Synthesis of Anisotropic Ag in DMF

From previous studies on the formation of Ag nanoparticles in DMF, we know that this organic solvent is a powerful reducing agent against Ag^+ ions. The chemical reaction that we proposed is the following:

$$HCONMe_2 + 2Ag^+ + H_2O \rightarrow 2Ag^0 + Me_2NCOOH + 2H^+, \quad (1)$$

where a carbamic acid is formed and an acidic environment is generated. Such an acidic medium was important for the polymerisation of SiO_2 during the deposition of silica shells on the formed silver nanoparticles [6]. Although this reaction can proceed at room temperature, the reaction rate dramatically increases as the temperature is raised, so that optimal conditions for the preparation of uniform nanoparticles were found at the boiling point of DMF (156 °C) [6]. In the case of using PVP as a stabilizer, it was also found [8] that the order of addition of the silver salt and polymer solutions affected the morphology of the obtained nanoparticles. In that work it was concluded that mixing of silver salt and PVP prior to addition onto boiling DMF yields more stable and monodisperse colloids. A typical TEM micrograph of PVP-stabilized Ag nanoparticles obtained by boiling in DMF is shown if Figure 1. Although a certain degree of polydispersity is observed, lower metal concentration or the use of microwave irradiation instead of boiling can lead to smaller and more uniform nanoparticles.

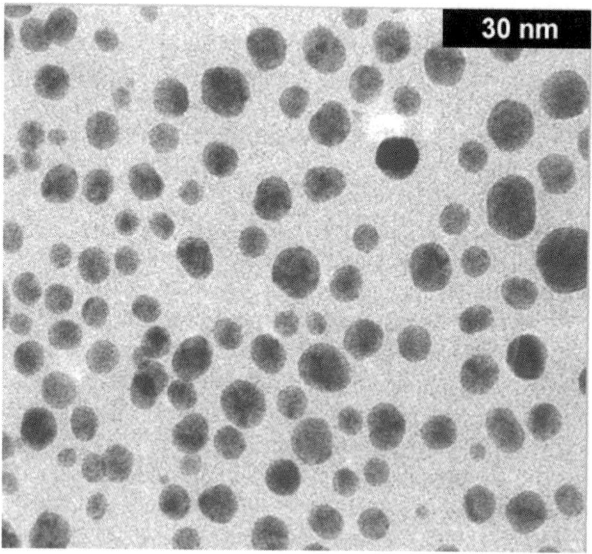

Figure 1. Typical TEM micrograph of Ag nanospheres obtained by reduction in DMF in the presence of PVP (0.76 mM), at low Ag^+ concentration (0.76 mM).

On the other hand, Xia and co-workers have recently reported [14] the synthesis of silver nanowires through the so-called polyol process [15], which is based on the use of ethylene glycol as a solvent and reductant at high temperature, in a similar process to the one described above using DMF. For the synthesis of nanowires, $PtCl_2$ was added prior to the silver salt and polymer, to form platinum nanoparticles that would then serve as seeds for the growth of the silver nanowires.

We adapted some of the experimental conditions used by Xia et al., soon realizing that the choice of a high silver concentration and high temperature were essential for the formation of anisotropic particles. Although nanowires were also formed under certain conditions (see below), they were never found to constitute a very high proportion within the sample. Oppositely, experimental conditions can be chosen so that a large proportion of (mainly) triangular, and in general polygonal nanoprisms were formed in solution. Initially, seed Pt nanoparticles were also used, though later they were substituted by Ag seeds, formed by reduction of $AgNO_3$ by PVP before addition to the boiling DMF. When the solution is added before the formation of Ag nuclei, only irregular and polydisperse particles are formed.

3. Formation of Rods and Polygons

As mentioned above, the shape (and size) of the obtained nanoparticles depends on several parameters, such as silver salt and PVP concentrations, temperature and reaction time. We have found that a silver concentration of 0.02 M was suitable for the formation of anisotropic particles, but again the concentration of PVP and the reaction temperature strongly influence the shape of the final particles. Two examples are shown below for the evolution of the particles as a function of boiling time in DMF.

The first example shows Ag nanorods, obtained for low PVP concentrations. Figure 2 shows that they are rather polydisperse, and coexist with nanoparticles of various shapes.

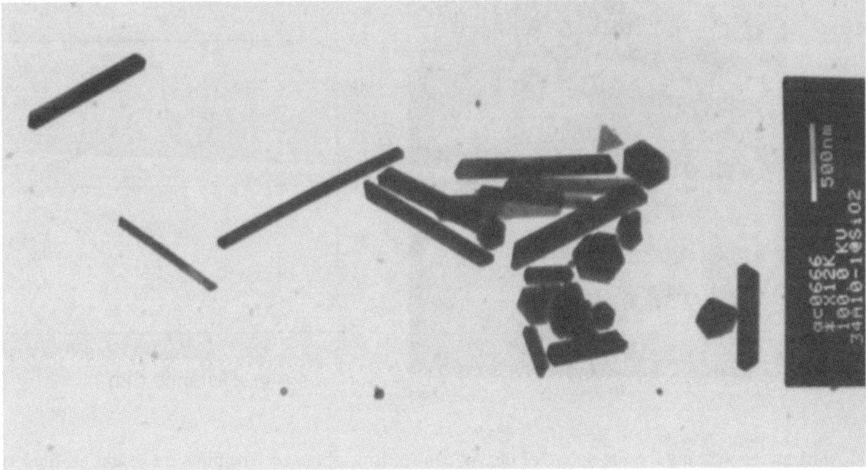

Figure 2. TEM micrograph showing Ag nanorods obtained by reduction in DMF in the presence of PVP (0.06 mM), at high Ag^+ concentration (0.02 M).

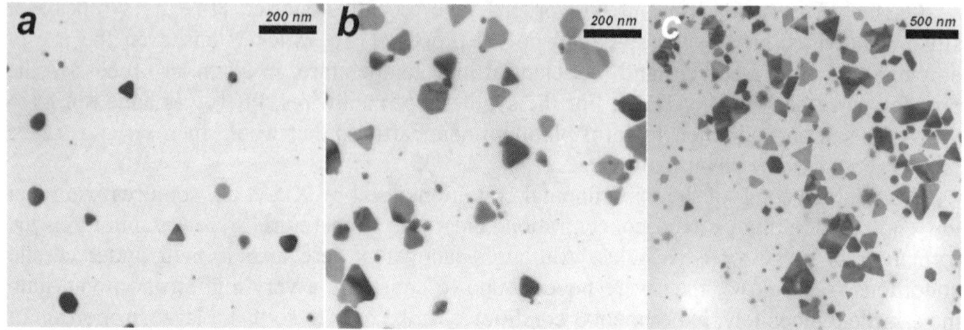

Figure 3. TEM micrographs showing the growth of Ag nanosprisms obtained by reduction in DMF in the presence of PVP (0.4 mM), at high Ag^+ concentration (0.02 M).

When higher PVP concentrations are used (0.4 mM), the formation of nanorods is suppressed, and mainly triangles and other polygonal shapes are observed. TEM observation indicates that initially, small spheres are formed which then assemble in certain shapes, and a melting-like process takes place, which leads to crystalline particles with well-defined shapes [16]. Additionally, in the case of nanoprisms, it was also observed that they become larger with time, and a wider variety of shapes are found for longer boiling times, as shown in Figure 3.

Atomic force microscopy (AFM) measurements demonstrated the flat geometry of the nanoprisms. An example is shown in Figure 4, indicating an average thickness of ca. 35 nm for lateral dimensions of the order of 200 nm. It can also be observed in this image that most of the nanoprisms are truncated triangles, which sometimes leads to other polygonal shapes.

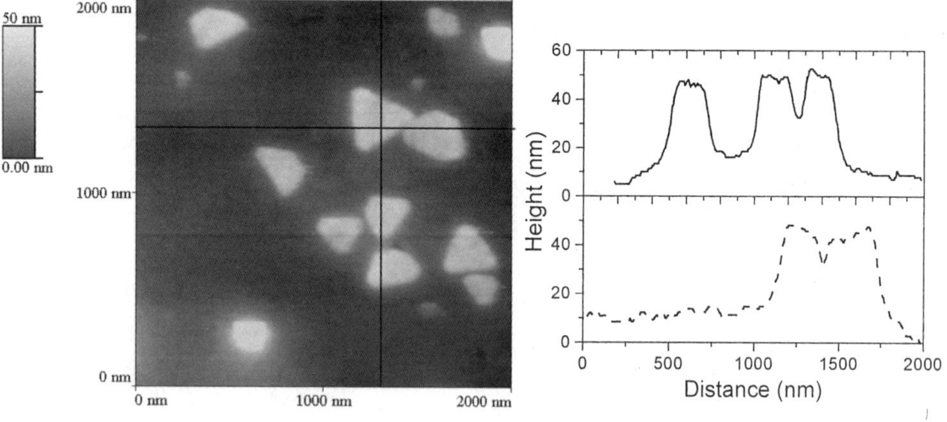

Figure 4. AFM image showing the flat nature of the Ag nanosprisms obtained. The plots are height profiles at the lines shown (solid: vertical; dashed: horizontal).

4. Optical Properties

4.1. OPTICAL SPECTRA IN SOLUTION

The study of the optical properties of these colloids is very interesting, since very few reports have been published on the formation of metal nanoprisms [12,13], which allow the experimental confirmation of theoretical calculations [13,17]. Such calculations are based on the discrete dipole approximation, so that the particle is divided into small elements which interact with each other through dipole-dipole interactions, and subsequent global evaluation of absorption and scattering. Using the discrete dipole approximation, Jin et al. [13] calculated the UV-visible spectrum of perfectly triangular nanoprisms (thickness 16 nm, lateral dimension 100nm) and of triangular nanoprisms truncated at each tip. Their results indicate that for the perfect triangles one main peak should be observed at high wavelengths (770 nm), corresponding to the in-plane dipole resonance, and two peaks with much lower intensity at lower wavelengths corresponding to the in-plane (470 nm) and out-of-plane (340) quadrupole resonances, while the out-of-plane dipole resonance is only observed as a shoulder at 410 nm. In the case of the truncated triangles, the main difference is a noticeable blue-shift of the in-plane dipole resonance down to 670 nm.

Figure 5 shows the experimental UV-visible spectra of the metal colloid during the formation of the nanoprisms. Initially, only one band is present, centred at 410 nm, corresponding to the dipole resonance of silver nanospheres. As the reaction proceeds, this band red-shifts, and several others gradually show up. After 20 min of boiling, four peaks are observed, which are similar to those predicted by Jin et al., except for an intense peak at 460 nm, which could be easily assigned to the presence of spherical particles within the final colloid, as observed by TEM.

Since the spheres present in the dispersion are in general noticeably smaller than the nanoprisms, they can be easily separated by centrifugation, with the advantage that the presence of PVP on the surface prevents aggregation during the process.

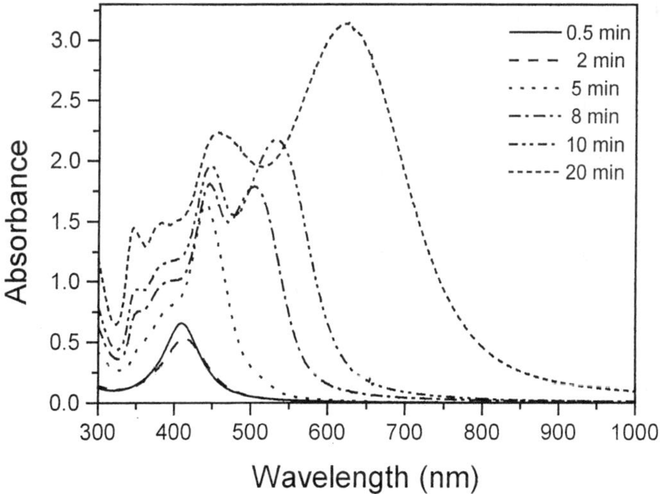

Figure 5. Time evolution of UV-visible spectra during the formation of Ag nanosprisms in DMF.

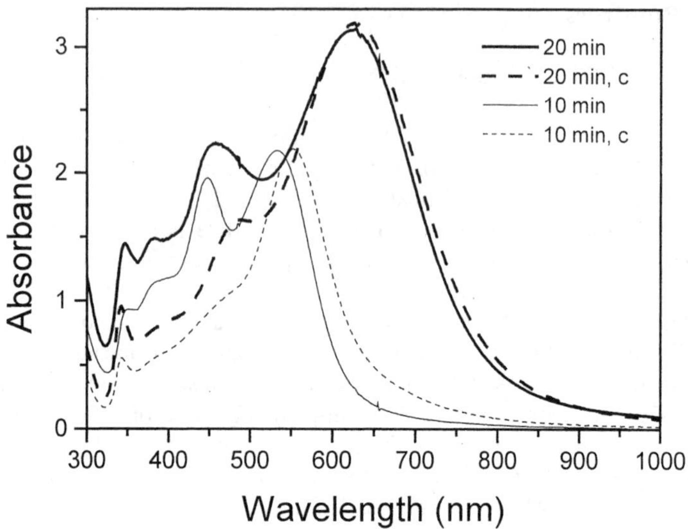

Figure 6. UV-visible spectra of Ag nanosprisms in DMF before (solid lines) and after (dashed lines) centrifugation and redispersion in ethanol. Boiling time is indicated.

Therefore, two samples prepared by boiling during 10 and 20 minutes, were carefully centrifuged (3000 rpm), the sediment was redispersed in ethanol and the spectrum of the resulting stable colloid was measured. The spectra before and after centrifugation/redispersion for both samples are shown in Figure 6. In both cases, the final spectrum basically coincides with the theoretical calculations, since the peak at 460 is strongly damped, as expected for a pure dispersion of nanoprisms.

It should also be mentioned that the in-plane resonance band is in general placed at lower wavelengths than expected, which can be explained on the basis of the deviations of the nanoparticles' shape with respect to perfect triangles (see Figure 3), especially at short boiling times.

4.2. REFRACTIVE INDEX EFFECTS

Once that the main features of the spectra of silver nanoprisms in solution are characterized, it is interesting to study how other parameters affect such spectra. One of the main factors influencing the plasmon resonance is the refractive index of the surrounding medium [3]. An increase in solvent refractive index leads to a red-shift of the plasmon band, as predicted by Mie theory. We have studied this effect on colloids of PVP-protected silver spheres [8], finding that when the solvent is changed from DMF (n = 1.426) to water (n = 1.333), the plasmon band is blue-shifted by just 5 nm (see inset of Figure 7), which agrees with theoretical calculations (see Figure 7 of ref. [8]). When a similar experiment is performed on PVP-protected Ag nanoprisms, a much stronger effect is observed, mainly on the in-plane dipole resonance, as shown in Figure 7. In the figure, spectra of two different nanoprism samples of different lateral dimensions (ca. 100 and 200 nm) have been plotted, showing the increasingly larger shift in the in-plane dipole plasmon resonance wavelength as the size of the nanoprisms is increased.

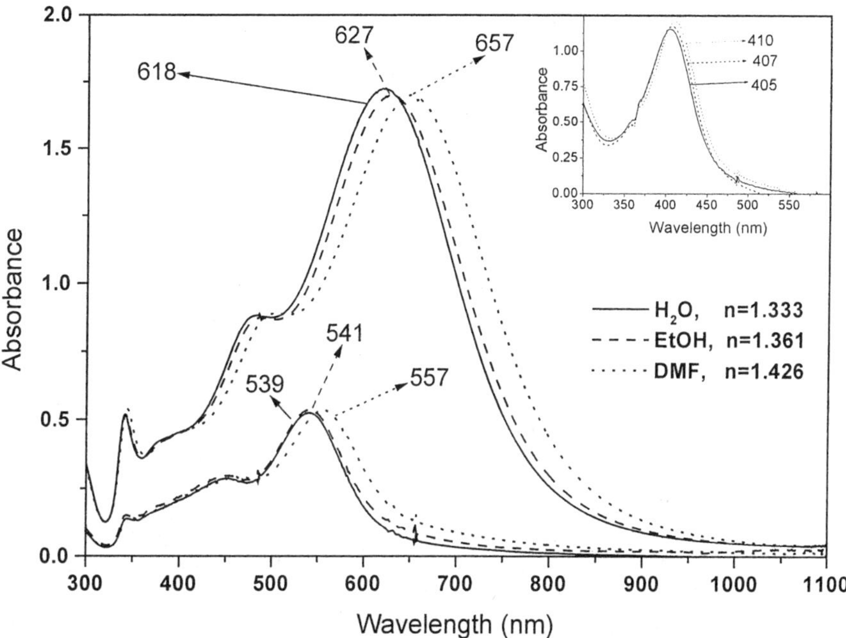

Figure 7. UV-visible spectra of Ag nanosprisms in DMF (solid lines), ethanol (dashed lines) and water (dotted lines). The upper curves correspond to average lateral dimensions of 200 nm, while the lower curves correspond to average lateral dimensions of ca. 100 nm. The inset shows the spectra for dispersions of Ag nanospheres.

For the larger particles, the red-shift associated to solvent exchange from water to DMF is as large as 39 nm, while for the smaller nanoprisms, the shift is of 18 nm. Interestingly, the peak at 340 nm, associated to the quadrupole out-of-plane resonance, remains almost constant (2 nm shift) during solvent exchange, which reflects the much higher sensitivity of the in-plane plasmon oscillation toward the nature of the surrounding medium. Such a sensitivity, which is expected to be of a similar magnitude toward adsorption of molecules to the nanoprism surface, can be exploited for the use of these systems as sensors.

4.3. EFFECTS OF CHEMICAL CHARGING

A second factor of interest that influences the plasmon resonance is the concentration of conduction electrons [2,3]. The effects of charging and discharging (spherical) silver nanoparticles in solution have been studied by electrochemical methods on polymer stabilized colloids [18], and by chemical charging-discharging on both polymer stabilized and silica-coated silver nanoparticles [19]. In these studies, it was observed that an increase in electron concentration promotes a blue-shift of the plasmon wavelength, while a red-shift is observed upon discharging. Because of its simplicity, we have chosen the chemical procedure, based on the monitoring of spectral changes upon addition of a strong reducing agent, such as $NaBH_4$.

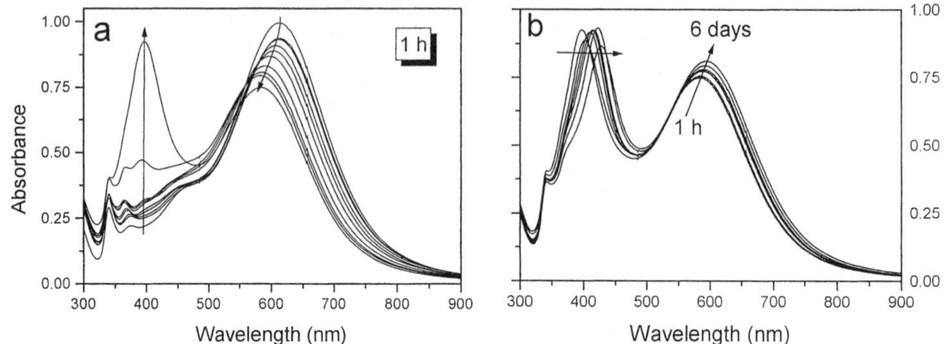

Figure 8. UV-visible spectra of Ag nanosprisms in water upon addition of $NaBH_4$. (a) Charging period; (b) Discharging period with associated red-shift.

Addition of borohydride, with a large negative redox potential for BH_4^- oxidation (E° = −1.2 V), leads to cathodic polarisation of the particles through the reaction

$$BH_4^- + Ag_n + 3H_2O \rightarrow BO_3^{3-} + Ag_n^{8-} + 10H^+ \qquad (2)$$

It was shown in ref. [19] that for 10 nm Ag nanoparticles electron injection is very fast (just a couple of minutes), while the discharging process is much slower (of the order of days). However, the results obtained for the addition of borohydride onto a dispersion of Ag nanoprisms are rather different, as can be observed in Figure 6. During the first hour after electron injection (Figure 8a) the in-plane dipole resonance band blue-shifts (as expected) but also drops in intensity, whilst the intensity of all other bands increases. Toward the end of this period, a new band quickly grows, which is centred at 398 nm. This is a very unexpected result, and therefore seems difficult to explain. Until now, the only plausible explanation for this (supported by TEM) is the formation of a large number of small silver nanoparticles, from the breakup of nanoprisms due to a large increase in surface tension upon charging.

Subsequently Figure 8b), both the new band and the in-plane dipole band slowly red-shift, during what can be termed the discharging process [19], which should affect both the nanoprisms and the newly formed, small spheres.

4.4. LAYER-BY-LAYER ASSEMBLED FILMS

The study of metal nanoparticles is not only interesting from the point of view of their single-particle behaviour, but also because of the complex collective properties when they are assembled close to each other. A number of studies have been made on assemblies of spherical nanoparticles [20-22], but very few on assemblies of anisotropic nanoparticles [12]. In this work, we made use of the layer-by-layer assembly (LBL) technique [23,24], based on sequential dipping of a substrate (a glass slide in our case) in an aqueous solution of a polyelectrolyte (here, poly-diallyldimethylammonium chloride, PDDA) and in an aqueous dispersion of the (negatively charged) nanoparticles, so that polyelectrolyte and nanoparticle monolayers are deposited during each dipping step, resulting in an alternate, sandwich-like structure.

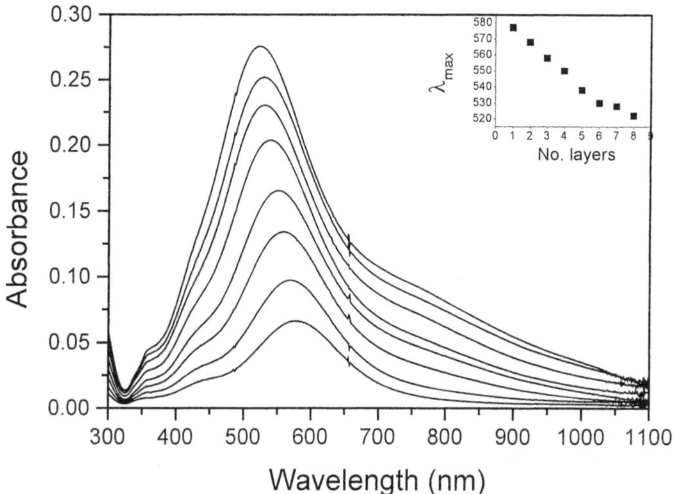

Figure 9. UV-visible spectra of layer-by-layer assembled films of Ag nanosprisms for the eight first monolayers. The inset shows the maximum position of the in-plane dipole as a function of layer number.

The LBL technique has been previously used for the assembly of gold nanospheres [20,25] and nanoprisms [12], and in both cases it was shown that the assembly of multilayers lead to a gradual red-shift of the plasmon band(s), provided that the particles/layers are close enough to each other. Such a red-shift is due to dipole-dipole interactions between neighbouring nanoparticles. The optical properties of multilayer films prepared through LBL assembly of the silver nanoprisms described above are shown in Figure 9, where the inset shows the position of the in plane dipole plasmon band vs. silver monolayer number.

It is clear that in this case, opposite to what was found for gold, the plasmon band gradually blue-shifts, with a change from 578 down to 521 after 8 dipping cycles! One could think that the reason for this could be an opposite behaviour for silver as compared to gold. However, similar experiments performed with 10 nm silver nanospheres yielded the expected trend of gradual red-shift when increasing the number of monolayers. Thus, a different sort of interactions seems to take place for this system, which at this moment we do not understand, and which need careful modelling.

5. Conclusions

We have demonstrated that silver nanorods and nanoprisms can be synthesized by thermal reduction of $AgNO_3$ in DMF, using PVP as a stabilizer. The spectra of the obtained nanoprisms agree with predictions from Mie theory for truncated triangles, and show large red-shifts upon transfer to a solvent with a higher refractive index. Thin films have been deposited through the layer-by-layer assembly method, and the optical spectra measured within the films behave differently to similar films deposited from spherical nanoparticles.

6. Acknowledgements

The authors are grateful to Ana Sánchez for her assistance in the experimental work. Financial support from the Spanish Xunta de Galicia (Project no. PGIDT01PXI30106PR) and Ministerio de Ciencia y Tecnología (Project no. BQU2001-3799) are acknowledged. I.P.-S. acknowledges the Spanish Ministry of Education and Culture for granting a FPI fellowship.

7. References

1. Faraday, M. (1857) Experimental relations of gold (and other metals) to light, *Philos. Trans. Roy. Soc. Lon.* **147**, 145-181.
2. Kreibig, U. and Vollmer, M. (1995) *Optical Properties Of Metal Clusters*, Springer Verlag, Berlin.
3. Mulvaney, P. (1996) Surface Plasmon spectroscopy of nanosized metal particles, *Langmuir* **12**, 788-800.
4. Pastoriza-Santos, I. and Liz-Marzán, L. M. (2002) Synthesis of Silver Nanoprisms in DMF, *Nano Lett.*, in press.
5. Pastoriza-Santos, I. and Liz-Marzán, L. M. (2000) Reduction of silver nanoparticles in DMF. Formation of monolayers and stable colloids, *Pure Appl. Chem.* **72**, 83-90.
6. Pastoriza-Santos, I. and Liz-Marzán, L. M. (1999) Formation and stabilization of silver nanoparticles through reduction by N,N-dimethylformamide, *Langmuir* **15**, 948-951.
7. Pastoriza-Santos, I., Koktysh, D., Mamedov, A. A., Giersig, M., Kotov, N. A., and Liz-Marzán, L. M. (2000) One-pot synthesis of Ag@TiO_2 core-shell nanoparticles and their layer-by-layer assembly, *Langmuir* **16**, 2731-2735.
8. Pastoriza-Santos, I. and Liz-Marzán, L. M. (2002) Preparation of PVP-protected metal nanoparticles in DMF, *Langmuir* **18**, 2888-2894.
9. van der Zande, B. M. I., Böhmer, M. R., Fokkink, L. G., and Schöneberger, C. (2000) Colloidal dispersions of gold rods : synthesis and optical properties, *Langmuir* **2000**, *16*, 451-458.
10. Link, S., Mohamed, M. B., and El-Sayed, M. A. (1999) Simulation of the optical absorption spectra of gold nanorods as a function of their aspect ratio and the effect of the medium dielectric constant, *J. Phys. Chem. B* **103**, 3073-3077.
11. Chang, S. S., Shih, C. W., Chen, C. D., Lai, W. C., and Wang, C. R. C. (1999) The shape transition of gold nanorods, *Langmuir* **15**, 701-709.
12. Malikova, N., Pastoriza-Santos, I., Schierhorn, M., Kotov, N. A., and Liz-Marzán, L. M. (2002) Layer-by-layer assembled mixed spherical and planar gold nanoparticles : control of interparticle interactions, *Langmuir* **18**, 3694-3697.
13. Jin, R.C., Cao, Y. W., Mirkin, C. A., Kelly, K. L., Schatz, G. C., and Zheng, J. G. (2001) Photoinduced conversion of silver nanospheres to nanoprisms, *Science* **294**, 1901-1903.
14 Sun, Y., Gates, B., Mayers, B., and Xia, Y. (2002) Crystalline silver nanowires by soft solution processing, *Nano Lett.* **2**, 165-168.
15. Fievet, F., Lagier, J. P., Figlarz, M. (1989) Preparing monodisperse metal powders in micrometer and submicrometer sizes by the polyol process, *MRS Bull*, December, 29-34.
16. Pastoriza-Santos, I., Giersig, M., and Liz-Marzán, L. M., unpublished results.
17. Schatz, G. C. (2001) Electrodynamics of nonspherical noble metal nanoparticles and nanoparticles aggregates, *J. Molec. Struct. (Theochem)* **573**, 73-80.
18. Ung, T., Giersig, M., Dunstan, D., Mulvaney, P. (1997) Spectroelectrochemistry of colloidal silver, *Langmuir* **13**, 1773-1782.
19. Ung, T., Liz-Marzán, L. M., Mulvaney, P. (1999) Redox catalysis using Ag@SiO_2 colloids, *J. Phys. Chem. B* **103**, 6770-6773.
20. Ung, T., Liz-Marzán, L. M., Mulvaney, P. (2001) Optical properties of thin films of Au@SiO_2 particles, *J. Phys. Chem. B* **105**, 3441-3452.
21. Henrichs, S., Collier, C. P., Saykally, R. J., Shen, Y. R., Heath, J. R. (2000) The dielectric function of silver nanoparticles Langmuir monolayers compressed through the metal insulator transition, J. Am. Chem. Soc., 122, 4077-4083.
22. Chumanov, G., Sokolov, K., Cotton, T. M. (1996) Unusual extinction spectra of nanometer-sized silver particles arranged in two-dimensional arrays, *J. Phys. Chem.* **100**, 5166-5168.

23. Kotov, N. A., Dekàny, I., and Fendler, J. H. (1995) Layer-by-layer self-assembly of polyelectrolyte-semiconductor nanoparticle composite films, *J. Phys. Chem.* **99**, 13065-13069.
24. Decher, G. (1997) Fuzzy nanoassemblies: toward layered polymeric multicomposites, *Science* **277**, 1232-1237.
25. Ung, T., Liz-Marzán, L. M., Mulvaney, P. (2002) Gold nanoparticle thin films, *Colloid Surf. A* **202**, 119-126.

MECHANICAL PROPERTIES OF SMALL METAL SPHERES AND RODS

PAUL MULVANEY,[1] JOHN SADER,[2] GREGORY V. HARTLAND,[3] AND MICHAEL GIERSIG[4]
[1] *School of Chemistry, The University of Melbourne, Victoria 3010, Australia*
[2] *Department of Mathematics and Statistics, The University of Melbourne, Victoria 3010, Australia*
[3] *Department of Chemistry and Biochemistry, University of Notre Dame, 251 Nieuwland Science Hall, Notre Dame, IN 46556-5670, USA*
[4] *Hahn-Meitner Institut, Abt. Photochemie, Glienickerstr.100, Berlin 15109, Germany*

1. Introduction

Over the last two decades, considerable interest has focussed on the synthesis of quantum dots and small metal crystallites. In particular, the optical, magnetic and electrical properties of small particles have been intensively investigated. In addition, an understanding of fundamental material properties such as melting point, Young's modulus and shear modulus will be essential if these particles are to be employed in nanoscale devices, motors or electronics. A vexed issue is establishing how one can probe mechanical properties such as rigidity or density, which appear to require manual contact with the materials. Recent work by several research groups has established that for metals such as gold and silver with pronounced surface plasmon resonances, laser excitation can cause rapid heating of the particle lattice [1,2]. The heating excites phonon modes, which can be monitored spectroscopically in the femto- to nanosecond time regime. In this report we summarise some of our recent results on relaxation in laser excited nanoparticles. A fuller account is given in the parent articles and references therein.

2. Theory of acoustic breathing modes

The zeroth order acoustic mode or breathing motion of a sphere was analysed quantitatively by Lamb [3]. He found that the breathing mode frequencies ω (radians/sec) for a homogeneous, elastic sphere with a free surface obeys

$$\omega = \frac{\tau\alpha}{a} \tag{1}$$

where a is the radius of the particle, and τ is the smallest positive root of the eigenvalue equation $\tau \cot \tau = 1 - \tau^2/4\kappa^2$ where $\kappa = \beta/\alpha$ is the ratio of the transverse β, and longitudinal α, speeds of sound in the material.

L.M. Liz-Marzán and M. Giersig (eds.), Low-Dimensional Systems: Theory, Preparation, and Some Applications, 77–86.

The acoustic breathing mode of a core-shell spherical particle, with core radius a_{core}, and shell thickness Δ is more complex [4]. Assuming that the core and shell can be modelled as continua, i.e., the molecular structure of the core and shell can be ignored, and are both composed of homogeneous, isotropic, linearly elastic materials, then enables an extension of Lamb's solution [3] to the case of a core-shell nanoparticle [5]. Unfortunately, the resulting expression is highly complex and therefore the complete expression is not presented here. Nonetheless, explicit analytical formulae can be obtained by performing an asymptotic expansion of the eigenvalue equation in cases where the shell thickness Δ is small or large in comparison to the core radius a_{core}. This results in the following explicit formulae for the natural resonant frequency ω:

$$\omega = \frac{\alpha_{core}\tau_{core}}{a_{core}}\left\{1 - A_{core}\left(\frac{\Delta}{a_{core}}\right) + O\left(\left(\frac{\Delta}{a_{core}}\right)^2\right)\right\} \quad , \quad \frac{\Delta}{a_{core}} << 1 \tag{2a}$$

$$\omega = \frac{\alpha_{shell}\tau_{shell}}{a_{core}+\Delta}\left\{1 + A_{shell}\left(\frac{a_{core}}{a_{core}+\Delta}\right)^3 + O\left(\left(\frac{a_{core}}{a_{core}+\Delta}\right)^4\right)\right\} \quad , \quad \frac{\Delta}{a_{core}} >> 1 \tag{2b}$$

where

$$A_{core} = \frac{16\kappa_{shell}^4 - 12\kappa_{shell}^2 + \left(\frac{\alpha_{core}}{\alpha_{shell}}\right)^2 \tau_{core}^2}{16\kappa_{core}^4 - 12\kappa_{core}^2 + \tau_{core}^2}\lambda \tag{3a}$$

$$A_{shell} = \frac{16(1+\tau_{shell}^2)\kappa_{shell}^4 - 8\kappa_{shell}^2\tau_{shell}^2 + \tau_{shell}^4}{3\left(16\kappa_{shell}^4 - 12\kappa_{shell}^2 + \tau_{shell}^2\right)} \cdot \frac{3 - 4\kappa_{core}^2 - \lambda\left(3 - 4\kappa_{shell}^2\right)}{3 - 4\kappa_{core}^2 + 4\lambda\kappa_{shell}^2} \tag{3b}$$

$$h_{core} = \frac{\omega}{\alpha_{core}} \quad , \quad h_{shell} = \frac{\omega}{\alpha_{shell}} \tag{3c}$$

$$\kappa = \frac{\beta}{\alpha} \tag{3d}$$

$$\lambda = \frac{\rho_{shell}\alpha_{shell}^2}{\rho_{core}\alpha_{core}^2} \tag{3e}$$

and τ satisfies the homogeneous eigenvalue equation:

$$\tau\cot\tau = 1 - \frac{\tau^2}{4\kappa^2} \tag{4}$$

where the subscripts "core" and "shell" henceforth refer to values in the core and shell, and ρ is the density of the material.

3. Results and discussion

We first briefly discuss the origin of the transient, oscillatory absorption curves observed following rapid laser excitation of the metal particles. The linear absorption coefficient of the metal particles in the visible regime is given accurately by the dipole formula

$$\alpha = \frac{18\pi n^3 \phi \varepsilon''(\omega)}{(\varepsilon'(\omega)+2n^2)^2 + \varepsilon''(\omega)^2)} \tag{5}$$

Here n is the refractive index of the solvent and $\varepsilon(\omega) = \varepsilon'(\omega) + i\,\varepsilon''(\omega)$ is the dielectric function of the metal. A peak occurs in the absorption spectrum when

$$\varepsilon'(\omega) = -2n^2 \tag{6}$$

For gold and silver, $\varepsilon'(\omega)$ obeys

$$\varepsilon'(\omega) = \varepsilon^{\infty} - \frac{\omega_p^2}{\omega^2} \tag{7}$$

Here, the plasma frequency, ω_p, is defined by

$$\omega_p^2 - Ne^2/m\varepsilon_o \tag{8}$$

where N is the electron density of the metal, m the electron mass, e the electron charge, ε_o the vacuum permittivity and ε_∞ is the high frequency contribution to the dielectric function. Combining Eqs. 6, 7 and 8, it is clear that the peak position depends on the electron density within the metal [6]. Heating causes a dilation of the lattice; this lowers the free electron density and results in a red-shift of the surface plasmon absorption band. Because of the high Q of the plasmon band, even a shift of a nanometre or two due to a dilation of less than 0.1% can result in pronounced bleaching or absorption increases. Interestingly then, changes in temperature of the metal nanocrystals are readily resolved by absorption spectroscopy. Furthermore, by monitoring the dynamics of the response to laser excitation, one gleans information on the conversion of electron energy into phonon modes, the mechanical oscillations of the particles, and the ultimate transfer of heat from the particles to the solvent bath. Hartland and colleagues have demonstrated that Eq. 1 accurately describes the oscillations observed during excitation of gold spheres of various sizes, and the interested reader is referred to that work [2,7].

Next, we move on to consider core-shell spheres. A shell layer around the metal core particle may serve several functions in a colloid system. It may be used to modify the surface plasmon oscillations of the particle, or the layer may provide colloid stability for the core, or it may isolate the particle from chemical perturbations. A shell will also perturb heat flow and mechanical oscillations of the core. As one may intuitively expect, the presence of a shell lowers the frequency of vibration. The amount depends on the relative thickness of the core and shell and composition, as may be expected [4,5]. A comparison of the exact solution with the analytical approximations is given for two types of core-shell particles in Figure 1. The first of these examples is for a silica shell over a gold core [8]. The silica shell is used to control particle packing in gold nanoparticle crystals, enhances the colloid stability of the gold sols facilitating solvent transfer, and allows high volume fraction sols in water to be prepared. The second case is for a lead coated gold particle. Lead is precipitated by underpotential deposition onto gold colloid surfaces under weakly reducing conditions [9]. The addition of a free electron metal with a higher plasma frequency than gold causes a blue shift of the plasmon band. Interestingly in this case, the metals have quite different Young's moduli, and the calculated mechanical breathing frequency is different to that for a gold particle of equal size.

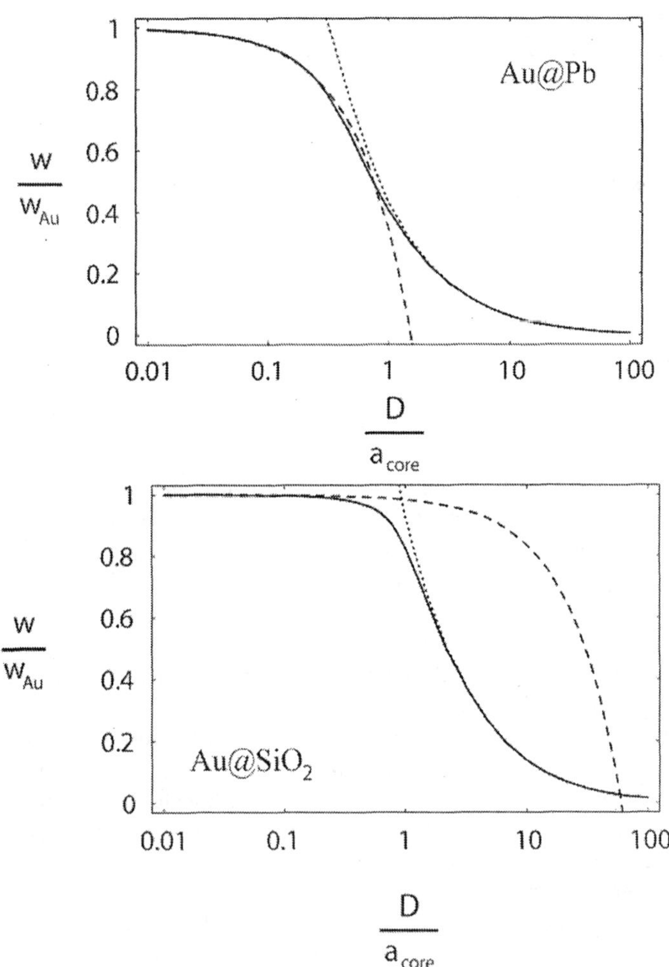

Figure 1. Ratio of breathing mode frequency ω for a core-shell sphere to that of pure Au sphere ω_{Au}, with identical core radius, as a function of normalized shell thickness, Δ. Exact solution (solid line); Asymptotic formulae given by dashed and dotted lines. Top: Au@Pb sphere; Bottom: Au@SiO_2 sphere [4].

In Figure 1, the effect of these two shells on the eigenfrequency of the core particle are shown using the asymptotic formulae, Eqs. 2a and 2b. By varying the material and geometric properties of the particles, in general it was found that the asymptotic formula Eq. 2a is accurate to within 5% if $\Delta / a_{core} < 0.3$, whereas Eq. 2b is accurate to within 5% if $\Delta / a_{core} > 1.5$. In Figure 2, we show the experimental data obtained for a series of Au @ Pb particles. The dashed line assumes complete deposition of lead ions onto the gold particles, while the full line allows for free lead nucleation, as predicted from the known rate constants for this system. There is remarkably good agreement, which extends down to a 1 nm lead layer. These results indicate that the vibrations of metallic core-shell nanospheres are accurately predicted by continuum models, even for very thin shell layers. Furthermore, the excellent agreement implies that (i) pulsed laser excitation of Au@Pb nanoparticles launches the breathing mode in the core-shell particle, and (ii) the Au core and Pb metal shell behave like bulk materials in the excitation of the breathing mode, despite their small dimensions.

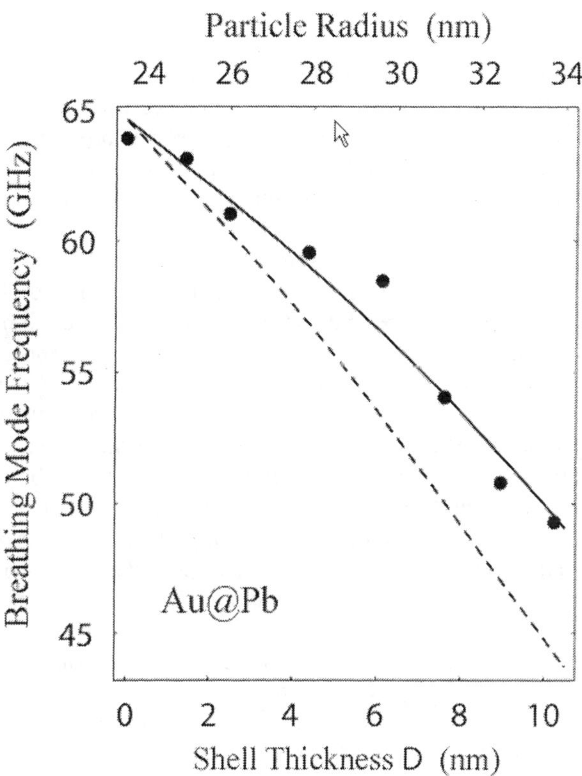

Figure 2. Comparison of experimental measurements (dots) to theoretical calculations (lines) of the breathing mode frequency of Au@Pb nanoparticles, for various Pb shell thicknesses. Au core radius is 23.5 nm. Calculations including Pb shell thickness correction (solid line), no thickness correction (dashed line) [4].

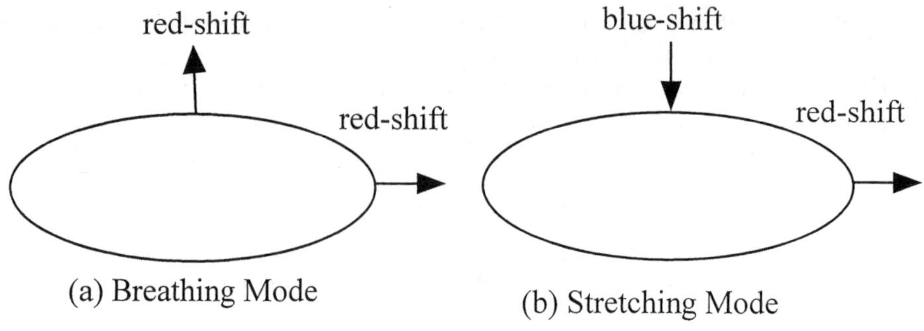

Figure 3: Cartoon illustrating the dependence of the absorption changes of the transverse and longitudinal mode on the eigenmode excited by the laser. For the breathing mode, the spectral shifts of each plasmon mode are in phase, whereas in 3(b), they are out of phase.

For non-spherical particles the situation is much more complex than for spheres and core-shell particles, where the symmetry of the particles reduces the number of likely modes which will be excited. For rods, the number and types of modes that can be excited is much more varied. It is not clear which vibrational modes will be excited, nor how the frequencies of these modes depend on the size, shape and elastic properties of the material.

For example, one must distinguish here between the breathing mode in which the particle undergoes simultaneous lattice expansion longitudinally and transversely and a normal longitudinal stretching mode. In this mode, stretching of the rod would be accompanied by contraction laterally, as determined by the Poisson ratio of the metal. In principle, the differences in these two modes shown in the cartoon of Fig. 3, could be distinguished from the optical response of the transverse and longitudinal modes. Breathing modes should result in in-phase red and blue shifts of both plasmon peaks, while the stretching mode would lead to out-of-phase optical changes in the two plasmon modes. Unfortunately at present, it is experimentally difficult to monitor the optical changes to the transverse mode, because they are quite weak. Changes in the transverse plasmon band are also more seriously affected by sample polydispersity, as will be shown below.

However, the length dependence of gold nanorod vibration frequencies has been probed. Figure 4 shows electron micrographs of typical gold nanorods produced using the protocol of Wang and colleagues [10]. The particle size depends strongly on reactor conditions. The particles are generally found to be single crystals as seen at higher magnifications.

Figure 5 presents transient absorption data for a gold nanorod sample with an average aspect ratio of 3.5. These experiments were performed with probe laser wavelengths of 650 nm and 780 nm. These two wavelengths lie on either side of the longitudinal plasmon band of this sample, which occurs at ~ 720 nm. There are several features to note about the transient absorption data. First, at early times (<20 ps) the signal at 650 nm is a bleach, whereas, the signal at 780 nm is an absorption. Second, both traces contain an initial decay, which has a time constant of several ps, followed by a slower modulated signal. The initial decay corresponds to cooling of the hot electrons created by the pump laser pulse and the modulations are due to the coherently excited vibrational modes of the rods. The relative magnitude of the modulations compared to the early signal from the hot electrons can be seen in the insert.

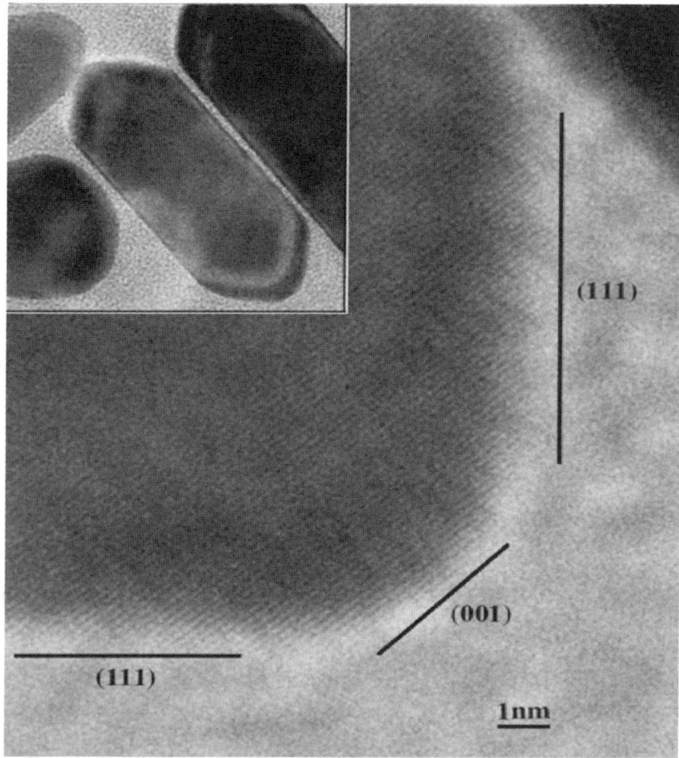

Figure 4: Electron micrograph of gold nanorods at lower and higher magnification.

The coherent vibrational motion is launched by the rapid heating of the electrons and lattice that follows ultrafast laser excitation. Both the period and the phase of the modulations are different for the 650 nm and 780 nm data. Fits to the experimental data using a damped cosine function yield a period and phase of 51 ps and 246° for the 650 nm data, compared to 65 ps and 184° for the 780 nm data. Note that a difference in the phase for transient absorption experiments performed on the red or blue sides of the plasmon band has been previously observed for spherical particles. However, for spheres the phase difference is 180°, i.e., the modulations are completely out-of-phase. Also note that the period of the modulations is much longer than that expected from Perner et al.'s results [11]. Specifically, from the length of the rods in our sample we would predict a period of $2L/c_l = 26$ ps if the observed modulations were due to expansion and contraction along the long axis of the rod. This is a factor of two faster than the experimental data in Figure 5.

In Figure 6 the period and phase are plotted against the probe laser wavelength for a series of transient absorption experiments with the same sample as in Figures 5. The period changes from 47 ps to 75 ps as the probe wavelength is shifted from the blue (600 nm) to the red (860 nm). For spherical particles the measured period does not change with the probe laser wavelength. This difference arises because the longitudinal plasmon resonance of the rods depends on their aspect ratio and, therefore, the length. Thus, different probe laser wavelengths interrogate different portions of the sample distribution. In contrast, for spherical particles the plasmon band does not shift with size (for particles < 50 nm diameter).

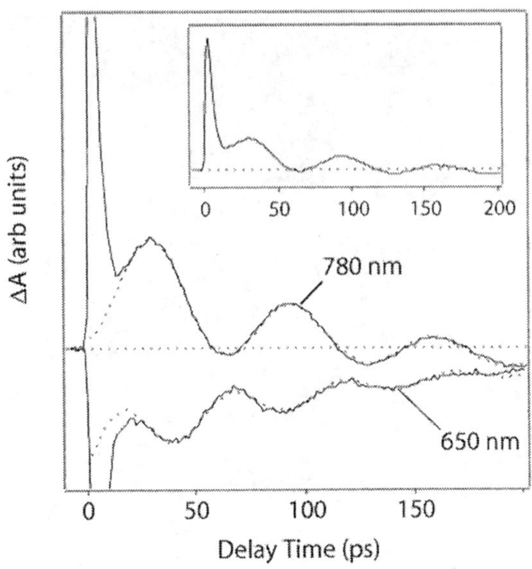

Figure 5: Transient absorption spectra for Gold nanorods with aspect ratio 3.5 following excitation at 400nm.

The relative contributions from different sized spheres in the sample does not change with the probe laser wavelength, and the main result of polydispersity is to damp the modulations. The phase of the modulations shows that the signal arises from a periodic red-shift in the position of the longitudinal plasmon band. This is consistent with either volume expansion (red-shift in the longitudinal plasmon band due to a decrease in electron density) or length expansion (red-shift due to an increase in the aspect-ratio).

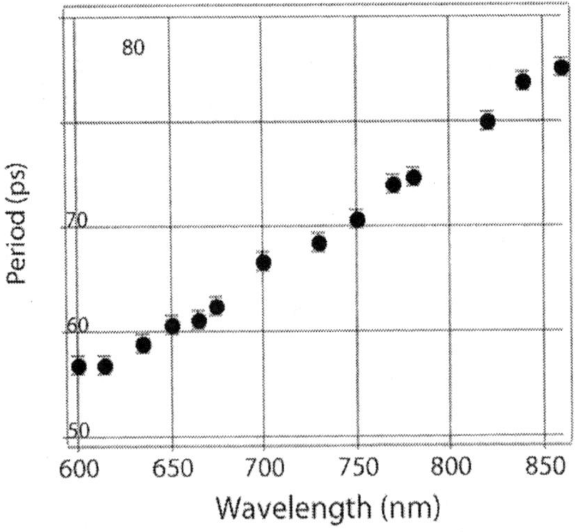

Figure 6. The period of the plasmon band oscillations versus probe laser wavelength following initial excitation at 400nm.

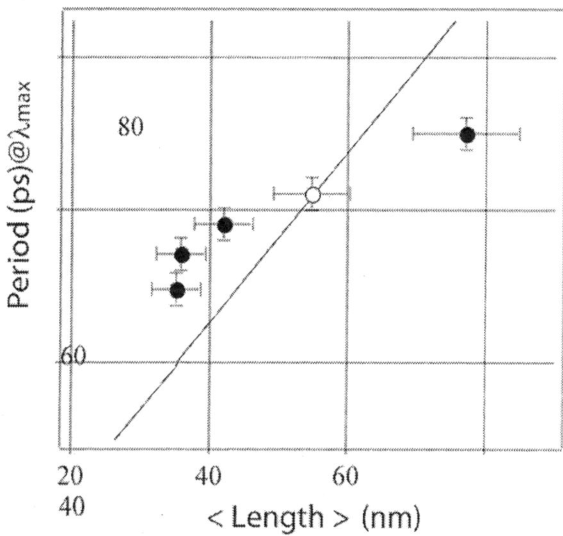

Figure 7: Relaxation time of the transient absorption signals for gold nanorods for several different aspect ratios.

The simulations of the signal in these experiments show that at probe laser wavelengths near the plasmon band maximum the effect of the sample polydispersity is minimized, i.e., the measured periods are close to the values expected from the average length of the rods. Note that this is true even if the vibrational period does not depend linearly on the rod length. Thus, to compare different samples, the probe laser should be tuned as close to the maximum of the longitudinal plasmon band as possible. Figure 7 shows results for different rod samples, where the period of the acoustic vibrational mode was determined from experiments with $\lambda_p \approx \lambda_{\max}$. The straight line is a fit to data assuming that the period is linearly proportional to the length of the rod, i.e., $T = 2L/\alpha$. The value of α obtained is $\alpha = 1800$ ms^{-1}, which is in between the longitudinal and transverse speeds of sound in Au: cl = 3240 ms^{-1} and ct = 1400 ms^{-1}, respectively.

Further experiments and detailed continuum mechanics calculations are currently being performed to elucidate the vibrational dynamics of these systems. In addition, we are developing a more sophisticated analysis of the transient absorption spectrum of the rods, to gain more insight into the homogeneous and inhomogeneous contributions to the longitudinal plasmon band.

4. Conclusions

We have attempted to provide a succinct account of our current experiments on heat and energy dissipation in nanoscale materials, and we have found that by and large, continuum models are able to successfully account for the spectroscopic changes that occur following picosecond laser excitation. In particular, the excitation of breathing modes in core-shell particles has been analysed successfully, and extends the previous analysis by Hartland of pure metal shells. In addition, we find that the vibrational modes of nanorods can be excited by laser excitation and the transient absorption spectra that results shows strong dependence on particle length. The signal decays rapidly due to polydispersity.

In summary, ultrafast laser excitation of Au nanorods coherently excites the low frequency acoustic vibrational modes. The measured vibrational periods are proportional to the length of the rods. Due to polydispersity, for a given sample the exact value of the period depends on the probe laser wavelength. Analysis of the period versus wavelength data provides information about the homogeneous linewidth of the longitudinal plasmon band. The results for a sample with an average length of 35 nm and a width of 10 nm show that the homogeneous linewidth is approximately two times larger than that calculated from the dielectric constant data for bulk Au.

5. Acknowledgements

The work described in this paper was supported by the National Science Foundation by Grant No. CHE98-16164 (GVH), and by the Particulate Fluids Processing Centre of the Australian Research Council and the Australian Research Council Grants Scheme (PM and JES). We thank NATO for sponsorship of the workshop and for assisting with manuscript publication costs.

6. References

1. Link, S., Mohamed, M. B., and El-Sayed, M. A. (1999) Simulation of the optical absorption spectra of gold nanorods as a function of their aspect ratio and the effect of the medium dielectric constant, *J.Phys. Chem.B* **103**, 3073-3077.
2. Hodak, J. H., Henglein, A., and Hartland, G. V. (2000) Photophysics of nanometer sized metal particles: electron-phonon coupling and coherent excitation of breathing vibrational modes, *J. Phys. Chem. B* **104**, 9954-9965.
3. Lamb, H. (1882) On the vibrations of an elastic sphere, *Proc. London Math. Soc.* **13**, 189-212.
4. Sader, J. E., Hartland, G. V., and Mulvaney, P. (2002) Vibrational modes of core-shell nanoparticles, *J.Phys.Chem.* **106**, 1399-1402.
5. Hodak, J. H., Henglein, A., and Hartland, G. V. (2000) Coherent excitation of acoustic phonon modes in bimetallic core-shell nanoparticles, *J. Phys. Chem. B* **104**, 5053-5055.
6. Mulvaney, P. (1996) Surface plasmon spectroscopy of nanosized metal particles, *Langmuir* **12**, 788-801.
7. Hodak, J. H., Henglein, A., and Hartland, G. V. (1999) Size dependent properties of Au particles: coherent excitation and dephasing of acoustic vibrational modes, *J. Chem. Phys.* **111**, 8613-8621.
8. Liz-Marzán, L. M., Giersig, M., and Mulvaney, P. (1996) Synthesis of nanosized gold-silica core-shell particles *Langmuir.* **12**, 4329-4336.
9. Mulvaney, P., Giersig, M., and Henglein, A. (1992) Surface chemistry of colloidal gold: deposition of lead atoms and accompanying optical effects, *J. Phys. Chem.* **96**, 10419-10424.
10. Yu, Y. Y., Chang, S. S., Lee, C. L., and Wang, C. R. C. (1997) Gold nanorods: electrochemical synthesis and optical properties, *J.Phys.Chem B* **101**, 6661-6662.
11. Perner, M., Gresillon, S., März, J., von Plessen, G., Feldmann, J., Porstendorfer, J., Berg, K.-J., and Berg, G. (2000) Observation of hot-electron pressure in the vibration dynamics of metal nanoparticles, *Phys.Rev. Lett.* **85**, 792-795.

POLYMER CONTROLLED CRYSTALLIZATION: SHAPE AND SIZE CONTROL OF ADVANCED INORGANIC NANOSTRUCTURED MATERIALS-1D, 2D NANOCRYSTALS AND MORE COMPLEX SUPERSTRUCTURES

SHU-HONG YU, HELMUT CÖLFEN
Max Planck Institute of Colloids and Interfaces,
Department of Colloid Chemistry,
MPI Research Campus Golm, D-14424, Potsdam, Germany

Abstract

Recently, exploration of mild environmentally friendly strategies for controlled fabrication of advanced inorganic materials with controlled shape, size, dimensionality, and structure has been heightened. Shape control and exploration of novel methods for self-assembling or surface-assembling molecules or colloids to generate materials with controlled morphologies and unique properties are among the hottest research subjects.

In this chapter, the latest advances in the polymer-controlled crystallization of various technically important inorganic crystals are summarized. The so-called double-hydrophilic block copolymers (DHBCs), which contain both a binding block and a solvating block for inorganic surfaces, are demonstrated to exert significant influence on the crystallization and morphology of various inorganic crystals such as $BaCrO_4$, $CaCO_3$, $BaCO_3$, $CdWO_4$, $BaSO_4$, ZnO, CdS under near natural conditions.

Cone-like bundles of $BaCrO_4$ nanofibers with diameter 10-20 nm and lengths up to 150 μm can be easily produced at room temperature in the presence of a phosphonated copolymer. A self-limited growth mechanism was proposed for the explanation of the high similarity of the $BaCrO_4$ nanofiber bundles. A controlled growth of $CaCO_3$, and $BaCO_3$ crystals with different sizes and surface strucutures was also addressed. In addition, a fine tuning of crystal morphology and crystal superstructures of 1D and 2D very thin $CdWO_4$ nanorods/nanobelts, and elongated nanosheets can be realized by a very simple aqueous route in the presence of block copolymers.

The results demonstrate that the integration of using DHBCs with taking advantages over the experimental conditions, such as the crystallization sites, temperature, pH value, reactant concentration, will provide very promising routes for controlling the shape, sizes, and microstructure of the hierachical inorganic crystals from nanoscale to macroscopic scale via a simple mineralization process. The materials with controllable shape, size, structure, and dimensionality are expected to find potential applications in the field of advanced materials.

L.M. Liz-Marzán and M. Giersig (eds.),
Low-Dimensional Systems: Theory, Preparation, and Some Applications, 87–105.

1. Introduction

Exploration of bioinspired morphosynthesis strategies, using self-assembled organic superstructures, organic additives, and/or templates with complex functionalization patterns, to template inorganic materials with controlled morphologies, has arisen a lot of attention [1-4]. Synthesis of inorganic crystals with specific size and morphology has arisen a lot of recent interest because of the potential to design new materials and devices in various fields such as catalysis, medicine, electronics, ceramics, pigments, and cosmetics [5-15].

Shape control has significant relevance in the fabrication of semiconductor nanocrystals [9,10], metal nanocrystals [11-15], and other inorganic materials [5-7], which may add additional variables in tailoring the properties of nanomaterials. Much effort has been made for fabricating one-dimensional nanowires or nanorods [16-22] by using hard templates such as carbon nanotubes [16,17] and porous aluminum templates [18-20] or by laser-assisted catalytic growth (LCG) [21,22], as well as controlled solution growth at room temperature or elevated temperature [9,10,23-25].

As an alternative strategy, bio-inspired morphosynthesis has been emerging as an important enviromentally friendly route to template inorganic materials with controlled morphologies by using self-assembled organic superstructures, organic additives, and/or templates with complex functionalization patterns [4-7]. The precipitation of inorganics with three-dimensional lipid and protein structures has been found to generate unusual morphologies such as oblong crystallites, mineralized organic tubules and disks, and microporous reticulated structures [26,27]. Reverse micelles and microemulsion techniques were widely used for the preparation of copper nanorods[13], higher-order structures such as $BaCrO_4$ chains and filaments [7], $BaSO_4$ filaments and cones [28], as well as $CaSO_4$ [29], and $BaCO_3$ nanowires [30] at room temperature.

Although various examples of shape control have been demonstrated, exploration of the facile synthesis of inorganic crystals with controllable size and shape in aqueous solutions at room temperature remains nonsystematic and challenging. Recently, it was shown that so-called double-hydrophilic block copolymers (DHBCs) [31-33] can exert a strong influence on the external morphology and/or crystalline structure of inorganic particles such as calcium carbonate [34-36], calcium phosphate [37], barium sulfate [38-40], and zinc oxide [41].

In the following parts, we will demonstrate the new advances of the morphosynthesis of $BaCrO_4$, $CaCO_3$, and $CdWO_4$ crystals by using DHBCs as crystal modifiers. The morphology of the particles can be systematically controlled by variation in functional group pattern and molecular structure of the DHBCs. Other effects such as the crystallization sites, temperature, pH value, reactant concentration, copolymer concentration on the morphology are also examined. The self-orginzation process of the superstructures under control of DHBCs will be discussed. The results demonstrate that the integration of using DHBCs, taking advantage of other experimental conditions, will provide more possibilities for the controlled synthesis of inorganic crystals with well-defined shape, size, and microstructures [42-44]. The powerful potential of DHBCs for rational design of inorganic nanocrystals and superstructures will be outlooked.

2. Double hydrophilic block copolymers (DHBCs) : New concept for controlled crystallization

It is well known that the adsoroption of ions or low molar mass additives onto the specific crystal surfaces can influence the crystal morphologies [45] as shown in Figure 1a (left). However, it is necessary to stablize the primary nanoparticles building blocks for further structure development to avoid the uncontrolled aggregation. This is only possible to a limited extent by electrostatic stabilization or low molar mass additives. The pure electrostatic stabilization is often not sufficient for the application of crystalline nanoparticles as building blocks for superstructures as the Debye length is usually in the range of the attractive van der Waals forces [46]. On the other hand, the stabilization by adsorption of polyelectrolytes is very effective electrosteric stabilization as shown for iron oxyhydroxide which was immediately stabilized by cellulose-sulfate and κ-carrageenan to avoid the precipitation of amorphous solid [47]. However, the limit for selective adsorption can be exceeded as the concentration of adsorpting groups is high in a polyelectrolyte of sufficient chain length, so that the non-selective adsorption occurs at all faces as shown in Figure 1b (left) leading to nondirected crystal growth [48]. Therefore, the ideal choice is to combine the requirement for stabilization and the selective adsorption functions together within one molecule, a block copolymer with a polyelectrolyte block-short enough for selective adsorption but long enough for sufficient interaction with the crystal surface- and a stabilizing block for steric stabilization as shown in Figure 1c (left). Block copolymers ususally consist of a hydrophilic and hydrophobic block so that they can be used as polymeric surfactants, which show the similar behavior to low molecular weight surfactants. Double-hydrophilic block copolymers (DHBCs) are a new class of polymers, which consist of two water-soluble blocks of different chemical nature. They polymers are typically rather small having block lengths between 10^3-10^4 g mol^{-1}. The solvating block shows good solublity in water. The binding block contains variable chemical patterns, which show strong affinity to minerals and have strong interaction with inorganic crystals as illustrated in Figure 1 (right part).

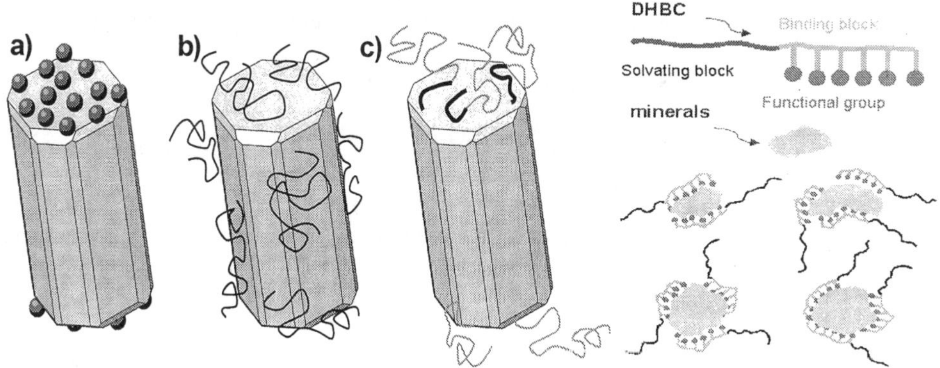

Figure 1 Left : (a) Specific adsorption of ions of low molar mass additives onto a crystal, (b) unspecific adsorption of polyelectrolytes and (c) specfic adsorption of block copolymers with a short polyelectrolyte block ; Right: The illustration scheme of the concept of DHBCs.

Different kinds of DHBCs with different funcational patterns were designed and used as crystal modifiers. Homopolymers poly(ethylene glycol) monomethyl ether (PEG, MW = 5000 g mol^{-1}) and poly(methacrylic acid, sodium salt) (PMAA, MW = 6500 g mol^{-1}) were purchased from Polysciences and Aldrich, respectively. A commercial block copolymer poly(ethylene glycol)-*block*-poly(methacrylic acid) (PEG-*b*-PMAA, PEG = 3000 g mol^{-1}, 68 monomer units, PMAA = 700 g mol^{-1}, 6 monomer units) was obtained from Th. Goldschmidt AG, Essen, Germany. The carboxylic acid groups of this copolymer were partially phosphonated (21%) to give a copolymer with carboxyl and phosphonated groups, PEG-*b*-PMAA-PO_3H_2, according to Ref. [34]. The degree of phosphonation was determined with quantitative ^{31}P-NMR-spectroscopy by comparing the signals of an internal standard (KH_2PO_4 signal at 3.1 ppm) with the signal of the polymer product (18.6 ppm) in D_2O [49].

A block copolymer containing a poly(ethylenediaminetetraacetic acid) (EDTA)-like carboxy-functionalized block, poly(ethylene glycol)-*block*-poly(ethylene imine)-poly(acetic acid) (PEG-*b*-PEI-$(CH_2CO_2H)_n$, PEG-*b*-PEIPA, PEG = 5000 g mol^{-1}, PEIPA = 1800 g mol^{-1}) was synthesized as described elsewhere [33].

The copolymers, which are based on PEG-*b*-PEI with various functional acidic groups – COOH, –PO_3H_2, –SO_3H, and –SH, were synthesized by further functionalization of the PEI block [33]. The ethyl phosphonic acid groups were added to the PEI block by the Michael type addition reaction of the amine group to the vinyl activated group of vinylphosphonic acid to give PEG-*b*-PEI-$(CH_2$-CH_2-$PO_3H_2)_n$(PEG-*b*-PEI-PEPA) [33]. PEG-*b*-PEI can also be modified by a simple standard acylation reaction of an amine group of PEI with acyl chloride such as lauroyl chloride ($C_{11}H_{23}COCl$) and will generate PEG-*b*-PEI(CH_2-CH_2-PO_3H_2)-$COC_{11}H_{23}$ (lauroyl) [33]. The reaction of the primary amino groups and also some secondary amino groups with methyl isothiocyanate produced PEG-*b*-PEI(HS=C-$NCH_3)_n$-$COC_{11}H_{23}$ [33]. The partially phosphorylated poly(hydroxy ethyl ethylene) block copolymer with PEG (PEG-*b*-PHEE–PO_4H_2(30%)) was synthesized as described in Ref. [50]. For the crystallization experiments, all copolymers were purified by exhaustive dialysis.

3. Morphosynthesis of various inorganic crystals and their hierachical structures

3.1 MORPHOSYNTHESIS OF BARIUM CHROMATE

A systematic morphosynthesis of barium chromate crystals with controlled morphology and novel superstructures by using double-hydrophilic block copolymers (DHBCs) with varying pattern of functional groups is described in this part. In addition, physico-chemical parameters such as the crystallization sites, temperature, the concentration of reactants and the copolymers were examined and also showed significant effects on the morphology of the resulting particles, demonstrating that the ability of the copolymer to interact with the inorganic crystals could be fine-tuned.

In the absence of polymeric additives, the precipitation process of $BaCrO_4$ occurs very quickly and the obtained products were found to be well-crystallized hashemite crystals confirmed by the corresponding XRD results. The dendritic X-shaped crystals with an average length of about 10 μm were obtained in the absence of additives.

However, Figure 2a (left) shows that nanofiber bundles were obtained in the presence of the phosphonated derivative PEG-*b*-PMAA-PO_3H_2 (21%, phosphonation degree). The particles obtained in the presence of PEG-*b*-PEIPA were found to be egg-shaped with an overall size of 100-180 nm. Egg-shaped particles with a size of 1 μm were also obtained with PEG-*b*-PMAA as a modifier, as shown in Figure 2b (left). The inhibition effect of PEG-*b*-PMAA for hashemite crystallization was found to be comparable to that of PEG-*b*-PEIPA, which is in coincidence with the results reported for barite crystallization [40]. As expected, a more pronounced partial hydrophobic modification of the functional polymer block results in mutual aggregation of the DHBCs and therefore an altered modification pattern, resulting also in altered mineralized superstructures. When the PEI backbone of PEG-*b*-PEI-(CH_2-CH_2-PO_3H_2) was additionally modified with lauroyl-$COC_{11}H_{23}$ moieties (PEG-*b*-PEI(CH_2CH_2-PO_3H_2)-$COC_{11}H_{23}$), the hybrid superstructures are composed of smaller aggregated spheres clustered together, and the rod-like structure obtained with the non hydrophobically modified polymer (Figure 2c, left) is lost. Uniform, well-defined lancet-like particles with a length of 20 μm and a width of 5 μm were obtained in the presence of PEG-PEI($CSHNCH_3$)$_n$-$COC_{11}H_{23}$ (Figure 2d, left). Apparently, in this case, the hydrophobic modification does not lead to clustered aggregated spheres as found for the phosphonated polymer analogue, suggesting a site selective block copolymer adsorption. The aspect ratio of *a*/*b* is as large as 5-5.4, which is much higher than that previously observed in the case of barite [51], an observation that supports the site selective block copolymer adsorption. When another phosphorus containing copolymer, the partially phosphorylated poly(hydroxy ethyl ethylene) block copolymer (PEG-*b*-PHEE-OPO_3H_2 (30%)) [29] was used, the morphology changed to smaller, trapezoidal aggregates with no clear expression of faces.

Figure 2 Left : Control of the morphologies of $BaCrO_4$ by DHBCs. The polymer concentration was kept at 1 g L^{-1} ([$BaCrO_4$] = 2 mM), pH 5: (a) PEG-*b*-PMAA-PO_3H_2 (21%, phosphonation degree); (b) PEG-*b*-PMAA; (c) PEG-*b*-PEI-(CH_2-CH_2-PO_3H_2)$_n$; (d) PEG-*b*-PEI(CSHN-CH_3)$_n$-$C_{11}H_{23}$; Right: SEM and TEM images of highly ordered $BaCrO_4$ nanofiber bundles obtained in the presence of PEG-*b*-PMAA-PO_3H_2 (21%, phosphonation degree) (1 g L^{-1}) ([$BaCrO_4$] = 2 mM), pH 5: (a) two fiber bundles with cone-like shape and length about 150 μm; (b) the very thin fiber bundles; (c) TEM image of the thin part of the fiber bundles, inserted electronic diffraction pattern taken along the <001> zone, showing the fiber bundles are well crystallized single crystals and elongated along [210]. Reprinted from Ref. 42 with permission from Wiley-VCH (Copyright 2002).

It is delineated that this polymer is already not purely hydrophilic anymore, but shows amphiphilic character and surface activity [50]. All DHBCs based on a bare PEI binding site give comparatively little control over the oriented growth of the crystals under the applied acidic conditions at pH 5, underlining that the growing crystals rather prefer to interact with the negatively charged $-PO_3H_2$ and $-COOH$ groups instead of the positively charged PEI block.

More complex morphologies of hashemite can be formed in the presence of PEG-*b*-PMAA-PO_3H_2. Figure 2a and 2b (right) show SEM images of fibrous superstructures with sharp edges composed of densely packed, highly ordered, parallel nanofibres of $BaCrO_4$. The TEM micrograph with higher resolution in Figure 2c (right) clearly shows the self-organized nature of the superstructure. Whereas the majority of the fibers appears to be aligned in a parallel fashion, gaps between the single fibers can form, but are also closed again. An electronic diffraction pattern taken from such an oriented planar bundle as shown in Figure 2d (right) confirmed that the whole structure scatters as a well crystallized single crystal where scattering is along the [001] direction and the fibers are elongated along [210].

The atomic surface structure modeling data for the surface cleavage of the hashemite crystal shows that the faces (1-22), (1-21), (1-20), (-120), which are parallel to [210] axis, contain slightly elevated barium ions, indicating that the negatively charged $-PO_3H_2$, $-COOH$ groups of PEG-*b*-PMAA-PO_3H_2 can preferentially adsorb on these faces by electrostatic attraction and block these faces from further growth. In contrast, the surface cleavage of the (210) face shows no attackable barium ions on the surface, suggesting that the functional polymer group does not favorably adsorb on this face, leading to a fast growth rate. This is favorable for the orientated growth along the [210] direction.

It is obvious that the fibers are grown along a single orientation direction and self-organized into a two or three dimensional superstructure, suggesting that in addition to the vectorially directed fiber formation, forces also act perpendicular to the fiber axis, which control their interspacing.

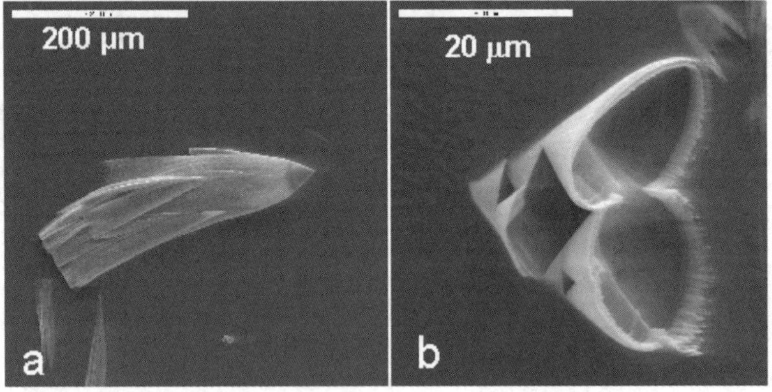

Figure 3 (a) Typical cone-like superstructure contained densely packed $BaCrO_4$ nanofiber bundles in the presence of PEG-*b*-PMAA-PO_3H_2 (21%, phosphonation degree) (1g L^{-1}) ([$BaCrO_4$] = 2 mM, pH 5), two days, 4 °C; (b) multi-funnel-like superstructures with remarkable self-similarity in the presence of PEG-*b*-PMAA-PO_3H_2 (1%, phosphonation degree) (1g L^{-1}) ([$BaCrO_4$] = 2 mM, pH 5). Reprinted from Ref. 42 with permission from Wiley-VCH (Copyright 2002).

All structures always grow from a single starting point. This may be due to the fiber growth against the glass wall or other substrates such as TEM grids which obviously provide the necessary heterogeneous nucleation sites. The growth front is always very smooth, suggesting the homogeneous joint growth of all single nanofilaments with the ability to cure occurring defects in line with the earlier findings for $BaSO_4$ [40].

The opening angle of the cones is always rather similar, which seems to depend on temperature, degree of phosphonation of the polymer, and polymer concentration [42].The control experiments show that the higher the temperature, the more linear the structures become. The superstructure developed more clearly, and to much larger size, when the mineralization temperature was lowered to 4 °C as shown in Figures 3a. A rather lower phosphonation degree (~ 1%) of a PEG-*b*-PMAA is already powerful enough to produce the fiber bundles and the superstructures as shown in Figure 3b. Interestingly, secondary cones can nucleate either from the rim or defects onto the cone, thus resulting in the tree like structures (Figure 3a). The fact that a cone stops growing once a second cone has nucleated at one spot on the rim shows that the growth presumably is slowing down with time, favoring the growth of the secondary cone. This is explained as the consequence of the influence of electrostatic dipole and multipole fields of the superstructure on approaching ions and colloidal building blocks, resulting in branching instead of further growth [48, 52-54]. The detailed structures of the cone-like bundles were found to be in the form of cone-in-a-cone "matrioshka" structure. The ability of the edges to attract each other and to fuse leads to the simultaneous occurrence of thin, very filigree fused cones (see Figures 3b). The conelike superstructures tend to grow further into a self-similar, multi-cone "tree" structure which was observed before for barite mineralized in the presence of polyacrylates, but only under very limited experimental conditions [40].

The initially formed particles are amorphous with sizes of up to 20 nm, which can aggregate to larger clusters. Evidently, this state of matter is the typical starting point for all types of highly inhibited reactions. The very low solubility product of barium chromate ($K_{sp} = 1.17 \times 10^{-10}$) shows that the superstructures do not really grow from a supersaturated ion solution but by aggregation/transformation of the primary clusters formed.

The formation mechanism of the development of the multicone-like superstructures is proposed as shown in Scheme 1: (i) At the beginning, amorphous particles are formed which are stabilized by the DHBCs (stage 1); (ii) heterogeneous nucleation of fibers occurs on glass substrates and the fibers grow under control of DHBCs, presumably by multipole field-directed aggregation of amorphous nanoparticles (stage 2, for a closer view on multipole field directed crystallization, see Ref. [40]); (iii) The growth is continuously slowed down until secondary nucleation or overgrowth becomes more probable than the continuation of the primary growth. This is a statistical observation and will lead to a distribution of cone sizes. The secondary cone will grow as the first ones have done; (iv) the secondary heterogeneous nucleation taking place on the rim can repeatedly occur depending on the mass capacity of amorphous nanoparticles in the system.

A plausible reason for the self-limiting growth may be seen in the crystal structure of hashemite/barite. These crystals have a mirror plane perpendicular to the *c*-axis, which means that a homogeneous nucleation will always result in crystals with identical charges of the opposite faces. Thus, no dipole crystals can be formed. For heterogeneous nucleation, however, this situation is different, a dipole crystal may be favored as one end of the crystal is determined by the heterogeneous surface, the other by the solution/dispersion. Accordingly, it has a dipole moment $\mu = Q \cdot l$ (Q = charge and l = length of the crystal) which increases while the crystal is growing.

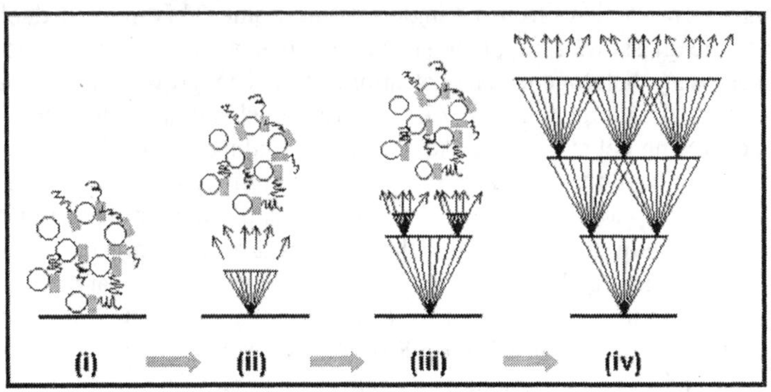

Scheme 1 Proposed mechanism for the formation of complex superstructure with self-similarity. Reprinted from Ref. 42 with permission from Wiley-VCH (Copyright 2002).

This implies self – limiting growth so that a new heterogeneous nucleation event on the rim should become favourable after the dipole moment has reached a critical value due to energy minimization. Such a self-limiting growth mechanism would also explain why the growth edge is so uniform.

The whole suggested mechanism relies on the absence of flow, which would disturb the directed aggregation according to local dipole/multipole fields. No fiber bundles or cones were obtained if the solution was stirred continuously at room temperature after mixing the reactants. Instead, only irregular and nearly spherical particles were obtained. This is again in agreement with a recently published mechanism where the fiber formation is due to the directed self-assembly of the primary particles [40], which is suppressed by stirring. The glass wall or other surface provides the necessary heterogeneous nucleation site for the birth of the fibers. In contrast, only spherical particles can be obtained when the same reaction was done in polypropylene (PP) or plastic bottles.

The concentration of the reactants, the ratio of cations and anions, pH, the copolymer concentration, and temperature have a strong influence on the precipitation of barium chromate. When the concentration of both ions was equimolar ($[Ba^{2+}]:[CrO_4^{2-}] = 1:1$) and was as low as 0.025 M (in presence of 1 g L^{-1} of copolymer), no precipitation occurred even after aging for three weeks at room temperature although the reference experiment without polymer showed crystallization. This means that the polymers are able to stabilize the precursor particles at this concentration level without the ability to undergo further structural transitions. When the concentration increased up to 0.05 M, fiber formtation occurred after one to four days. Further increase of the ion concentrations up to 0.2 M resulted in very quick precipitation. TEM shows that in this case, only ellipsoids or boulder-like crystals with sizes of about 200-600 nm were produced. If the $[Ba^{2+}]:[CrO_4^{2-}]$ molar ratio was changed to 1:5 or 2.5:1, the solution becomes turbid as soon as adding Na_2CrO_4 into the system. Again, no strong control by the polymer was exerted, and only ellipsoidal nanoparticles or aggregates were obtained.

These results show that the polymer essentially interacts with "neutral" crystals that show no overall surface enrichment of either one or the other ionic species. This mode of interaction is clearly different from the previous observation from a microemulsion system, in which the higher aspect ratios of $BaCrO_4$ nanofilaments were found only at higher $[Ba^{2+}]:[CrO_4^{2-}]$ molar ratio (= 5:1), that is there the interaction with the anionic AOT surfactants of this particular system is mediated by the excess of barium ions [7].

When the concentration of the polymer was lowered to 0.2 g L^{-1}, the solution quickly becomes turbid, and the precipitate consisted of nanoparticles of mostly spherical shape. This result suggests that the interaction capacity of the phosphonated copolymer at lower concentration is not high enough to stabilize the fibers with their high internal surface, but still shows strong interaction with all of the crystal faces. When the concentration of the polymer increased up to 1 g L^{-1}, only fiber bundles are obtained as discussed before. Upon increase of the polymer concentration to 2 g L^{-1}, the solution stays transparent and yellow and a very loose flocculate is obtained after two or three days, which was found to be composed of aggregated, but not oriented, and therefore more dispersed fiber bundles.

This ties in well with the idea of a strong interaction of the copolymer with the crystals, where the increased concentration leads to a higher binding and a higher polymer content in the precipitate, making the hybrid morphology more dispersed. These results indicate that only a limited concentration window exists for the formation of 3-dimensionally highly orientated fiber bundles. Apparently, the force, which is responsible for the 3-D packing can be over-compensated by a high polymer surface coverage and the resulting steric and possible electrostatic repulsion.

When the pH value was varied from 5.5 to 3.5, 2 respectively, no precipitate was found, even after aging for two weeks. The suitable pH range for the formation of $BaCrO_4$ fibers was within 5-7. In contrast, when the pH was changed to pH 10, the precipitate was found to be composed of nearly spherical particles in the μm range that formed a superstructure or aggregates of nanoparticles. This suggests that a selective adsorption of the block copolymer can only take place in a limited pH range whereas beyond the pH limits, the polymer adsorbs in a non-selective fashion.

Increasing the temperature from 4 °C (cone-like fiber bundles with large opening angle) over 25 °C (more linear and extended fiber bundles) to 50 °C results in faster nucleation and less specificity of the growth process. Small twinned or peanut-shaped nanoparticles were found at 50 °C. Further increase of the temperature to 100 °C shows a continuation of this trend. Again, only peanut-shaped aggregates were found.

This observation holds for both, glass and polypropylene bottles, that is higher temperatures indeed allow homogeneous nucleation to prevail over the heterogeneous nucleation at lower temperatures, which leads to fibers. Here, a different species, the peanuts, where a nucleus can grow in two opposite directions, were found. Heterogeneous nucleation, presumably, has not stopped, but lost importance because of the acceleration of the homogeneous nucleation rate and the dominance of this stage for the crystallization process. Peanuts are formed at a much higher overall rate than fibers under this condition.

It should be pointed out that the growth of fiber hybrid superstructures is promoted by the addition and recrystallization of primary colloidal particles rather than the build-up by single ions. The reason that the accelerated homogeneous nucleation by applying temperature resulted in peanut or dumbell-like particles, which stoppted growth, could be related with the characteristics of the primary particles. These particles carry the character of an electrostatic multipole field which is perpetuated and amplified throughout the growth process organized by the polymer. The detailed mechanism, however, needs further elucidation. The further modification of the glassware (negative and positive charge) or by addition of highly polar colloidal nuclei could result in better control of the heterogeneous nucleation. In addition, the distribution of the polymer within the larger superstructures and the interface between the crystals and the polymer could be followed by confocal Raman microscopy and neutron scattering.

3.2 MORPHOSYNTHESIS OF CALCIUM CARBONATE WITH CONTROLLED SIZE AND SURFACE STRUCTURES

Calcium carbonate ($CaCO_3$) has been widely used as model system for studying the biomimetic process due to the abundance in nature and also its important industrial application in paints, plastics, rubber, or paper [55]. Biomimetic synthesis of $CaCO_3$ crystals in the presence of organic templates and/or additives has been intensively investigated in recent years [56, 57].

Rhombohedral shaped calcite crystals consisting of six (104) faces were easily produced under equilibrium conditions in the absence of the polymers, which was believed to be the macroscopic expression of the unit cell as shown in Figure 4.

Biomimetic crystallization of large $CaCO_3$ spherules with controlled surface structures and sizes has been recently realized by a slow gas-liquid diffusion reaction of the CO_2 released from ammonia carbonate in a closed desiccator in the presence of double-hydrophilic block copolymers (DHBCs) [43]. Contrary to previous work where calcite composites were prepared by a rapid supersaturation/precipitation technique, the double jet injection [34,38,39], the gas-diffusion reaction technique is much slower and closer to the speed of calcium carbonate morphogenesis in biomineralization which can range from a few hours for eggshells to many days [58].

The slow precipitation process finally leads to uniform big spherules in the presence of poly(ethylene glycol)-*block*-poly(ethylene imine)-poly(acetic acid)(PEG-*b*-PEIPA), which after three days still contain a variety of intermediates with shapes such as rods, small spheres, peanuts, dumbbells, and overgrown bigger spherules (insert part as shown in Figure 5a, left). Big $CaCO_3$ calcite spherules with a size of about 90 μm can be obtained. Such comparably big spherules with complex surface structures of calcite rhombohedra and resulting cavities cannot be easily generated by conventional solution growth methods. The high-magnification SEM image shows those structures in more detail, which imply that the big spherules result from a complex growth mechanism as first found for fluoroapatite by Busch and Kniep [52,53].

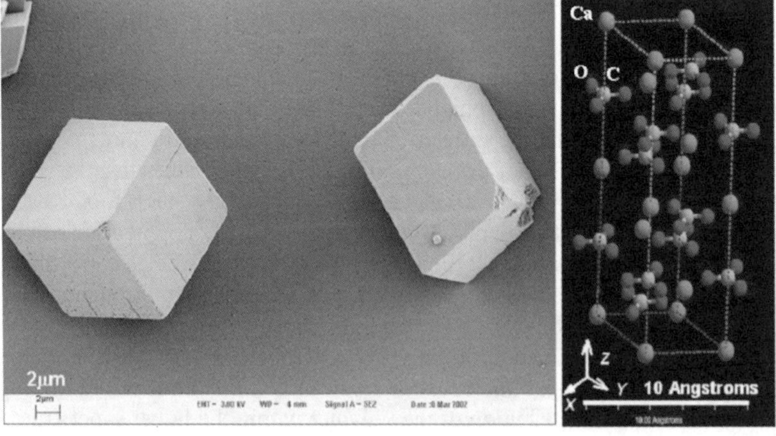

Figure 4 Left: SEM image of the $CaCO_3$ particles grown in the absence of block copolymer after two weeks; Right: cell structure of calcite crystals by *Cerius*^{2} software. Blue ball: Ca; gray: C; red: O.

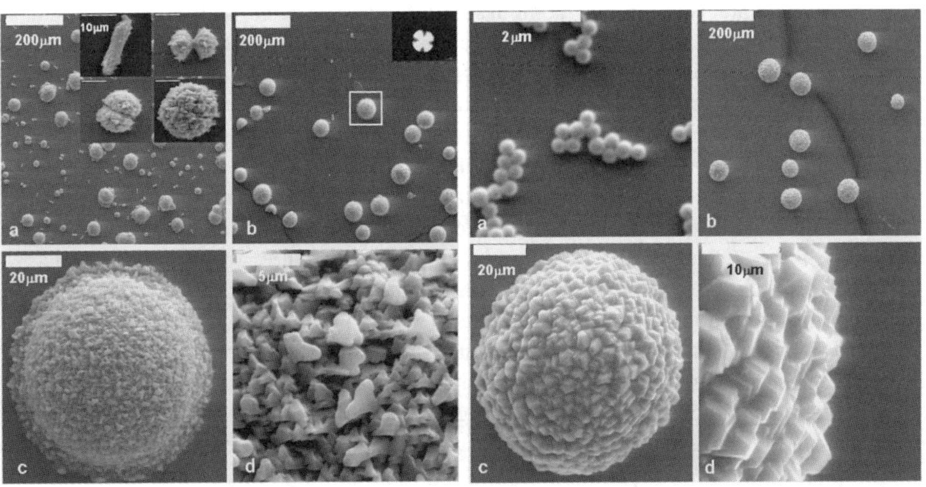

Figure 5 Left : (a) SEM images of the $CaCO_3$ particles with different morphology, showing that the morphology evolution process was nicely "*captured*" by stopping the reaction after 3 days, in the presence of PEG-*b*-PEIPA (1 g L^{-1}); The insert shows that big spherules are developed starting from small nanoparticles and grow further to form elongated rods, peanuts, dumbbells, and finally assembled into complete spherules ; (b) a full view of uniform spherules with average size of 90 μm, the insert shows a typical polarized optical microscope image showing the perfect spherule characteristics; (c) a typical big spherule formed after two weeks; (d) the surface structure of the spherule. Right: SEM images of $CaCO_3$ particles grown on a glass substrate in the presence of PEG-*b*-PMAA (1 g L^{-1}): (a) $CaCO_3$ nanoparticles with uniform sizes (310 nm) grown after 3 hours; (b) big $CaCO_3$ particles grown after 3 days ; (c) a typical big spherule with size of 90 μm formed after two weeks; (d) enlarged surface structure of the big spherule. Reprinted from Ref. 43 with permission from Wiley-VCH (Copyright 2002).

At the first stage, rod-like particles are formed which can growth at their ends resulting in dumbbell-like particles. These dumbbells further grow into spherical particles consistent with earlier findings on the same polymer, which resulted in the formation of hollow spheres [34] and are consistent with a particle aggregation along electric field lines [52,54]. The fact that we can simultaneously observe particles in different stages of their development indicates that nucleation and subsequent aggregation are slow. This is typical for all slow evaporation techniques, opposite to the double jet procedure.

After two weeks of ripening and equilibration, all particles display a spherical superstructure with uniform size of 80 μm, as shown in Figure 5b (left). The polarized optical microscope (POM) image in Figure 5b (left) shows the perfect spherulitic structure of the composed spherules. Figure 5c (left) shows a typical spherule with a size of 80 μm. The surface of the superstructure is very rough and is composed of smaller truncated and randomly oriented calcite "*cornerstones*" with average size of about 3-5 μm in diameter, and a lot of cavities among them, as shown in Figure 5d (left) which is similar to the formation of snow-ball like cobalt oxide superstructures [59]. This implies a multi-step growth mechanism where at first the rod to sphere transition occurs with subsequent overgrowth of the spheres once they are formed.

Similar looking spherules with different sizes and different structural elementary units can also be formed using PEG-*b*-PMAA as crystal growth modifier. Figure 5a (right) shows the spherical $CaCO_3$ nanoparticles with a uniform diameter *ca.* 310 nm, which were obtained in the very early stages after 3 hours reaction. With the time prolonged up to three days, the precipitate was already composed of bigger particles with both spherical and twinned superstructure with size in the range of 30 μm to 60 μm, respectively. Again, the intermediates of the big spherules are overgrown. The surface is again found to be very rough, with 5–10 μm sized calcite crystals establishing the superstructure. Further prolongation of the reaction time up to two weeks leads again to big spherules with sizes in the range of 66 μm to 93 μm, as shown in Figure 5b (right).

The high-magnification SEM image (Figure 5c, right) shows a typical spherule with an overall diameter of 93 μm, its surface is composed of calcite rhombohedral subunits. The inner part of the crystal is different from the surface structure: within, one can detect a clearly layered radial growth structure [43], whereas on the outside, the same crystals adopt a more equilibrated rhombohedral appearance as shown in Figure 5d (right).

Figure 5 (right) suggests a growth mechanism of nanoparticle aggregation as already observed earlier for metastable vaterite spheres, with subsequent aggregation of these spheres as already suggested in Figure 5a (right). Later, the transformation to stable calcite beginning on the surface may lead to the calcite rhombohedra on the particle surface. Further overgrowth by the process of Ostwald ripening can lead to the observed radial outgrowth to big spherules with sizes up to 93 μm (Figure 5c, right) at the expense of the smaller particles thus sharpening the particle size distribution. The surface structure is very similar to that reported by Küther *et al*, where Au nanocolloids were used as nuclei for crystallization of calcite spherules with sizes about 10-15 μm [60,61]. The crystals were reported to organize radially and grow radiating out from the center with the direction of growth along the crystallographic [001] direction [60,61]. This is consistent with the WAXS results which shows no or only little exposition of the 006 face. The radial outgrowth of the inner part of the crystals we observed here is also consistent with that reported by Jennings *et al* [62], where such crystals were grown only in the presence of both Mg^{2+} and SO_4^{2-} ions in the solution.

3.3 FABRICATION OF 1D AND 2D CADIUM TUNGSTATE NANOCRYSTALS

Tungstate materials have attracted much research interest because of their important luminescence behavior, structural properties, and potential applications [63]. Cadmium tungstate ($CdWO_4$) nanocrystals with a monoclinic wolframite structure are considered to be highly interesting due to their high average refractive index, low radiation damage, low afterglow, and high X-ray absorption coefficient [64]. They can be used, for instance, as an X-ray scintillator [65, 66]. Other tungstates with Scheelite structure MWO_4 (M = Ca, Ba, Pb) also display interesting excitonic luminescence, thermoluminescence, and stimulated Raman scattering (SRS) behavior.

$CdWO_4$ nanorods were previously synthesized by a hydrothermal process at 130 °C [67]. The nanorods of different compositions reported so far generally have rather small aspect ratios (length-to-diameter) of 2–10 [68], resulting in weak symmetry breaking surface effects. It must also be mentioned that a reverse micelle templating method has recently been used to synthesize uniform $BaWO_4$ nanorods with diameters of 5 nm and aspect ratios of about 150 by using barium bis(2-ethylhexyl)sulfosuccinate [Ba(AOT)$_2$] micelles, which were reacted with NaAOT microemulsion droplets containing sodium tungstate (Na_2WO_4) [68]. The large excess of surfactants as well as the very low concentration throughout synthesis, however, could restrict the applicability of this procedure.

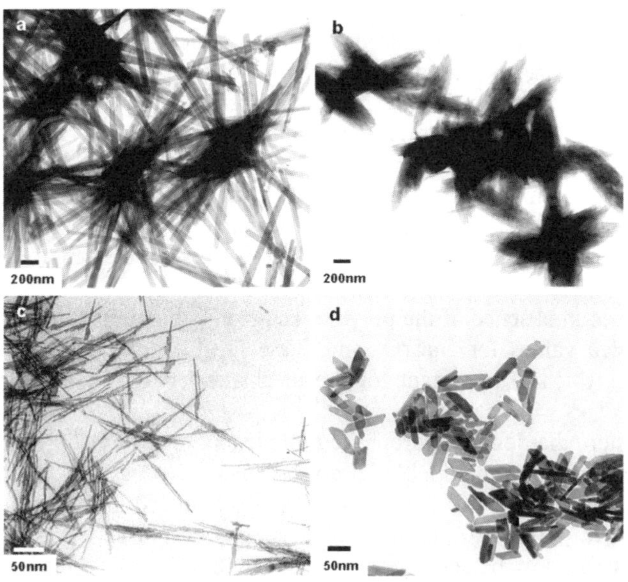

Figure 6 TEM images of the samples obtained under different conditions: (a)-(b) No additives: (a) pH = 5.3, double jet, $[Cd^{2+}]$: $[WO_4^{2-}] = 8.3 \times 10^{-3}$ M (final solution), at room temperature; (b) pH = 5.3, double jet, $[Cd^{2+}]$: $[WO_4^{2-}] = 8.3 \times 10^{-3}$ M (final solution), then hydrothermal crystallization: 80 °C, 6 h; (c) pH = 5.3, in the presence of PEG-*b*-PMAA (1 g L^{-1}), 20 mL, double jet, $[Cd^{2+}]$:$[WO_4^{2-}] = 8.3 \times 10^{-3}$ M (final solution), then hydrothermal crystallization: 80 °C, 6 h. (d) direct hydrothermal treatment of 20 mL solution containing equal molar $[Cd^{2+}]$:$[WO_4^{2-}] = 8.3 \times 10^{-2}$ M, pH = 5.3, at 130 °C, 6 h, in the presence of 1 g L^{-1} PEG-*b*-PMAA-PO_3H_2 (21%). Reprinted from Ref. 44 with permission from Wiley-VCH (Copyright 2002).

In this part, we present a facile aqueous solution route for the synthesis of extremely thin 1D and 2D $CdWO_4$ nanoparticles with controlled sizes (length, width, thickness) using a combination of the double jet injection of simple inorganic reactants and double-hydrophilic block copolymers (DHBCs) as crystal growth modifiers.

The precipitation of $CdWO_4$ was carried out in a double jet reactor thermostated at 25 °C [44]. Well-crystallized $CdWO_4$ nanorods/nanobelts can be easily synthesized at room temperature in the absence of additives at room temperature. TEM image in Figure 6a shows very thin, uniform $CdWO_4$ nanorods/nanobelts with lengths in the range of 1 to 2 μm and a uniform width of 70 nm along their entire length (aspect ratio of about 30). The slow and controlled reactant addition by the doube-jet technique under stirring mainains formation of intermediate amorphous nanoparticles at the jets [34] so that nanoparticles are the precursors for the further particle growth rather than ionic species. This growth mechanism is crucial to obtain the observed belt structure.

Successive hydrothermal ripening after the double jet reaction leads to a rearrangement of the rods into 2D lense-shaped, raft-like superstructures with a resulting lower aspect ratio as shown in Figure 6b. The thickness of the rods and the raft-like superstructures could be evaluated by scanning force microscopy (SFM), revealing about 6-7 nm based on topography height profiles. The structure modeling data clearly shows the W octahedra chain within the wolframite structure. The thin nature of the elongated nanoparticles could be related with the chain structure of W octahedra in the wolframite type structure [69].

The self-organization of the nanorods into 2D structures induced by hydrothermal treatment is very similar to that of a recent report described for $BaWO_4$ and $BaCrO_4$ nanorods [70], but differs significantly from the assembly of the short $BaCrO_4$ and CdSe nanorods where ribbon-like and vertical rectangular/hexagonal superstructures are favored [7,9]. There are two reasons for the rods to align in parallel [70], first, to maximize the entropy of the self-assembled structure of rod like or nematic objects by minimizing the excluded volume per particle in the array, as first suggested by Onsager [71], and second, because of the higher sum of van der Waals forces along the length of a nanorod as compared to its tip [72].

The aspect ratio in absence of the polymer is already about 30, which is higher than the previously reported values for $BaCrO_4$ and CdSe [7,9]. Additionally, these particles are comparably thin, which is important for the mechanical performance in nanocomposites [73].

In a second step, the double hydrophilic block copolymer PEG-*b*-PMAA was added to the solvent reservoir before the double-jet crystallization process and the mixture was then hydrothermally ripened at 80 °C. Figure 6c shows that uniform nanofibers with a diameter of 2.5 nm, a length of 100-210 nm, and an aspect ratio of 40-85 can be readily obtained. These nanofibers can now be regarded as "real" 1D objects, since the number of surface atoms is comparable with those embedded within the structure. In addition, it is seen that the single fibers are well separated, indicating a sufficient steric stabilization brought in by the adsorbed DHBCs. Hydrothermal ripening at 80 °C for longer time or at higher temperatures (120 °C) results in this case again in a co-alignment of the rods along their axis to form similar raft-like and very thin 2D-superstructures, as described above.

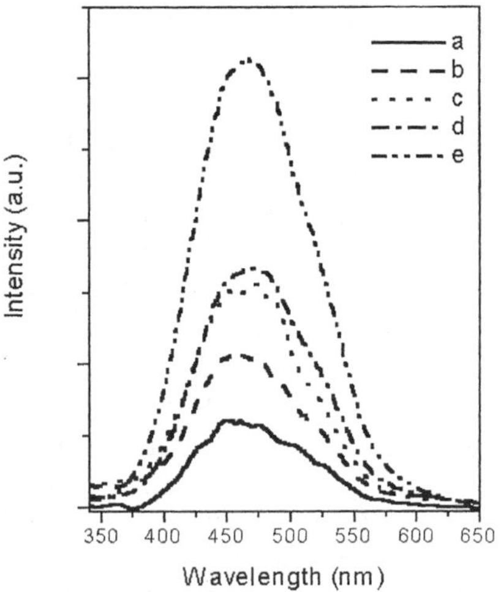

Figure 7 Room temperature photoluminescence spectra of the samples obtained under different conditions: (a) pH = 5.26, double jet, the final solution $[Cd^{2+}]/[WO_4^{2-}] = 8.3 \times 10^{-3}$ M; b) and c) pH 5.26, double jet, the final solution $[Cd^{2+}]:[WO_4^{2-}] = 8.3 \times 10^{-3}$ M, then hydrothermal crystallization: b) 80 °C, 6 h; (c) 120 °C, 6 h; d) pH 5.3, in the presence of PEG-*b*-PMAA (1 g L^{-1}), 20 mL, double jet, the final solution $[Cd^{2+}]/[WO_4^{2-}] = 8.3 \times 10^{-3}$ M, then hydrothermal crystallization: 80 °C, 6 h. e) direct hydrothermal treatment of 20 mL solution containing equal molar $[Cd^{2+}]/[WO_4^{2-}] = 8.3 \times 10^{-4}$ M, pH 5.3, at 130 °C, 6 h, in the presence of 1 g L^{-1} PEG-*b*-PMAA-PO_3H_2 (21%). Reprinted from Ref. 44 with permission from Wiley-VCH (Copyright 2002).

In addition, a new polymer-driven morphology arises when the partly phosphonated hydrophilic block copolymer PEG-b-PMAA-PO_3H_2 (21%) (1 g L^{-1}) is added at an elevated temperature of 130 °C even without using the double jets but at higher concentrations ($[Cd^{2+}]/[WO_4^{2-}] = 8.3 \times 10^{-2}$ M) and coupled supersaturation. Figure 6d shows that very thin platelet-like particles with a width of 17-28 nm, a length of 55-110 nm, and an aspect ratio of 2-4 are obtained by a direct hydrothermal process.

Figure 7 shows the luminescence spectra of the different $CdWO_4$ nanostructures obtained under different conditions and with different polymers, as compared on the base of similar concentrations. The spectrum characteristics were very similar to that of the other reported scheelite tungstate crystals (AWO_4, A = Pb, Ca, Ba, Sr) [74]. It is seen that the absolute luminescence intensity increases with increasing hydrothermal crystallization temperature (Figures 7a-7c), indicating the perfection of the crystals. Quite unexpectedly, the luminescence efficiency is further increased for the species crystallized in the presence of the different DHBCs (at lower crystallization temperature), where the best performing system increases the efficiency by a factor of two. This is explained by a highly perfected structure where quenching surface defects are suppressed or blocked by the polymers. The increase of luminescence efficiency by blocking of surface states was already observed for DHBC-stabilized CdS quantum dots [75], whereas the mode of perfecting of crystal surfaces by surface active polymers was only described very recently [76].

The $CdWO_4$ nanostructures obtained in absence of polymer exhibit a blue emission band in the range of 400-550 nm centered around 460 nm when excited at 253 nm, which agrees well with the data for the single crystals obtained at high temperatures [65], but is blue-shifted compared to the previously reported "intrinsic luminescence" at 480-490 nm [77]. The emission band shape might be explained considering Jahn-Teller active vibrational modes of t_2 symmetry influencing the WO_4^{2-} complex anion of slightly distorted tetrahedral symmetry to lead to a structured absorption band for the $A_1 \rightarrow T_{1(2)}$ transitions [78,79].

4. Conclusions and outlook

In summary, we demonstrated that polymer controlled crystallization is an very powerful route for morphosynthesis of advanced inorganic materials with controlled shape, size, dimensionality, and length scales by using double-hydrophilic block copolymers (DHBCs) as crystal modifiers. The results demonstrated that the combination of using DHBCs, taking advantage of other cooperative effects of experimental conditions such as the crystallization sites, temperature, pH value, and reactant concentration, will extend the possibilities for the controlled synthesis of inorganic crystals with well-defined shape, size, and superstructures.

$BaCrO_4$ nanofibers and nanfiber bundles with large aspect ratios can be easily fabricated at room temperature by using the phosphonated block copolymers. The produced calcium carbonate particles with well-defined morphologies, sizes, and surface structures could have some interesting applications, e.g. as site specific absorber and chromatographic materials. A very simple aqueous route to prepare uniform and very thin $CdWO_4$ nanorods/nanobelts, and elongated nanosheets at room temperature starting from simple inorganic reactants by the double jet crystallization technique is also presented. This technique provides nanoparticulate precursors for further crystallization which is in contrast to all previously reported techniques. Additionally, application of two different double hydrophilic block copolymers throughout a hydrothermal ripening process allowed a fine tuning of both crystal morphology and crystal superstructure. The prepared structures display a very strong blue/green luminescence at room temperature, where the quantum efficiency is highly improved by addition of the DHBCs.

It is expected that DHBC copolymers with more complex patterns of chemical groups may provide even more effective tools for the controlled synthesis of inorganic crystals with well-defined architectures from nanoscale to macroscopic scale, and this possibility is currently being further extended also for other mineral systems. The materials with well-defined shape, sizes, and structures could find interesting applications in the fields of advanced materials.

5. Acknowledgement

We acknowledge financial support by the Max Planck Society and the DFG (SFB 448). S.-H. Yu thanks the Alexander von Humboldt Foundation for granting a research fellowship. H. C. thanks the Dr. Hermann Schnell foundation for financial support. We thank the director of the Max Planck Institute of Colloids and Interfaces, Prof. Dr. Markus Antonietti, for his sound advices and interests on this project. Prof. Dr. M. Sedlak and Dr. Jan Rudloff are acknowledged for the synthesis of the phosphonated and phosphorylated polymers. We also thank Prof. Dr. Michael Giersig for HRTEM work, and Dr. Mastai Yitzhak for helpful discussion.

6. References

1. Mann, S., Ozin, G. A. (1996) Synthesis of inorganic materials with complex form, *Nature* **382**, 313-318.
2. Mann, S. (2000) The chemistry of form, *Angew. Chem. Int. Ed.* **39**, 3393-3406.
3. Dujardin, E., Mann, S. (2002), Bio-inspired materials chemistry, *Adv. Mater.* **14**, 775-788.
4. Estroff, L. A., Hamilton, A. D. (2001) At the Interface of Organic and Inorganic Chemistry: Bioinspired Synthesis of Composite Materials, *Chem. Mater.* **13**, 3227-3235.
5. Archibald, D. D., Mann, S. (1993) Template mineralization of self-assembled anisotropic lipid microstructures, *Nature* **364**, 430-433.
6. Yang, H., Coombs, N., Ozin, G. A. (1997) Morphogenesis of shapes and surface patterns in mesoporous silica, *Nature* **386**, 692-695.
7. Li, M., Schnablegger, H., Mann, S. (1999) Coupled synthesis and self-assembly of nanoparticles to give structures with controlled organization, *Nature* **402**, 393-395.
8. Matijević, E. (1996) Controlled colloid formation, *Curr. Opin. Colloid Interface Sci.* **1**, 176-183.
9. Peng, X. G., Manna, L., Yang, W. D., Wickham, J., Scher, E., Kadavanich, A., Alivisatos, A. P. (2000) Shape control of CdSe nanocrystals, *Nature* **404**, 59-61.
10. Manna, L., Scher, E. C., Alivisatos, A. P. (2000) Synthesis of Soluble and Processable Rod-, Arrow-, Teardrop-, and Tetrapod-Shaped CdSe Nanocrystals, *J. Am. Chem. Soc.* **122**, 12700-12706.
11. Ahmadi, T. S., Wang, Z. L., Green, T. C., Henglein, A., El-Sayed, M. A. (1996) Shape-controlled synthesis of colloidal platinum nanoparticles, *Science* **272**, 1924-1926.
12. Gibson, C. P., Putzer, K. (1995) Synthesis and characterization of anisometric cobalt nanoclusters, *Science* **267**, 1338-1340.
13. Pileni, M. P., Ninham, B. W., Gulik-Krzywicki, T., Tanori, J., Lisiecki, I., Filankembo, A. (1999) Direct relationship between shape and size of template and synthesis of copper metal particles, *Adv. Mater.* **11**, 1358-1362.
14. Zhou, Y., Yu, S. H., Wang, C. Y., Li, X. G., Zhu Y. R., Chen, Z. Y. (1999) A novel ultraviolet irradiation photoreduction technique for the preparation of single-crystal Ag nanorods and Ag dendrites, *Adv. Mater.* **11**, 850-852.
15. Park, S.-J., Kim, S., Lee, S., Khim, Z. G., Char, K., Hyeon, T. (2000) Synthesis and magnetic studies of uniform iron nanorods and nanospheres, *J. Am. Chem. Soc.* **122**, 8581-8582.

16. Lieber, C. M. (1998) One-dimensional nanostructures: Chemistry, physics & applications, *Solid State Commun.* **107**, 607-616.
17. Hu, J. T., Odom, T. W., Lieber, C. M. (1999) Chemistry and physics in one dimension: Synthesis and properties of nanowires and nanotubes, *Acc. Chem. Res.* **32**, 435-445.
18. Klein, J. D., Herrick, R. D., Palmer, D., Sailor, M. J., Brumlik, C. J., Martin, C. R. (1993) Electrochemical fabrication of cadmium chalcogenide microdiode arrays, *Chem. Mater.* **5**, 902-904.
19. Martin, C. R. (1994) Nanomaterials-a membrane-based synthetic approach, *Science* **266**, 1961-1966.
20. Routkevitch, D., Bigioni, T., Moskovits, M., Xu, J. M. (1996) Electrochemical fabrication of CdS nanowire arrays in porous anodic aluminum oxide templates, *J. Phys. Chem.* **100**, 14037-14047.
21. Duan, X. F., Lieber, C. M. (2000) General synthesis of compound semiconductor nanowires, *Adv. Mater.* **12**, 298-302.
22. Duan, X. F., Lieber, C. M. (2000) Laser-assisted catalytic growth of single crystal GaN nanowires, *J. Am. Chem. Soc.* **122**, 188-189.
23. Stein, A., Keller, S. W., Mallouk, T. E. (1993) Turning down the heat-design and mechanism in solid-state synthesis, *Science* **259**, 1558-1564.
24. Trentler, T. J., Hickman, K. M., Goel, S. C., Viano, A. M., Gibbons, P. C., Buhro, W. E. (1995) Solution-liquid-solid growth of crystalline III-V semiconductors-an analogy to vapor-liquid-solid growth, *Science* **270**, 1791-1794.
25. Yu, S. H., Yoshimura, M. (2002) Shape and phase control of ZnS nanocrystals: Template fabrication of wurtzite ZnS single-crystal nanosheets and ZnO flake-like dendrites from a lamellar molecular precursor ZnS-$(NH_2CH_2CH_2NH_2)_{0.5}$, *Adv. Mater.* **14**, 296-300.
26. Friberg, S. E., Wang, J. F. (1991) A new method to prepare uniform particles-precipitation from lyotropic liquid-crystals, *J. Dispersion Sci. Technol.* **12**, 387-402.
27. Mann, S., Archibald, D. D., Didymus, J. M., Douglas, T., Heywood, B. R., Meldrum, F. C., Reeves, N. J. (1993) Crystallization at inorgainc-orgainc interfaces-biominerals and biomimetic synthesis, *Science* **261**, 1286-1292.
28. Hopwood, J. D., Mann, S. (1997) Synthesis of barium sulfate nanoparticles and nanofilaments in reverse micelles and microemulsions, *Chem. Mater.* **9**, 1819-1828.
29. Rees, G. D., Evans-Gowing, R., Hammond, S. J., Robinson, B. H. (1999), Formation and morphology of calcium sulfate nanoparticles and nanowires in water-in-oil microemulsions, *Langmuir* **15**, 1993-2002.
30. Qi, L. M., Ma, J., Cheng, H., Zhao, Z. (1997) Reverse micelle based formation of $BaCO_3$ nanowires, *J. Phys. Chem. B* **101**, 3460-3463.
31. Cölfen, H. (2001) Double-hydrophilic block copolymers: Synthesis and application as novel surfactants and crystal growth modifiers, *Macromol. Rapid Commun.*, *22*, 219-252.
32. Sedlak, M. Antonietti, M. Cölfen, H. (1998) Synthesis of a new class of double hydrophilic block copolymers with calcium binding capacity as builders and for biomimetic structure control of minerals, *Macromol. Chem. Phys.* **199**, 247-254.
33. Sedlak, M. Cölfen, H. (2001) Synthesis of double-hydrophilic block copolymers with hydrophobic moieties for the controlled crystallization of minerals, *Macromol. Chem. Phys.* **202**, 587-597.
34. Cölfen, H., Antonietti, M. (1998) Crystal design of calcium carbonate microparticles using double-hydrophilic block copolymers, *Langmuir* **14**, 582-589.
35. Cölfen, H., Qi, L. M. (2001) A systematic examination of the morphogenesis of calcium carbonate in the presence of a double-hydrophilic block copolymer, *Chem. Eur. J.* **7**, 106-116.
36. Marentette, J. M., Norwig, J., Stockelmann, E., Meyer, W. H., Wegner, G. (1997) Crystallization of $CaCO_3$ in the presence of PEO-block-PMAA copolymers, *Adv. Mater.* **9**, 647-651.
37. Antonietti, M., Breulmann, M., Göltner, C., Cölfen, H., Wong, K. K., Walsh, D., Mann, S. (1998) Inorganic/organic mesostructures with complex architectures: Precipitation of calcium phosphate in the presence of double-hydrophilic block copolymers, *Chem. Eur. J.* **4**, 2493-2500.
38. Qi, L. M., Cölfen, H., Antonietti, M. (2000) Crystal design of barium sulfate using double-hydrophilic block copolymers, *Angew. Chem. Int. Ed.* **39**, 604-607.
39. Qi, L. M., Cölfen, H., Antonietti, M. (2000) Control of barite morphology by double-hydrophilic block copolymers, *Chem. Mater.* **12**, 2392-2403.
40. Qi, L. M., Cölfen, H., Antonietti, M., Lei, M., Hopwood, J. D., Ashley, A. J., Mann, S. (2001) Formation of $BaSO_4$ fibers with morphological complexity in aqueous polymer solutions, *Chem. Eur. J.* **7**, 3526-3532.
41. Öner, M., Norwig, J., Meyer, W. H., Wegner, G. (1998) Control of ZnO Crystallization by a PEO-b-PMAA Diblock Copolymer, *Chem. Mater.* **10**, 460-463.
42. Yu, S. H., Cölfen, H., Antonietti, M. (2002) Control of the morphologenesis of barium chromate by using double-hydrophilic block copolymers (DHBCs) as crystal growth modifiers, *Chem. Eur.J.* **8**, 2937-2345.

43. Yu, S. H., Cölfen, H., Hartmann, J., Antonietti, M. (2002) Biomimetic crystallization of calcium carbonate spherules with controlled surface structures and sizes by double-hydrophilic block copolymers (DHBC), *Adv. Funct. Mater.* **12**, 541-545.
44. Yu, S. H., Antonietti, M., Cölfen, H., Giersig M. (2002) Synthesis of very thin 1D and 2D $CdWO_4$ nanoparticles with improved fluorescence behavior by polymer control crystallization, *Angew. Chem. Int. Ed.* **41**, 2356-2360.
45. Adair, J. H., Suvaci, E. (2000) Morphological control of particles, *Curr. Opinion Colloid Inter. Sci.* **5**, 160-167.
46. Napper, D. H. (1983) *Polymeric stabilization of colloidal dispersions*, Academic press London, New York, Paris, San Diego, San Francisco, Sao Paulo, Sydney, Tokyo, Toronto.
47. Jones, F., Cölfen, H., Antonietti, M. (2000) Iron oxyhydroxide colloids stabilized with polysaccharides, *Colloid Polym. Sci.* **278**, 491-501.
48. Cölfen, H., (2001) *Biomimetic mineralization using hydrophilic copolymers: Synthesis of hybrid colloids with complex form and pathways towards their analysis in solution*, Habilitation thesis, Potsdam.
49. Kieczykowski, G. R., Ronald, R. B., Melillo, D. G., Reinhold, D. F., Grenda, V. J., Shinkai, I. (1995) Preparation of (4-Amino-1-Hydroxybutylidene)bisphosphonic Acid Sodium Salt, MK-217 (Alendronate Sodium). An Improved Procedure for the Preparation of 1-Hydroxy-1,1-bisphosphonic Acids, *J. Org. Chem.* **60**, 8310-8312.
50. Rudloff, J., Antonietti, M., Cölfen, H., Pretula, J., Kaluzynski, K., Penczek, S. (2002) Double-hydrophilic block copolymers with monophosphate ester moieties as crystal growth modifiers of $CaCO_3$, *Macromol. Chem. Phys.* **203**, 627-635.
51. Bromley, L. A., Cottier, D., Davey, R. J., Dobbs, B., Smith, S., Heywood, B. R. (1993) Interactions at the organic/inorganic interface: molecular design of crystallization inhibitors for barite, *Langmuir* **9**, 3594-3599.
52. Kniep, R., Busch, S. (1996) Biomimetic growth and self-assembly of fluorapatite aggregates by diffusion into denatured collagen matrices, *Angew. Chem. Intl. Ed.* **35**, 2624-2626.
53. Busch, S., Dolhaine, H., DuChesne, A., Heinz, S., Hochrein, O., Laeri, F., Podebrad, O., Vietze, U., Weiland, T., Kniep, R. (1999) Biomimetic morphogenesis of fluorapatite-gelatin composites: Fractal growth, the question of intrinsic electric fields, core/shell assemblies, hollow spheres and reorganization of denatured collagen, *Eur. J. Inorg. Chem.* **10**, 1643-1653.
54. Cölfen, H., Qi, L. M. (2001) The mechanism of the morphogenesis of $CaCO_3$ in the presence of poly(ethylene glycol-b-poly(methacrylic acid), *Progr. Colloid Polym. Sci.* **117**, 200-203.
55. Dalas, E., Klepetsanis, P., Koutsoukos, P. G. (1999) The overgrowth of calcium carbonate on poly(vinyl chloride-co-vinyl acetate-co-maleic acid), *Langmuir* **15**, 8322-8327.
56. Falini, G. Albeck, S. Weiner, S. Addadi, L. (1996) Control of aragonite or calcite polymorphism by mollusk shell macromolecules, *Science* **271**, 67-69.
57. DeOliveira, D. B., Laursen, R. A. (1997) Control of calcite crystal morphology by a peptide designed to bind to a specific surface, *J. Am. Chem. Soc.* **119**, 10627-10631.
58. Mann, S. (2001) *Biomineralization, Principles and Concepts in bioinorganic materials chemistry*, Oxford University Press.
59. Jones, F., Cölfen, H., Antonietti, M. (2000) Interaction of kappa-carrageenan with nickel, cobalt, and iron hydroxides, *Biomacromolecues* **1**, 556-563.
60. Küther, J., Seshadri, R., Tremel, W. (1998) Crystallization of calcite spherules around designer nuclei, *Angew. Chem. Int. Ed.* **37**, 3044-3047.
61. Küther, J., Seshadri, R., Nelles, G., Assenmacher, W., Butt, H.-J., Mader, W., Tremel, W. (1999) Mercaptophenol-protected gold colloids as nuclei for the crystallization of inorganic minerals: Templated crystallization on curved surfaces, *Chem. Mater.* **5**, 1317-1325.
62. Tracy, S. L., Francois, C. J. P., Jennings, H. M. (1998) The growth of calcite spherulites from solution I. Experimental design techniques, *J. Cryst. Growth* **193**, 374-381.
63. Saito, N., Sonoyama, N., Sakata, T. (1996) Analysis of the excitation and emission spectra of tungstates and molybdate, *Bull. Chem. Soc. Jpn.* **69**, 2191-2194.
64. Lotem, H., Burshtein, Z. (1987) Method for complete determination of a refractive-index tensor by bireflectance-application to $CdWO_4$, *Opt. Lett.* **12**, 561-563.
65. Pustovarov, V. A., Krymov, A. L., Shulgin, B. (1992) Some peculiarities of the luminescence of inorganic scintillators under excitation by high-intensity synchrotron radiation, *Rev. Sci. Instrum.* **63**, 3521-3522.
66. Tanaka, K., Miyajima, T., Shirai, N., Zhang, Q., Nakata, R. (1995) Laser photochemical ablation of CdWO4 studied with the time-of flight mass spectrometrometric technique, *J. Appl. Phys.* **77**, 6581-6587.
67. Liao, H. W., Wang, Y. F., Liu, X. M., Li, Y. D., Qian, Y. T. (2000) Hydrothermal Preparation and Characterization of Luminescent $CdWO_4$ Nanorods, *Chem. Mater.* **12**, 2819-2821.
68. Kwan, S., Kim, F., Akana, J., Yang, P. D. (2001) Synthesis and assembly of $BaWO_4$ nanorods, *Chem. Commun.* 447-448.

69. Chichagov, A. P., Ilyukhin, V. V., Belov, N. V. (1966) Crystal structure of $CdWO_4$, *Soviet Physics-Doklady* **11**, 11-13.
70. Kim, F., Kwan, S., Akana, J., Yang, P. D. (2001) Langmuir-Blodgett nanorod assembly, *J. Am. Chem. Soc.* **123**, 4360-4361.
71. Onsager, L. (1949) The effect of shape on the interaction of colloidal particles, *Ann. (N.Y.) Acad. Sci.* **51**, 627-659.
72. Nikoobakht, B., Wang, Z. L., El-Sayed M. A. (2000) Self-assembly of gold nanorods, *J. Phys. Chem. B* **104**, 8635-8640.
73. Giannelis, E. P. (1996) Polymer layered silicate nanocomposites, *Adv. Mater.* **8**, 29-35.
74. Nikl, M., Bohacek, P., Mihokova, E., Kobayashi, M., Ishii, M., Usuki, Y., Babin, V., Stolovich, A., Zazubovich, S., Bacci, M. (2000) Excitonic emission of scheelite tungstates AWO_4 (A = Pb, Ca, Ba, Sr), *J. Lumin.* **87**, 1136-1139.
75. Qi, L. M., Cölfen, H., Antonietti, M. (2001) Synthesis and characterization of CdS nanoparticles stabilized by double-hydrophilic block copolymers, *Nano. Lett.* **1**, 61-65.
76. Peytcheva, A., Antonietti, M. "Carving on the nanoscale": Polymers for the site-specific dissolution of calcium phosphate, (2001) *Angew. Chem. Int. Ed.* **40**, 3380-3383.
77. Chirila, M. M., Stevens, K. T., Murphy, H. J, Giles, N. C. (2000) Photoluminescence study of cadmium tungstate crystals, *J. Phys. Chem. Solids* **61**, 675-681.
78. Polak, K., Nikl, M., Nitsch, K., Kobayashi, M., Ishii, M., Usuki, Y., Jarolimek, O. (1997) The blue luminescence of $PbWO_4$ single crystals, *J. Lumin.* **72-74**, 781-783.
79. Toyozawa, T., Inoue, M. (1966) Dynamical Jahn-Teller effect in alkali halide phosphors containing heavy metal ions, *J. Phys. Soc. Japan* **21**, 1663.

NANOCRYSTALLINE Er^{III}/Si^{IV}@ZnO MULTILAYERS: A DETAILED OPTICAL AND STRUCTURAL STUDY

M.KOHLS[1], G. MÜLLER[1], L. SPANHEL[2], C. URLACHER-LELUYER[3], J. MUGNIER[3], J. DUMAS[4], J.C. PLENET[4], G. MCMAHON[5], D. SU[6], AND M. GIERSIG[6]

[1] *Lehrstuhl für Silicatchemie, Bayerische Julius-Maximilians-Universität Würzburg, Röntgenring 11, D-97070 Würzburg, Germany*

[2]*Laboratoire Verres et Céramiques, Université de Rennes 1, CNRS UMR 6512, Institute de Chimie de Rennes, CS 74205, 35042 Rennes Cedex, France*

[3]*Laboratoire de Physico-Chimie des Matériaux Luminescents, Université Claude Bernard Lyon 1, CNRS UMR 5620, 43 Bd. du 11 Novembre,69622 Villeurbanne Cedex, France*

[4]*Département de Physique des Matériaux, Université Claude Bernard Lyon 1, CNRS UMR 5586, 43 Bd. Du 11 Novembre, 69622 Villeurbanne Cedex, France*

[5]*CANMET, Materials Technology Laboratory, 568 Booth Street, Ottawa, Ontario K1A0G1, Canada*

[6]*Hahn-Meitner Institut, Abt. Physikalische Chemie, Glienickerstr. 100, 15109 Berlin, Germany*

1. Introduction

Structural and optical studies on erbium-doped nanomaterials are of broad interest due to their large potential in advanced photonic applications [1-8]. Their wet chemical processing (sol-gel, solution combustion synthesis, colloidal routes etc.) has been largely explored using different erbium hosts such as titania [1], zirconia [2], II-VI-semiconductors [3,4], aluminosilicates [5], silica [6,7] as well as yttria [8]. Of particular interest to us are the Er^{III}-doped planar wave guides prepared from so-called "polymeric sols" containing weakly branched molecular clusters. Previous optogeometrical studies [1,2] on these sol-gel nanostructures have indicated that thermal annealing of wet gel layers seems to be the most crucial processing step. Critically:

1. elimination of high energy vibrations (liberating organics and OH-groups),
2. blocking of Er-clustering as the result of phase separation/transformation processes
3. suppression of nanocrystal formation/growth which give rise to increased scattering are needed in order to achieve the optimal active waveguide performance (reduced damping losses and a long life time of the photoexcited Er^{III} ions).

Following our recent brief communication [4], we present in this paper a detailed optical and structural study of multilayers prepared from "particulate sols" composed of 5-6 nm large ZnO nanocrystals. We will show that co-doping (inclusion of Si^{IV} ions in addition to Er^{III}) of nanocrystals plays a decisive role in controlling the lifetime of the NIR-Er^{III} emission ($^4I_{13/2} \rightarrow {}^4I_{15/2}$ transition centred around 1540 nm) and the resulting film quality.

L.M. Liz-Marzán and M. Giersig (eds.), Low-Dimensional Systems: Theory, Preparation, and Some Applications, 107–120.

2. Experimental Section

2.1 SYNTHESIS AND CO-DOPING OF ZnO COLLOIDS

21.95 g $Zn(Ac)_2$ dihydrate were dispersed in 155 mL 1-propanol and refluxed in oil bath at 120-130 °C for 10 minutes. During this time, 20-30 mL of the solvent phase were distilled away. Subsequently, 45 mL TMAH base were rapidly added to the still hot and strongly stirred precursor solution. At that time, one could observe a sudden formation of a white gel-like precipitate which disintegrates within a few minutes yielding an optically clear sol. The volume of the sol was reduced to 50 mL by rotary evaporation resulting in a 2M ZnO nanocolloid, which has been finally purified by membrane microfiltration.

To this pre-cleaned sol, dehydrated erbium acetate powder was added which readily dissolves within a few minutes under vigorous stirring. The Er-concentrations were varied between 0.5 and 10 at.% (with respect to Zn). Subsequently, TEOS (tetraethoxy-silane) was added to the Er@ZnO sol. Typically, atomic Zn/Si^{IV} ratios between 5 and 40 were used. The resulting colloids were stirred for 1 h prior to coatings.

2.2 FILM PREPARATIONS

The dust free sols (pressed through 0.2 μm Millipore filters prior to coatings) were used to prepare thin films on pre-cleaned Herasil-1® fused silica substrates via dip coating techniques. The dynamic viscosity of the coating solutions ranging between 2.1 and 2.9 mPa s was adjusted by diluting the 2 M sols with an appropriate amount of ethanol. The substrate withdrawal speed was adjusted to 2 - 13 cm min^{-1}. The wet films were directly transferred into a sintering oven and thermally cured between 400 °C and 900 °C.

2.3 OPTICAL SPECTROSCOPY

Optical absorption spectra were collected with a *Perkin Elmer Lambda* spectrophoto-meter. The time resolved NIR-fluorescence investigations were performed at room temperature with a pulsed Nd/YAG-OPO-laser (pump power = 10-40 mJ, pulse width = 3-5 ns) in combination with a *Jobin-Yvon* 270M monochromator connected to an InGaAs-detector and a *Le Croy 9361* oscilloscope. For the film characterisation under wave guiding conditions, a coupling prism method was used as depicted in Figure 1.

Either rutile (n = 2.584 at 632 nm) or a LaSF35-prism (n = 2.015) were utilised to launch the light waves into the ZnO films while the propagating TE- and TM-modes were selected with a polarization filter. Three different wave guide techniques were employed to characterise the ZnO films:

(1) *NIR-fluorescence measurements* (part a, Figure 1). Ar^+- laser pumped Ti-Sapphire laser (300 mW) served to excite Er^{III} ($^4I_{15/2} \rightarrow {}^4I_{9/2}$ transition at 800 nm) within the ZnO waveguide. The 1.5 μm fluorescence spectra were collected with a Jobin Yvon U1000 monochromator and a nitrogen cooled Ge-detector in combination with a Le Croy 9400 Dual 125 MHz-oscilloscope.

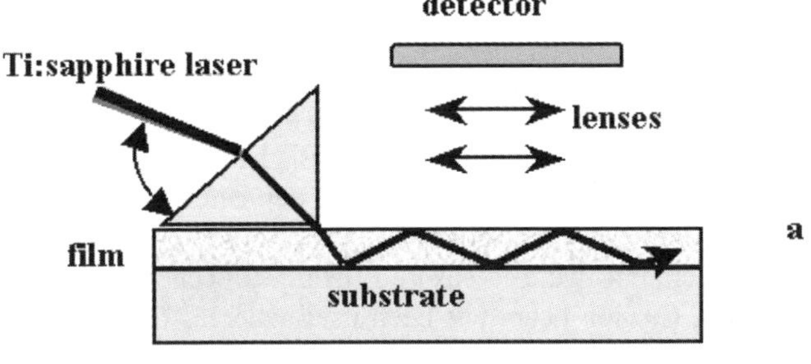

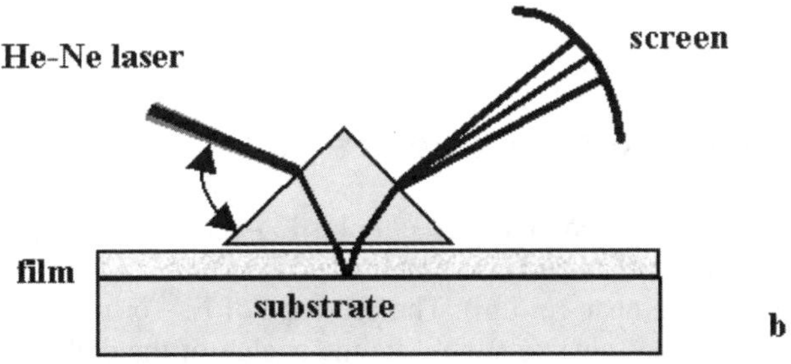

Figure 1. Experimental set-up of waveguide spectroscopic techniques [1,2].

(2) *Light damping experiments* (part a, Figure 1) were conducted using a red He-Ne laser light (632.8 nm, 10 mW) coupled into the ZnO film while the propagation distance was recorded with a CCD camera. For propagation distances below 0.5 cm, the damping coefficient α was approximated with the well known Tien-formula

$$\alpha = 27/L \ [\mathrm{dB\ cm^{-1}}] \tag{1}$$

where L is the experimentally observed propagation distance. Alternatively, due to the distance - L dependent scattering light intensity - I_s of the propagating modes, the value of α was calculated according to

$$-\alpha \times \Delta L = 10 \times \log [I_s (L + \Delta L) / I_s] \tag{2}$$

(3) *"m- line spectroscopy"* (part b, Figure 1) was used to determine the thickness and the effective refractive index of the films. For this purpose, HeNe-laser light (543.5 nm, 7 mW) was coupled into the films using a symmetric prism that allows one to observe, under certain incident angle conditions, the so-called "black line" on a screen. With this technique and using the following formula

$$N_m = \sin\delta_m \times \cos\sigma + (n_p{}^2 - \sin^2\delta_m)^{1/2} \times \sin\sigma \tag{3}$$

the effective refractive index N_m of the m-th mode was calculated where n_p is the refractive index of the prism, δ_m is the incident angle of the mode and σ is the prism angle.

2.4. STRUCTURAL AND ANALYTICAL MEASUREMENTS

X-ray investigations on sols and films were carried out at room temperature with a STOE STADI P diffractometer (CuKα_1-radiation, λ = 1.5406 nm, Bragg-Brentano-geometry).

Secondary Ion Mass Spectrometry (SIMS) was used to determine the elemental distribution of the elements in co-doped ZnO films deposited on Si-wafers. All analyses were performed using a Cameca double-focusing magnetic sector SIMS. After the analyses, depths of the sputter craters were measured using a Tencor profilometer. For the determination of oxygen, Cs^+ ion beam has been used whereas oxygen ion beam served to record the atomic profiles of Zn, Er, Si and B. To visualise the interface regions within multiple layers, focused ion beam (FIB) and secondary electron beam (SEB) images were recorded.

3. Results

3.1. CO-DOPING OF PARTICULATE ZnO SOLS

In our previous studies on alcoholic ZnO colloids [3,4] it has been shown that dissolution of erbium acetate in particulate ZnO sols activates the characteristic Er^{III} NIR-fluorescence ($^4I_{13/2} \rightarrow {}^4I_{15/2}$ transition near 1.5 μm). The presence of Er^{III} or other trivalent cations (Al^{III}, In^{III}, B^{III}) in ZnO particles also partly blocks the ageing of the colloids [9]. This process of slow nanocrystal growth can be completely suppressed by adding TEOS to the sols which can be monitored using optical absorption spectroscopy. Figure 2 displays the changes in optical spectra of three different ZnO sols (0%, 5 at.% and 20 at.% TEOS) taken after 3 hours, 2 days and 2 weeks. Without TEOS, a red-shift of the absorption spectra is noticed (crystal growth within the quantum confinement regime) which coincides with the observations generally made with these undoped samples (see for example [10]).

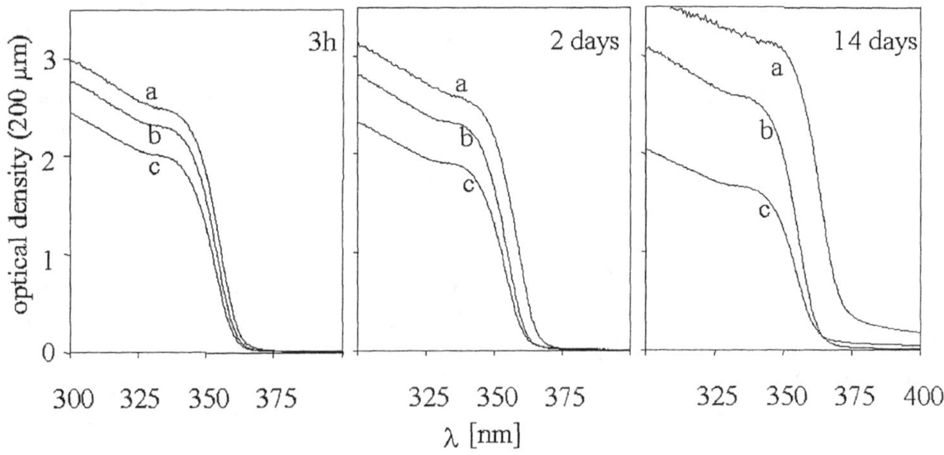

Figure 2. Effect of TEOS on the growth of 0.5 M ZnO nanocolloids in 1-propanol under ambient laboratory conditions. a: without TEOS, the crystallite size increased from 3 nm to 5 nm after 2 weeks; b: 5 at.% TEOS, c: 20 at.% TEOS (with respect to Zn).

As 20 at.% TEOS are added to the ZnO sol, the spectral red-shift is no longer detected and after 4 weeks, nanocolloid gelation is observed. The suppressed growth in these slightly alkaline alcoholic sols (pH ~ 8-9) can be understood assuming a deposition of TEOS onto hydroxylated $(ZnO)_n$ particles according to

$$(ZnO)_n\text{-}OH + EtO\text{-}Si(OEt)_3 \rightarrow EtOH + (ZnO)_n\text{-}O\text{-}Si(OEt)_3 \quad (4)$$

whereas the subsequent very slow hydrolysis and condensation of chemisorbed $Si(OEt)_3$ ligands (abbreviated further below as SiR_3) might produce silica-bridged $(ZnO)_n$ particles and thus, a sol-gel transition after one month:

$$(ZnO)_n\text{-}O\text{-}SiR_3 + H_2O \rightarrow EtOH + (ZnO)_n\text{-}O\text{-}SiR_2\text{-}OH \quad (5)$$

$$(ZnO)_n\text{-}OSiR_3 + (ZnO)_n\text{-}OSiR_2\text{-}OH \rightarrow EtOH + (ZnO)_n\text{-}OSiR_2\text{-}O\text{-}SiR_2O\text{-}(ZnO)_n \quad (6)$$

or

$$(ZnO)_n\text{-}OH + (ZnO)_n\text{-}O\text{-}SiR_2\text{-}OH \rightarrow EtOH + (ZnO)_n\text{-}O\text{-}SiR_2\text{-}O\text{-}(ZnO)_n \quad (7)$$

The minor drop in optical density after 2 weeks (sample c, Figure 2) could stem from a partial decomposition of the ZnO particles. For example, a slow uptake of Zn ions by the present Si-OH groups could take place:

$$(ZnO)_n + HO\text{-}SiR_2\text{-}O\text{-}(ZnO)_n \rightarrow (ZnO)_{n-1} + HOZnO\text{-}SiR_2\text{-}O\text{-}(ZnO)_n \quad (8)$$

To examine the above proposed interfacial reactions (4)-(8), we performed time resolved ^{29}Si-NMR measurements. In these experiments we firstly dissolved 2 at.% Er^{III} salt in a 2M ZnO sol and then, after adding 20 at.% TEOS, we collected the ^{29}Si-NMR spectra after different times. The result is shown in Figure 3 below, displaying two representative spectra of this study:

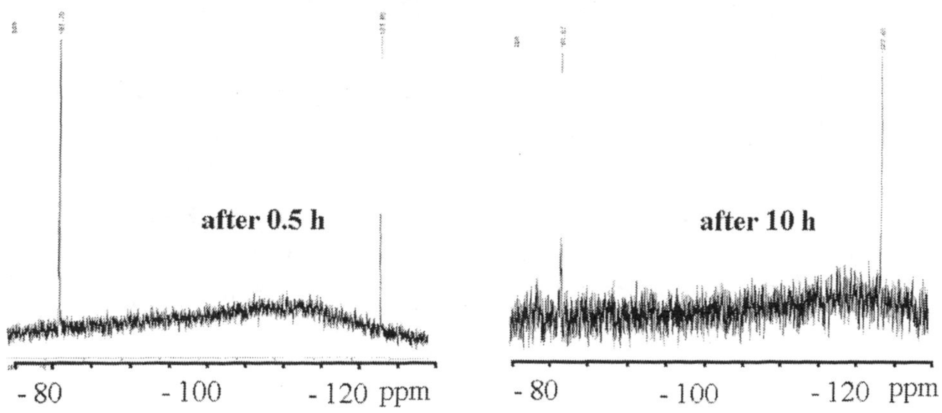

Figure 3. ^{29}Si-NMR spectra of 1-propanolic Er,Si:ZnO colloid (2M ZnO, 2 at.% Er^{III}, 20 at.% TEOS) 0.5 hours and 10 hours after TEOS addition (TMS was the reference). The increased signal to noise ratio in the right part is due to shorter acquisition time chosen for 10 hours old sols.

One recognises a slow disappearance of TEOS signal at –81.7 ppm and concomitant appearance of one single peak at –123.7 ppm. After 20 hours, the TEOS was completely consumed (the –81.7 ppm signal disappeared) while the peak at –123.7 ppm reached its maximum intensity. Furthermore, in the absence of Er^{III}, the single product peak appeared at lower fields nearby –106 ppm (not shown in the Figure). From these results we note: Firstly, the observed high field shift of the product peak in the presence of Er^{III} is indicative of Si-O-Er sequences. That these sequences play an important role in blocking the phase separations and ZnO crystallisation will be shown further below in Section 3.2.

Secondly, the unexpected observation of just one high-field shifted peak indicates a relatively rapid reaction producing one single product. The expected intermediary Q^{ij} species (as generally detected during hydrolysis and condensation of TEOS [11]) are missing. Here, Q denotes the tetravalent Si atom, "i" represents the number siloxanes $Si(-O-Si)_i$ or $Si(-O-Zn)_i$ bridges attached to Si and "j" denotes the number of silanol groups $Si(-OH)_j$. At present, only a tentative explanation can be given: It appears that the observed products are probably Q^{40} or Q^{50} states situated inside or between the ZnO particles. The *inside-location* could be understood taking into account the previously detected mass fractality [12,13], e.g., the presence of differently sized interstitial cavities within the porous ZnO particles. There also probably exist a minor concentration of undetectable and condensable Q^{22} or Q^{20} sites responsible for the aforementioned gelation after several weeks (overall interconnection of ZnO particles).

To visualise the ZnO nanocrystals, we performed HRTEM measurements before and after the co-doping. Without Er^{III} and Si^{IV}, the as-prepared ZnO particles generally are nearly spherical, as already reported by several groups. Surprisingly, this is not the case if one observes the electron microscopic images of co-doped ZnO nanocrystals in Figure 4. Most of the ZnO nanocrystals are nearly triangular in shape (obviously tetrahedral in 3D space, edge lengths around 6 nm) and often stick together either at the vertexes or along the edges. This is also seen if we use tin alkoxides instead of TEOS (part c in Figure 4).

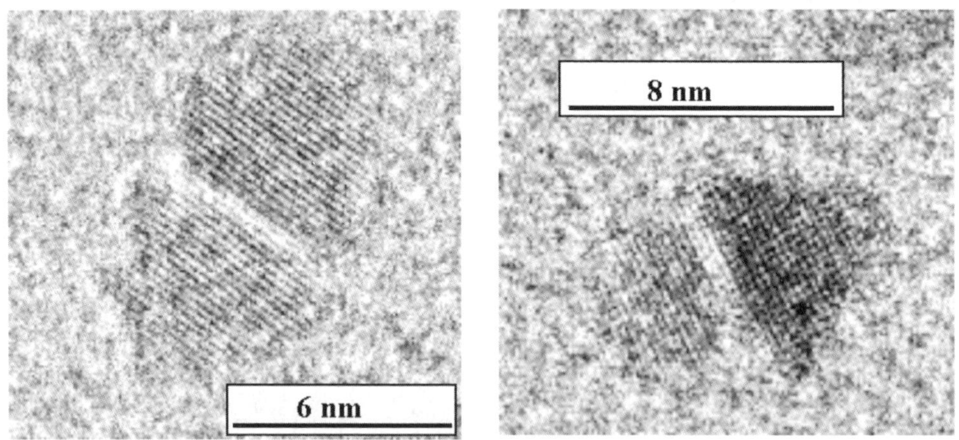

Figure 4. HRTEM images of highly diluted co-doped ZnO sols. Part (a): 2 at.% Er^{III}, 5 at.% Si^{IV}, part (b): 2 at.% Er^{III}, 5 at.% Sn^{IV}

The observed lattice planes reflect the presence of hexagonal zincite which coincides with the XRD data (see following Section). To conclude, the above results suggest tetrahedrons to be the primary growth products which become spherical later on if co-doping is not performed. Indeed, the mechanism of the nucleation and growth appears to be more complicated than that based on classical Ostwald theories.

3.2. STRUCTURAL STUDY OF THERMALLY ANNEALED Er,Si@ZnO-LAYERS

In this section, we discuss structural changes observed during sintering of the "co-doped" ZnO nanocrystals. We begin with the X-ray diffraction data (Figure 5) collected on Si@ZnO-, Er@ZnO- and Si,Er@ZnO films. The JCPDS powder data are also included along with the calculated average crystallite sizes. Here, an uncertainty of ± 15 % exists for all given sizes calculated from the FWHM values of the diffraction peaks using the well known Debye-Scherrer formula [14].

Part (a) of Figure 5 compares the XRD fingerprints of sintered ZnO layers (at 800 °C in air) containing in each case 2 at. % Er^{III} but different Si^{IV} amounts. We observe the same tendency of Si^{IV} to block the growth as already discussed above for the colloidal ZnO samples. Namely, with increasing Si^{IV} content, the diffraction peaks of the sintered nanophases (hexagonal zincite) become increasingly less intense and broader. The initially 6 nm large crystals grow up to 35 nm if no TEOS is used whereas they only double their size to 12 nm in the presence of 20 at.% Si^{IV}. Furthermore, in the diffraction patterns of films with 20 at.% Si^{IV}, orthorhombic zinc *ortho*-silicate nano-domains show up. Their concentration is rather low with respect to the present ZnO particles.

Now we focus on part (b) of Figure 5 displaying the evolution of Si^{IV}-free ZnO XRD-patterns as a function of Er^{III} concentration. Again, blocking of further crystallisation and growth takes place that was already noted in earlier studies on $Er:Ti^{IV}$ systems [15]. Without co-doping, the ZnO nanocrystals grow from 6 nm to 85 nm while they remain small with increasing Er^{III} content. Above 4 at.% Er^{III}, they grow to 25 nm and in addition, a simultaneous formation of cubic erbium oxide phases is detected. We do not want discuss the origin of the preferred (100) orientation of the ZnO nanocrystals (recognised by comparing the film patterns with the JCPDS powder data) that has been already reported in [9]. Particularly interesting is the diffractogram of a film sample with 5 at.% Si^{IV} and 10 at.% Er^{III} (most upper fingerprint in part (b) of *Figure 5*). The absent phase separation shows that Si^{IV} blocks not only the ZnO crystal growth but it also suppresses the Er_2O_3 formation. We can only speculate what kind of nanostructures exist in these "overdoped" nanocrystalline ZnO films free of foreigner crystalline nanophases. Perhaps, some sort of alloys are formed composed of interconnected ZnO particles with interpenetrated molecular -Si-O-Er-O-Si- sequences. Their presence was already suggested from the ^{29}Si-NMR data of the co-doped ZnO colloids (Figure 3) and also discussed by the interpretation of the FTIR film spectra [16]. We exclude the case where Er^{III} is entrapped inside small silica clusters without structural contribution of ZnO, since from silica glass technologies it is well known that the dominant Er_2O_3-SiO_2 phase separation is only blocked if one uses additionally Al^{III} [17].

To further explore the sintering induced ZnO growth, we collected crystal size data on films with 5 at.% Si^{IV} and 2 at.% Er^{III} as function of sintering temperature. The result is shown in part (c) of Figure 5 along with size data of other Er^{III}/Si^{IV} concentration conditions.

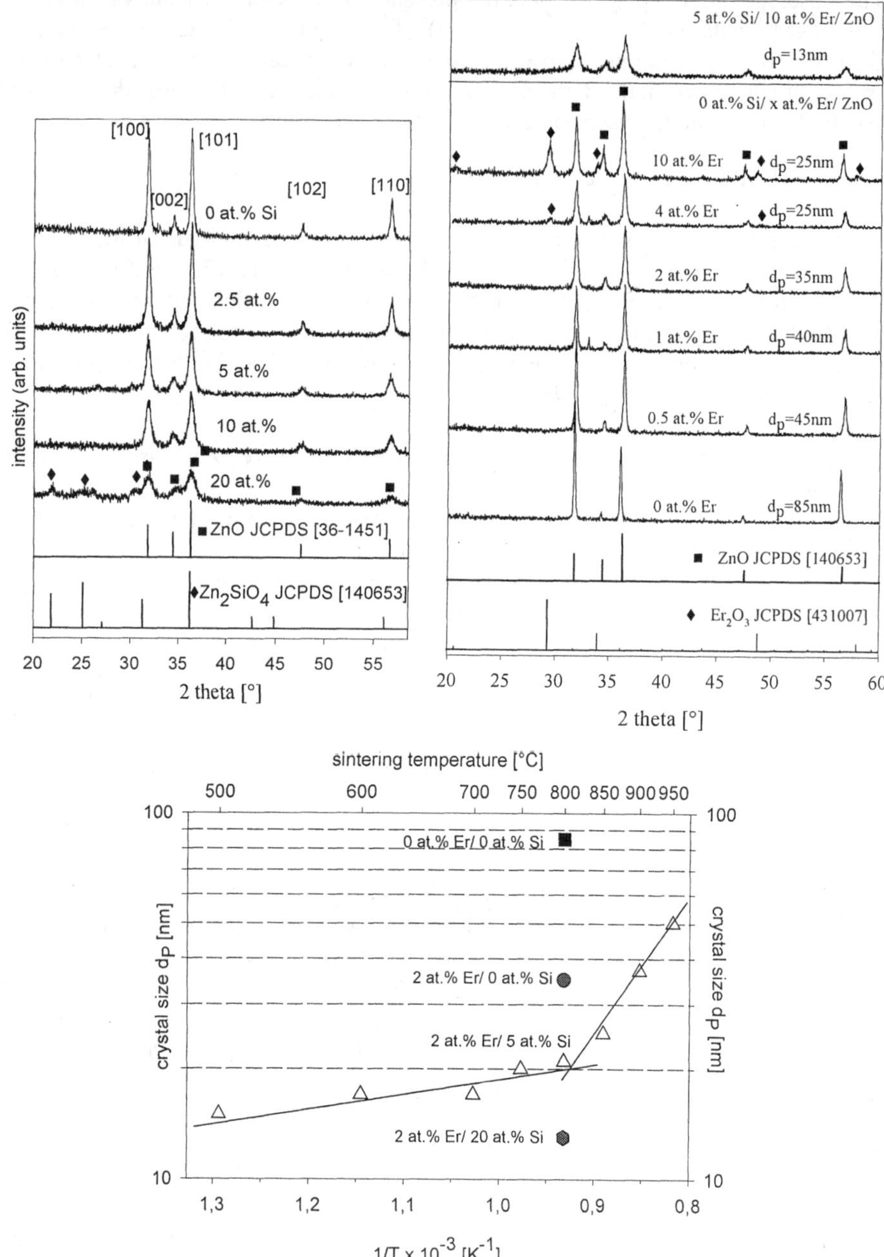

Figure 5. XRD spectra of ZnO films of different chemical compositions thermally annealed at 800°C in air. Part (a): 2 at.% Er^{III} and x at.% Si^{IV} (0 < x < 20); Part (b): x at.% Er^{III} (0< x < 10) without Si^{IV} (0 < x < 20), for comparison, a difffractogram of a [10 at.% Er^{III} + 5 at.% Si^{IV}]- composition is also included; part (c): Semilogarithmic plot of the ZnO crystal size (2 at.% Er^{III} and 5 at.% Si^{IV}) plotted against the reciprocal sintering temperature. For details see text.

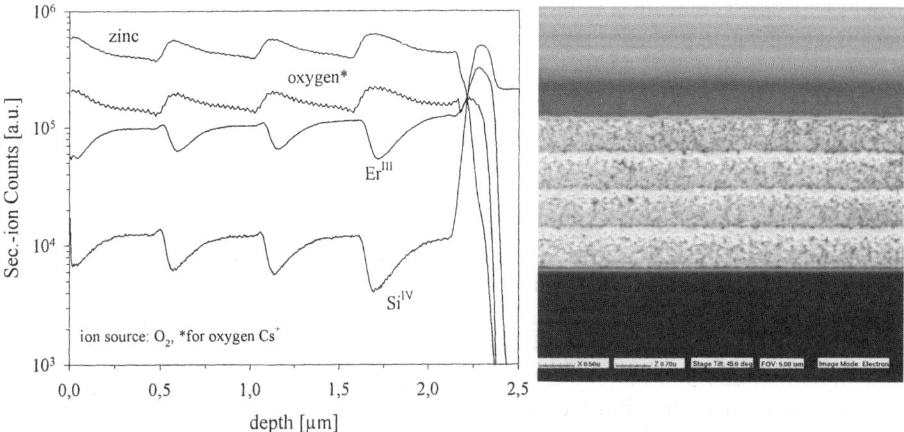

Figure 6. SIMS-depth profile (at the left) and FIB-image (at the right) of 4-fold coated ZnO multilayer (5 at.% Si^{IV} and 2 at.% Er^{III}) deposited on a Si-wafer. For experimental details see text and Section 2.

Two different growth mechanisms exist between 500 °C and 950 °C reflected by two linear regions of the log (size) vs. 1/T plot, from which we can determine the corresponding activation energies E_a.The E_a value of the low T-regime is 21 kJ/mol suggesting a surface diffusion process where the crystals grow from ~12 nm at 500°C to ~21 nm at 800°C. This process is coupled to the liberation of organics and disappearance of OH-groups as seen in complementary FTIR measurements [16]. On the other side, the E_a value of the high T-regime was determined to be 195 kJ/mol reflecting a reaction controlled process during which the ZnO crystals further double their size from 21 nm to 45 nm. We believe that it is the begining of the zinc silicate formation that explains the significantly higher activation barrier for crystal growth above 800°C.

We close this chapter by presenting some Secondary-ion-mass spectroscopy (SIMS) results and focused-ion-beam (FIB) image of co-doped ZnO multilayers. Each successive ZnO layer (with 5 at.% Si^{IV} and 2 at.% Er^{III}) was first preheated at 400 °C under constant flow of dry air and after the 4-fold coating, the sample was finally annealed at 800°C in ambient air. Part (a) of Figure 6 shows the SIMS depth profile of this multilayer deposited on a Si-wafer (part a).

The periodicity of SIMS atom concentration profiles correlates with the number of coatings observed in the FIB-image (see Figure 6) and additionally indicates that within each of the four layers, a concentration gradient exists. Particularly interesting is the asynchronicity of the detected Er/Si- and Zn/O - profils within each of the 4 successive layers. This trend correlates with the XRD data of Figure 5 where it is shown that Si^{IV} blocks the formation of separated Er_2O_3 and ZnO nanophases. Indeed, the proposed existence of Si-O-Er-O-Si-O-sequences eliminating the Er^{III} clustering within the ZnO nanocrystals also explains the synchronic periodicity of the Er^{III}/Si^{IV} patterns. The clearly visible interfacial regions between the successive layers (see the FIB- picture) deserve a few final comments. These regions are probably formed via partial infiltration of the rough upper parts of pre-cured layers with incoming ZnO nanoparticles of the successive coatings. Then, in the finally cured multilayers, the nanomaterial is more dense between, than within, the layers. This and the SIMS-asynchronicity might create refractive index gradients, a kind of Bragg reflector similar to a photonic cavity, which might explain the recently reported noticeable optical gain of these multilayers [4].

3.3. OPTICAL PROPERTIES OF SINTERED Er^{III},Si^{IV}-ZnO MULTILAYERS

3.3.1 Light propagation losses and refractive index measurements

That Er^{III} inhibits the crystallisation in zirconia and titania waveguides elaborated from polymeric cluster solutions is well known [2,15]. In our film samples prepared from particulate Si^{IV},Er^{III}@ZnO sols, the Si^{IV} or Sn^{IV} play a similar role, preventing the already existing ZnO nanocrystals from further growth. This co-doping step enables light propagation and refractive index adjustments in the resulting multilayers. Up to 3 TE- and TM- modes could be coupled into the 1 μm thick films whereas at least one mode was seen by thickness between 300 nm and 600 nm. It should be mentioned that the quality of the final film depends upon the film preparation conditions. Intrinsic morphological defects, the number of coatings as well as the sintering temperature all strongly influence the light propagation distance. This is seen in the camera images of a propagating TE_0-mode inside five differently prepared Er^{III} free 20%Si^{IV}@ZnO films, shown in Figure 7.

The presence of micro-cracks in multiply coated films causes shortening of the propagation distance and dispersion of light as can be seen in part (a) of the above Figure. In crack-free samples, the propagation distance increases with the number of coatings, as easily recognised by comparing images (b) and (c). Furthermore, the light propagation shortens by raising the sintering temperature from 350°C to 750°C (compare c,d,e). To gain further quantitative optical and structural information, we employed waveguiding spectroscopies (see Section 2.3) to determine the damping losses and refractive indices of different multilayers. From the experimental refractive indices we additionally calculated the corresponding porosity values. The results of these investigations are summarised in Table 1.

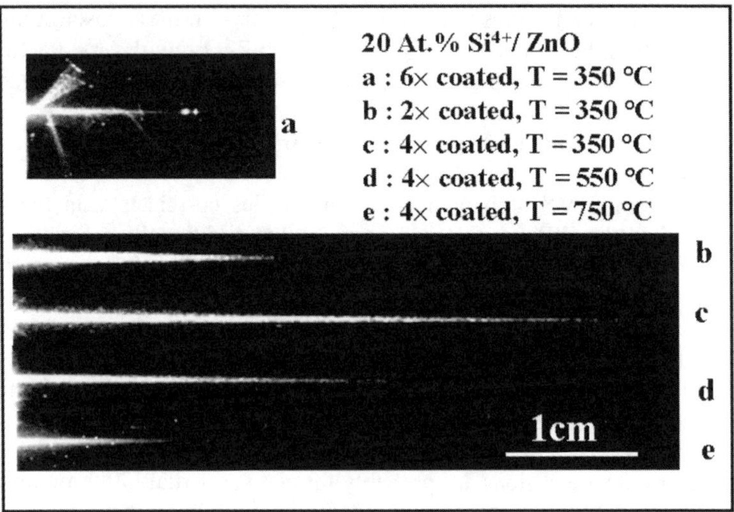

Figure 7. Camera images of a propagating HeNe-laser light (10 mW) inside five differently prepared 20%Si^{IV}@ZnO films. Here, Si/SiO_2-wafers were used as the substrates.

TABLE I. A comparison of optical/structural parameters of selected ZnO - film samples elaborated under different preparation conditions

doping conditions & sintering temperature	refractive index (λ = 543.5nm)	Damping [dB/cm]	porosity [%]
20 At.% Si^{IV}, 350°C 550°C 750°C	1.533	2.81 6.39 9.25	46.02
5 At.% Si^{IV}, 350°C	1.602		44.93
5 At.% Si^{IV} 2 At.% Er^{III}, 350°C	1.621		42.78
5 At.% Si^{IV} 2 At.% Er^{III}, 750°C	1.639		40.72

From the above table it is evident that with increasing Si^{IV} content the refractive index decreases while it increases on the addition of Er^{III}. The sintering at higher temperatures makes the multilayers more dense and less porous, which raises the refractive index. Consequently, the multilayer collapses, which causes a reduction in thickness from 630 nm at 350 °C to 542 nm at 750 °C. There are several sources of increased damping at higher sintering temperatures. For example, the nanocrystal boundary scattering and internal reflections due to asynchronic atom concentration gradients or inhomoge-neously aggregated particles forming pores of different sizes. Sol and film preparations in clean dust free rooms would also improve the light propagation.

3.3.2 Er^{III}-NIR-Fluorescence

Due to the low absorption coefficient of electronic 4f-transitions, the characteristic Er^{III}-1.54 μm fluorescence is only observed in bulk materials or in thin films of rather higher optical quality. The sensitivity of these measurements is largely increased by employing waveguide techniques. Figure 8 displays the characteristic NIR-fluorescence data of Er^{III},Si^{IV}@ZnO waveguides sintered between 400°C and 950 °C. On the left, the broad emission spectra excited at 800 nm ($^4I_{9/2}$ state) show Stark splitting and slightly shift to higher energies with increasing sintering temperature. This indicates structural changes in the local Er^{III} environment. Such observations were not possible by exciting the samples under traditional front face conditions. Even if excited at 520 nm ($^2H_{11/2}$ state), where Er^{III} has a much higher absorption coefficient, front face techniques were unable to deliver data with films sintered below 600 °C.

The inset of the right part of Figure 8 shows how the mean lifetime of the 1.5 μm fluorescence ($^4I_{13/2}$ state) increased with rising sintering temperature, which correlates with the disappearance of OH-groups and organic acetates ligands as seen in FTIR spectra. The thermal removal of these high energy vibration sources additionally accompanied from the partial elimination of -Er-O-Er- clustering is needed to suppress the rapid nonradiative deactivation that has been already noted in studies of several other research groups. The presence of clustered and nonclustered Er^{III} can be checked by recording the fluorescence decay curves shown in part (b) of Figure 8. One recognises a complex decay containing a fast (< 50 μs) and a slow millisecond time component. These two time components reflect the presence of different Er^{3+} coordination states.

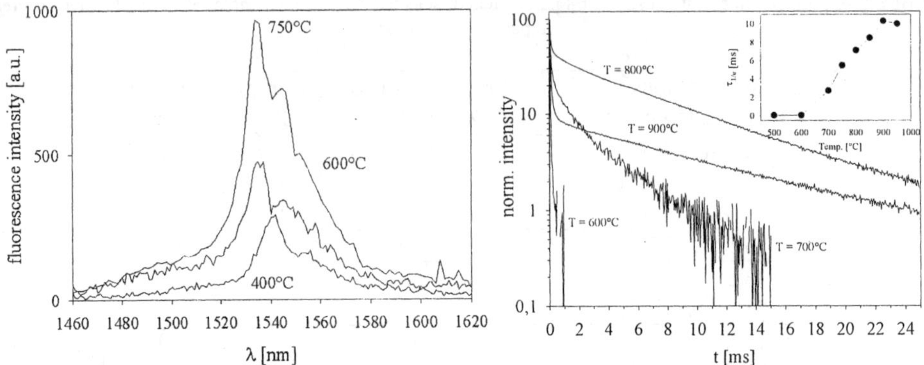

Figure 8. NIR-fluorescence data of 2At.%Er^{III},5At.%Si^{IV}@ZnO-films pre-heated between 400 °C and 950 °C. The emission spectra were taken in waveguiding modus while the lifetime data were collected via front face excitation. Both studies were performed under ambient laboratory conditions. For experimental details see Section 2.3.

We attribute the fast decay component to the presence of fast energy relaxation along the clustered Er^{3+} ions. The slow ms-decay component comes from homogeneously distributed non-clustered Er^{3+} ions which is the situation required to produce an efficient amplifier. The ratio of the both components is strongly dependent on sintering temperature and chemical post-treatments of the sols and films. Below 600 °C, the fast decay component dominates the NIR-fluorescence decay. Between 600 and 800 °C, the lifetime rises from < 20 μs up to 4-5 ms, due to the removal of OH-groups and the breakage of the Er-O-Er-clusters. A co-doping of Er/ZnO sols with TEOS rises the mean fluorescence lifetime further to about 7 ms after sintering at 800 °C. At 900 °C however, the concentration of non-clustered Er^{III} drops down due to the beginning phase separation (zinc silicate formation). Concomitantly, the fast decay component starts to dominate the NIR fluorescence as clearly seen in Figure 8.

Aided by their amphoteric nature, the Si^{IV},Er^{III}@ZnO films can be micropatterned employing conventional photolitographic techniques combined with a wet chemical acidic etching [4,18]. The resulting stripe wave guides exhibit fluorescence life times about 10 ms with a significantly decreased contribution of the fast decay component. This can be seen in Figure 9. at left. Such micropatterns exhibit a high internal optical gain ranging between 60 and 120 cm^{-1} at room temperature. This corresponds to approximately 12 dB at a pumping distance of only 500 μm. In comparison, conventional several meters long Erbium-doped glass fiber amplifiers exhibit a net gain of about 10-15 dB/cm.

As can be seen on the right side in Figure 9, the mean lifetime $\tau_{1/e}$ and the intensity of the NIR-fluorescence are also strongly dependent on the Er^{III} concentration. At concentrations above 2 at.%, both the life time and the intensity of the 1.54 μm fluorescence decline due to ongoing Dexter energy transfer processes in increasingly clustered Er^{III} . Perhaps, future multiple doping of the ZnO nanocrystals with three or four foreigner cations might further improve the quality of the above discussed nanomaterial.

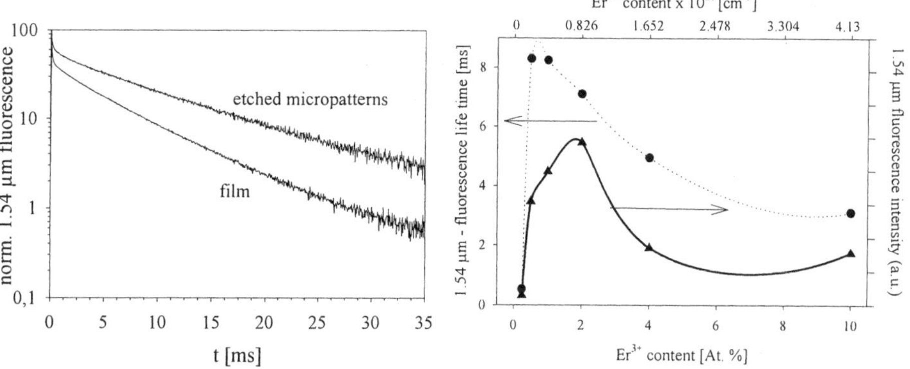

Figure 9. left: effect of micropatterning via a wet acidic etching on the NIR-fluorescence life time of 2 at.%Er^{III}, 5 at.%Si^{IV}@ZnO layers sintered in air at 750 °C; right: the mean life time and NIR-fluorescence intensity as a function of the Er^{III} content.

4. Acknowledgement

This work was supported by the BMBF/Siemens AG (project 03N1007C9). We are thankful to Dr. R. Bertermann for NMR measurements. We thank Prof. P. Mulvaney for critical reading of the manuscript and the valuable comments and suggestions.

5. References

1. Bahtat, A., Bouazaoui, M, Bahtat, M., Garapon, C., Jacquier, B. and Mugnier J. (1996) Up-conversion fluorescence spectroscopy in Er^{3+} :TiO_2 planar waveguides prepared by a sol-gel process, *J. Non-Cryst. Solids* **202**, 16-22.
2. Urlacher, C., Marco de Lucas, C. and Mugnier, J. (1997) Chemical and physical aspects of sol-gel process for planar waveguides elaboration: application to zirconia waveguides, *Synthetic Metals* **90**, 199-204.
3. Schmidt, T., Spanhel, L., Müller, G., Kerkel, K. and Forchel, A. (1998) Activation of 1,54 µm – Er^{3+}-Fluorescence in Concentrated II-VI-Semiconductor Cluster Environments, *Chem. Mater.* **10**(1), 65-74.
4. Kohls, M., Schmidt, T., Katschorek, H., Spanhel, L., Müller, G., Mais, N., Wolf, A. and Forchel (1999) A Simple Colloidal Route to Planar Micropatterned Er@ZnO – Amplifiers, *Adv. Mater.* **11**(4), 288-292.
5. Orignac, X., Barbier, D., Du, X., Almeida, R.M., McCarthy, O. and Yeatman, E. (1999) Er-Alumosilica via sol-gel, *Opt. Mater.* **12**, 1-12.
6. Sloof, L.H., de Dood, M.J.A., van Blaaderen, A. and Polman, A. (2000) Erbium-implanted silica colloids with 80% luminescence efficiency, *Appl. Phys. Lett.* **76**, 3682-3684.
7. Lipson, M. and Kimerling, L.C. (2000) Er^{3+} in strong light-confining microcavity, *Appl. Phys. Lett.* 77, 1150-1152.
8. Capobianco, J.A., Vetrone, F., Boyer, J.Ch., Speghini, A. and Bettinelli, M. (2002) Enhancement of Red Emission via Upconversion in Bulk and Nanocrystalline Cubic Y_2O_3:Er^{3+}, *J. Phys. Chem. B* **106**, 1181-1187.
9. Hilgendorff, M., Spanhel, L., Rothenhäusler, Ch. and Müller, G. (1998) From ZnO Colloids to Nanocrystalline Highly Conductive Films, *J. Electrochem. Soc.* **145**, 3632-3638.
10. Van Dijken, A., Meulenkamp, E.A., Vanmaekelbergh, D. and Meijerink, A. (2000) The kinetics of the radiative and nonradiative processes in nanocrystalline ZnO particles upon photoexcitation, *J. Phys. Chem. B* **104**, 1715-1723.
11. Lee, K., Look, J.L., Harris, M.T. and McCormick, A.V. (1997) Assessing Extreme Models of the Stöber Synthesis Using Transients under a Range of Initial Compo-sition, *J. Colloid Interface Sci.* **194**, 78-88.
12. Lorenz, C., Emmerling, A., Fricke, J., Schmidt, T., Hilgendorff, M. and Spanhel, L. (1998) Aerogels containing strongly luminescing ZnO nanocrystals, *J. Non.-Cryst. Solids* **238**, 1-5.
13. Tokumoto, M.S., Pulcinelli, S.H., Santilli, C.V. and Craievich, A.F. (1999) SAXS study of the kinetics of formation of ZnO colloidal suspensions, *J. Non-Cryst. Solids* **247**, 176-182.

14. West, A.R. (1984) *Solid State Chemistry and its Applications*, John Wiley & Sons, New York.
15. Bahtat, A., Bouazaoui, M., Bahtat, M., Mugnier, J. (1994) Fluorescence of Er^{3+} ions in TiO_2 planar waveguides prepared by a sol-gel method, *Opt. Comm.* **111**, 55-60.
16. Kohls, M. (2000) *Dissertation Thesis*, University of Würzburg.
17. Desurvire, E. (1994) *Erbium-Doped Fiber Amplifiers - Principles and Applications*, John-Wiley & Sons, New York.

LASER PHOTOETCHING IN NANOPARTICLES PREPARATION AND STUDY OF THEIR PHYSICAL PROPERTIES

I.DMITRUK[1,2] , YU.BARNAKOV[2], AND A.KASUYA[2]
[1] *Kyiv National Taras Shevchenko University, Kyiv, Ukraine*
[2] *Center for Interdisciplinary Research, Tohoku University, Sendai, Japan*

Abstract. Size-selective photochemical etching is useful for obtaining semiconductor nanoparticles with desired size and narrow size distribution. Photoetching can be used not only for particles preparation but also for study of their physical properties. It helps to resolve excited states of electron–hole pair in nanoparticles and to estimate the homogenous broadening of exciton absorption peak. Photoetching can increase the photoluminescence quantum yield by increasing the lifetime of charge carriers in nanoparticles. A computer model of the photoetching process has been elaborated, which helps to understand the process and to determine the optical properties of the nanoparticles.

1. Introduction

Chemical methods for the preparation of semiconductor nanoparticles have been the subject of constant interest for many years, because they are promising materials for the production of large amounts of material and are rather flexible for production of particles with different size and shape. Much attention was paid to obtaining the smallest possible nanoparticles and with narrow size distributions [1-4]. However, if obtaining particles of different sizes is not a problem, the size-distribution of as-grown particles often is far from desired. Techniques such as liquid-phase chromatography, size-selective precipitation and some others utilize the same idea to pick up particles of desired size from a broad as-grown ensemble. Such techniques improve size distribution but the practical realization is not always so easy for different materials and for different particle sizes.

Another approach toward an active control of nanoparticle size and size distribution has been exploited in much less extent. The idea is to decrease the size of the bigger particles in the ensemble to that of the smallest ones by means of size-selective photochemical etching. While photosensitivity of colloid solutions has been known long time ago, the first comprehensive study of the chemistry of the etching process and its possible application for controlled decrease of the size of semiconductor nanoparticles was presented in [5]. Later several groups of researches demonstrated the possibility not only to decrease particle size, but also to obtain narrower size distribution for CdS [6,7]

L.M. Liz-Marzán and M. Giersig (eds.),
Low-Dimensional Systems: Theory, Preparation, and Some Applications, 121–131.

and some other semiconductors [8]. The expected advantage of size-selective photoetching is that it is easy to obtain particles of any desired size while the total number of particles in solution is preserved.

In the present paper we demonstrate the validity of this method for CdSe nanoparticles, the possibility to obtain extremely small particles of perfect structure and also to present interesting findings on the understanding of the photoetching process and its application not only to produce nanoparticles but also to study their fundamental properties.

2. Samples

Experiments were performed with CdSe nanoparticles with diameters in the range of 12 - 30 Å and size distribution about 10 - 20 % prepared by a two-phase protocol reaction. Sodium selenosulfite, Na_2SeSO_3, was chosen to release slowly Se-ions. The Cd-ion was in the form of $[CdN(CH_2COO)_3]^-$. The surfactant used was $CH_3(CH_2)_9NH_2$. With these ions and surfactant introduced in a water phase with pH=10, particle growth starts after addition of toluene. It takes place likely in the water–toluene interface. Particles of different sizes can be produced depending on temperature, growth time, and concentration of cadmium ions. Particles covered by surfactant molecules were suspended in toluene, hexane or ethanol. This colloid solution was stable for several weeks or months depending on particle size. Details on particle preparation will be reported elsewhere.

Encountering difficulties for the microscopic imaging of such small nanoparticles covered with the layer of surfactant we relayed on optical measurements for real-time monitoring of particle size and size distribution. The method is based on the well-known quantum size-effect [9] and is sensitive enough for such small particles with diameters $d \ll a_B$ (a_B – exciton Bohr radius). A typical absorption spectrum of CdSe colloid solution is presented in Figure1 (curve a). It consists of a narrow absorption band around 420 nm and background increasing to the short wavelength side. This background demonstrates more or less pronounced structure and can be interpreted probably as absorption into excited states of electron-hole pair in CdSe nanoparticle. Sometimes the absorption background tails to energies much lower than the main peak which reveals a not so good size distribution i.e. the presence of considerably larger particles.

3. Etching

The idea of size-selective photoetching of nanoparticles can be illustrated by the possibility of size-selective excitation of their photoluminescence. The luminescence spectrum of the sample (Figure1, curve b) consists of a narrow band-edge luminescence band slightly shifted to the low-energy side from the absorption peak and a broad band with maximum around 550 nm. The half-width of the band-edge luminescence band corresponds approximately to the half-width of the absorption band and is determined also mainly by particle size distribution.

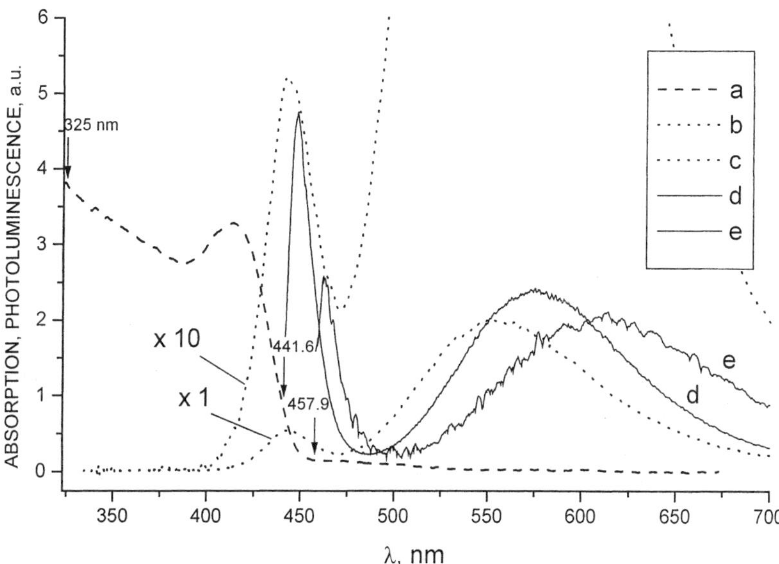

Figure 1. Photoluminescence spectra of CdSe nanoparticles under nonselective UV 325 nm excitation (b, c), size-selective 441.6 nm and 457.9 nm excitation (d, e), and absorption spectrum of the sample (a). T=77 K. Arrows indicate excitation wavelength positions.

Such hypotheses about size-dependent inhomogeneous broadening of absorption and band-edge luminescence was proven by experiments with resonant size-selective excitation. Using an excitation wavelength close to the low energy side of the absorption peak we can excite only nanoparticles of the certain size. In this case band-edge photoluminescence appears as a narrow asymmetric line at the long-wavelength side of the nonselective luminescence band with constant Stokes shift from excitation (curves d, e). Its asymmetry may be due to phonon-assisted transitions not resolved at this temperature. Under selective excitation, the long-wavelength broad band also changes its shape and position. Details of this behavior and possible interpretations will be discussed in another report.

In the present paper we focus on attempts to change particle size and size distribution by photochemical etching. Such process of photoetching of colloid nanoparticles with the presence of oxygen is well known for semiconductor sulfides [5, 6]. Assuming the same behavior for CdSe nanoparticles we studied changes in the absorption spectra under 325 nm and 442 nm irradiation by He-Cd laser. The UV line at 325 nm provides nonselective excitation of all particles while the line at 442 nm corresponds to the long wavelength wing of the absorption peak and can interact only with the biggest nanoparticles. In both cases we observed a decrease of intensity of the absorption peak and its shift to high-energy side (Figure 2). This means a decrease of nanoparticle size due to photochemical etching. The details of the photoetching process are not completely understood yet but it is obvious that oxygen present in solution plays a main role.

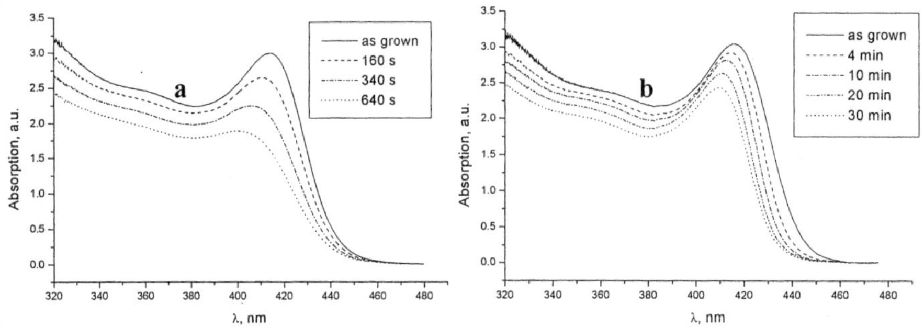

Figure 2. Photo-chemical etching of CdSe nanoparticles with 325 nm (a) and 442 nm (b) laser irradiation.

$$CdSe + 2O_2 + \text{light} \longrightarrow Cd^{2+} + SeO_4^{2-} \tag{1}$$

or

$$2CdSe + 3O_2 + \text{light} \longrightarrow 2Cd^{2+} + 2SeO_3^{2-}. \tag{2}$$

The process starts from the absorption of a quantum of light by CdSe nanoparticles, then charge separation takes place. A subsequent electrolysis–like process leads to removal of one CdSe molecule from the particle. Another explanation can be the excitation energy transfer from a CdSe nanoparticle to an oxygen molecule followed by nanoparticle etching by exited oxygen.

As it is seen from the figures, the etching efficiency for UV irradiation is much higher despite the lower power of this line. The probable reason is the higher absorption coefficient at 325 nm (Figure 1,a). A more careful study of Figure 2,a and Figure 2,b reveals two more differences. In the case of 325 nm irradiation the absorption peak shifts as a whole and becomes broader while under 442 nm irradiation mostly the long-wavelength wing of the absorption peak shifts to higher energy side and peak narrows. To obtain numerical data we have performed multi-peak analysis of the absorption spectra. The results for peak position (representing average size of nanoparticles) are presented in Figure 3. In both cases there is energy increase, but while for UV irradiation it is approximately linear, for 442 nm irradiation it shows some saturation. The band-edge peak in photoluminescence spectra shows the same shift. The decrease of absorption intensity in the process of etching can be explained simply by a decrease of the total amount of CdSe in solution. The most drastic difference between photoetching with UV irradiation and with 442 nm light was found in half-width of absorption peak. It broadens in the first case and narrows in the latter (Figure 3). The observed decrease of peak half-width means a narrowing of size distribution. An improvement is clearly seen in the case of 442 nm irradiation. The reason of such a difference in the results of photoetching by UV and 442 nm light relies in the fact that the UV line is absorbed by all nanoparticles while the 442 nm line excites only nanoparticles at one side of the size distribution. These nanoparticles with the bigger sizes are being etched, i.e. their average size is decreased, leading to an improvement in size distribution. This also explains the differences in etching kinetics (Figure 2).

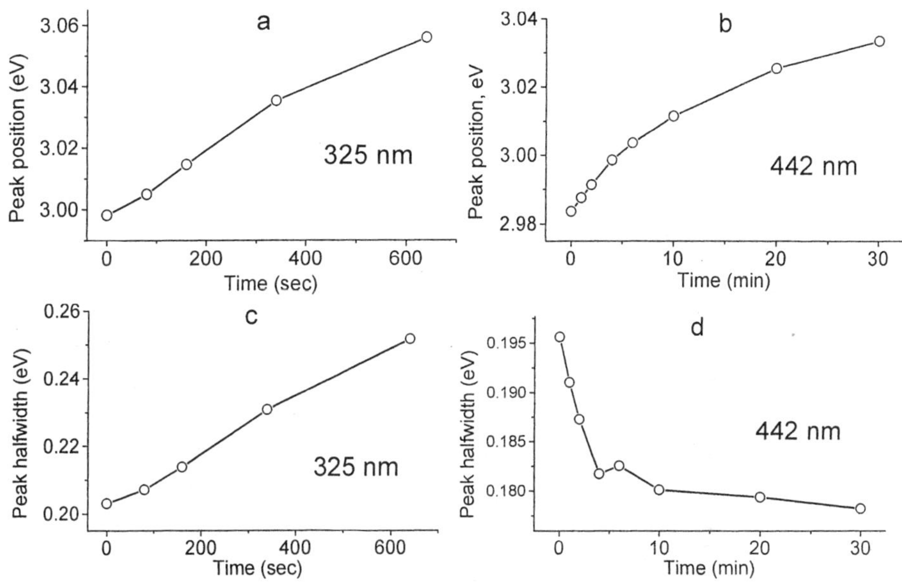

Figure 3. Photo-chemical etching of CdSe nanoparticles with 325 nm and 442 nm laser irradiation.

4. Differential Absorption Spectra Obtained by Short Time Photoetching

For a more precise comparison between selective and non-selective photoetching, let us refer to Figure 4. Removal of the long wavelength wing of the absorption peak under 442 nm irradiation is clearly seen in comparison with UV etching.

In this figure we have also presented etching efficiency curves obtained as the difference in absorption spectra before and after etching, divided by initial absorption (curves d and e for 325 nm and 442 nm irradiation respectively). For UV light, the etching efficiency curve is flat and broad but for 442 nm it is a much narrower band with a maximum at 442 nm. There must be no broadening of this etching efficiency curve due to particle size distribution because only the particles with their excitation energy equal to the energy of laser quantum are excited and etched. Thus the halfwidth of the etching efficiency curve is determined by homogenous broadening similar to experiments on hole burning [10]. It provides the possibility to estimate the lifetime of this state of initially excited electron-hole pair: $\tau = h/\Delta E = 22$ fs.

Then the electron-hole pair looses coherence due to its interaction with LO-phonons and relaxes to a lower-energy state. The transition from this state is observed in the luminescence spectrum.

The above described method for obtaining differential absorption spectra and etching efficiency curves yields some more interesting results. For example, it helps to resolve excited states of the electron-hole pair in the nanoparticles (Figure 5). The absorption background is not sensitive to size-selective photoetching with long wavelength light and excited states are observed more clearly in the differential absorption spectrum.

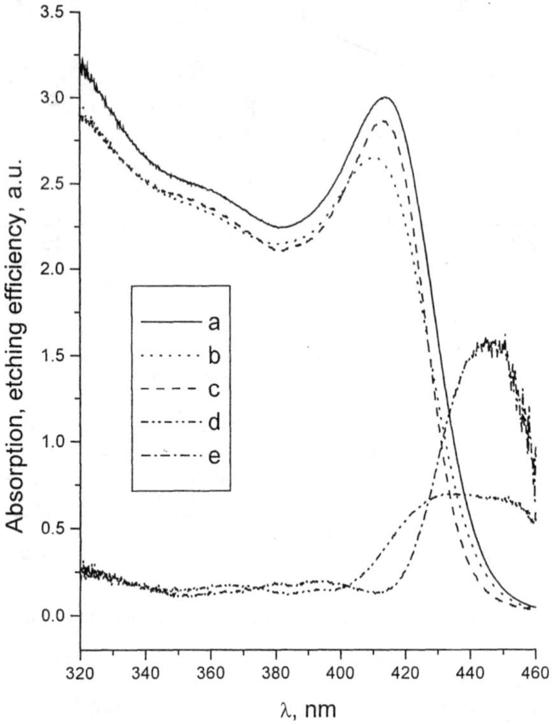

Figure 4. Absorption spectra of initial CdSe nanoparticle sample (a), after 160 s etching with 325 nm irradiation (b), after 5 min etching with 442 nm irradiation (c) and etching efficiencies for 325 nm (d) and 442 nm (e) irradiation.

Selective and non-selective photoetching with UV light also helps to check if specific absorption peaks really belong to the nanoparticles. If some features in the absorption spectrum do not shift to shorter wavelengths and do not decrease in amplitude even under UV irradiation absorbed by any CdSe nanoparticles, it means that they correspond to some impurity or byproduct of particle preparation.

5. Obtaining of Small Particles with Narrow Size Distribution

Taking into account the above described behavior of CdSe nanoparticles in photoetching under different conditions we can attempt to obtain small nanoparticles with narrow size distribution starting from rather big and not monodisperse ones. Figure 6 presents the results of such an experiment together with size estimates based on absorption peak position [11]. The initial sample had a broad, two-peak size distribution. After etching for 8 h with 70 mW 442 nm line of He-Cd laser we obtained a distinct peak at 410 nm corresponding to a particle size of 1.25 nm.

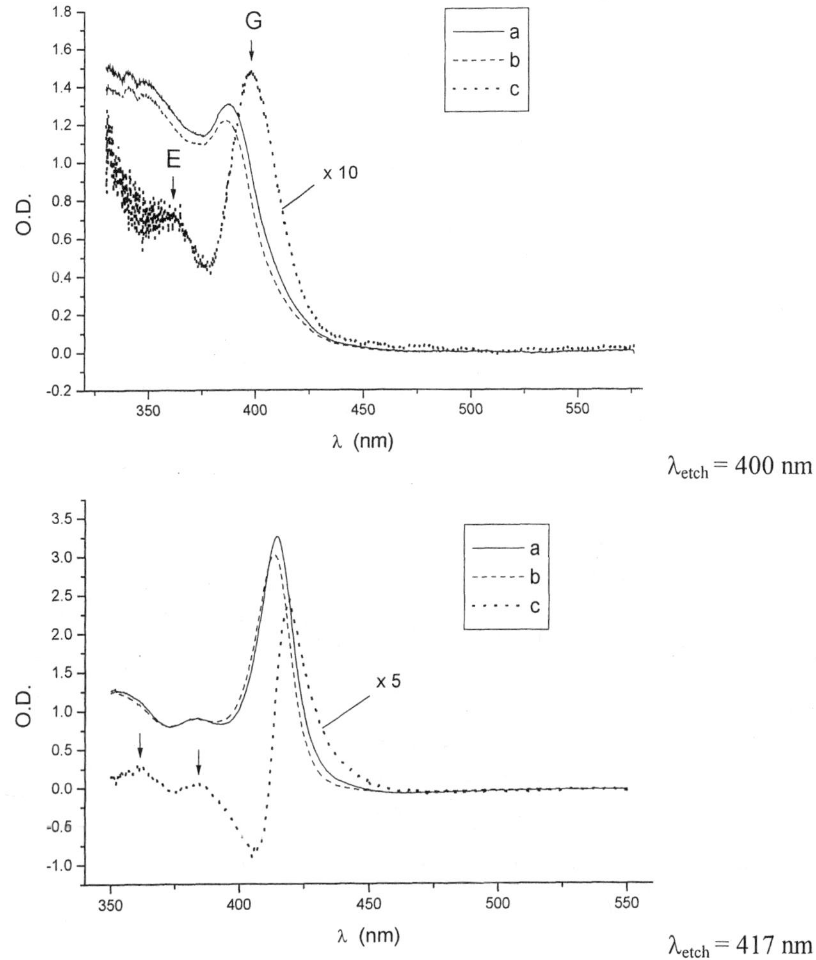

Figure 5. Absorption spectra of the initial (a) and photoetched (b) samples and their difference (c). Arrows indicate ground and excited state of electron-hole pair in nanoparticle.

At this stage, the photoetching process reached saturation because the obtained particles did not absorb 442 nm radiation. To proceed with photoetching we changed wavelength to 410 nm and in 20 min we obtained a peak at 400 nm. Then the process slowed down and we had to change wavelength to 400 nm. The absorption peak shifted to 388 nm and then stopped. Further irradiation did not shift the absorption peak but only decreased its intensity. Such a behavior means that for particles of that size (estimated as 0.9 nm) further photoetching leads to complete destruction of the particles. It seems that the obtained particles are the smallest CdSe nanoparticles stable in solution.

6. Change of Photoluminescence Spectra

Assuming that defect areas of nanocrystals which act as charge traps will be removed during the first in photoetching process, we can expect a considerable improvement of the luminescent properties of nanoparticles. Such a behavior was observed in the experiments. In all studied samples photoetching leads to a considerable increase of total photoluminescence quantum yield, in some samples up to one order of magnitude. But not all the samples demonstrated a considerable improvement of band-edge luminescence (Figure 7b). In others the intensity of band-edge luminescence remains the same and only long-wavelength B-band considerably increases (Figure 7a).

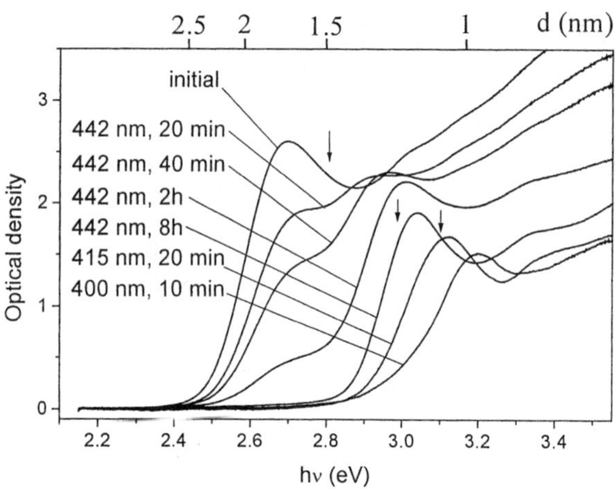

Figure 6. Obtaining of small CdSe clusters by consequent size-selective photoetching. Arrows indicate positions of laser lines used.

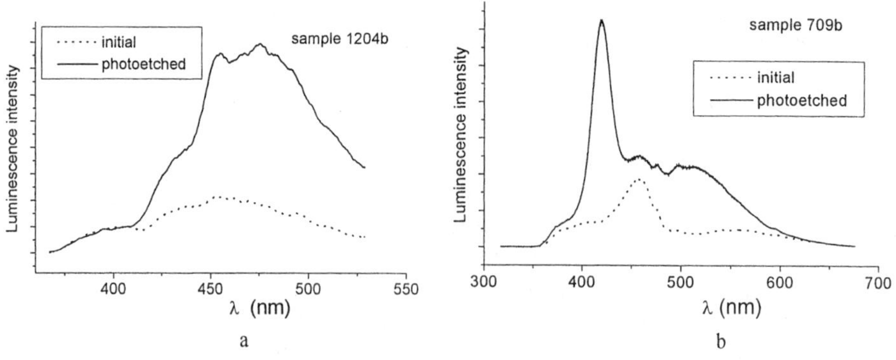

Figure 7. Change of photoluminescence spectra of CdSe nanoparticle samples prepared in different conditions.

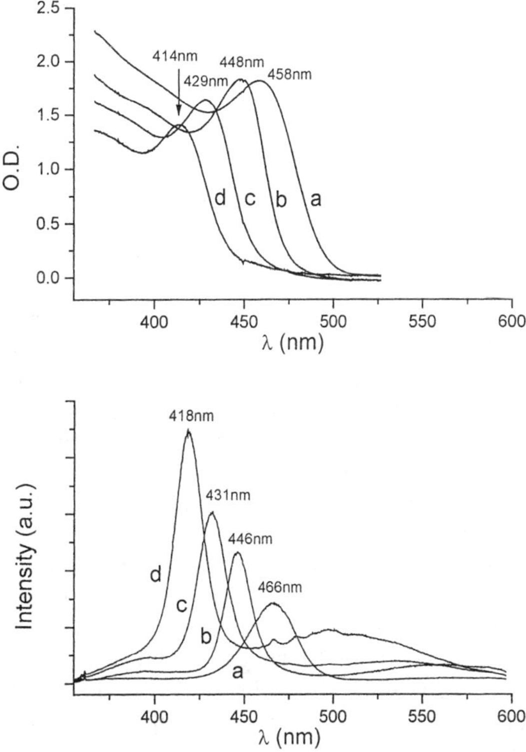

Figure 8. Change of absorption and photoluminescence spectra of CdSe nanoparticles with perfect crystalline structure.

The difference can be attributed to the different conditions on the surface of particles or to different origin of the defects responsible for B-band. The observed increase of quantum yield appears to be due to an increase of lifetime of charge carriers in the nanoparticles. Improvements in particles preparation made it possible to obtain as-grown samples with strong band-edge luminescence. Probably these nanoparticles have much less defects, and they demonstrate the expected behavior in photoetching experiments (Figure 8). Narrowing of particle size distribution and increase of photoluminescence quantum yield are clearly seen.

7. Computer Simulation

A computer model of the photoetching process has been elaborated. It takes into account direct and phonon-assisted transitions into the ground state of electron-hole pair in nanoparticles. Direct transition is supposed to be homogeneously broadened with halfwidth δE. Phonon-assisted transitions, both with emission and with absorption of phonon, have been included to describe the increase of absorption to high-energy side.

For simplicity only one LO-phonon 25 meV, so-called actual phonon, is taken into account as the major contributor to electron-phonon processes, and we neglect the dependence of its frequency on particle size. Electron-phonon coupling constant is the only fitting parameter of the model. It is chosen for the best similarity of calculated and experimental absorption spectra. Of course it can be determined experimentally or estimated but it is not the subject of the present study. Initial size distribution of particles in the sample is supposed to be Gaussian with halfwidth Δd. Size dependence of the energy of electronic transition in the particles is taken from an interpolation of empirical data for CdSe [11]. A huge number of particles (10^4-10^6) has been taken to constitute the ensemble in order to obtain smooth, statistically reliable results. The absorption spectrum of each particle is calculated as described above and the absorption of the sample is obtained by summing up the absorption of all particles. Photoetching is simulated as a removal of material from the particle atom by atom with the probability proportional to absorption of the given particle at the laser wavelength. Each photoetching event shifts a given particle in size distribution to the smaller size and as a result changes the absorption spectrum of that particle and of the whole ensemble. Resembling experimental conditions, the simulation is performed in subsequent stages with a certain number of photons irradiating the sample in each stage. Results of simulation of consequent change of absorption spectra are presented in the following figures with parameter t, 2t, 3t, ... denoting etching stage similar to irradiation time in real experiments.

Initial size distribution and homogenous broadening of electronic transition both contribute to the width of the absorption peak. But they change in different way during the photoetching process. If the laser wavelength corresponds to the low-energy wing of the absorption peak, narrowing of size distribution is expected (Figure 9), very similar to the behavior observed in the experiment. The final width of the absorption peak (and the size distribution correspondingly) is determined mostly by homogeneous broadening of the electronic transition. Differential absorption spectra also look similar to experimental ones, Figure 10. The half-width of the peak in differential spectra is almost equal to the homogeneous width of the electronic transition.

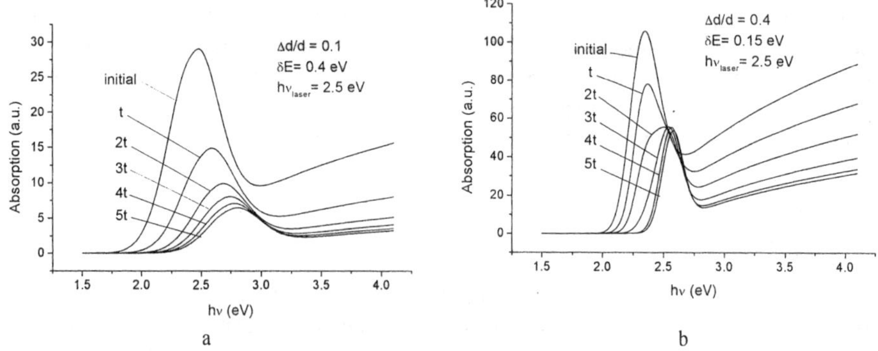

Figure 9. Computer simulation of photoetching process of the samples with different size distribution Δd/d and different homogeneous broadening δE.

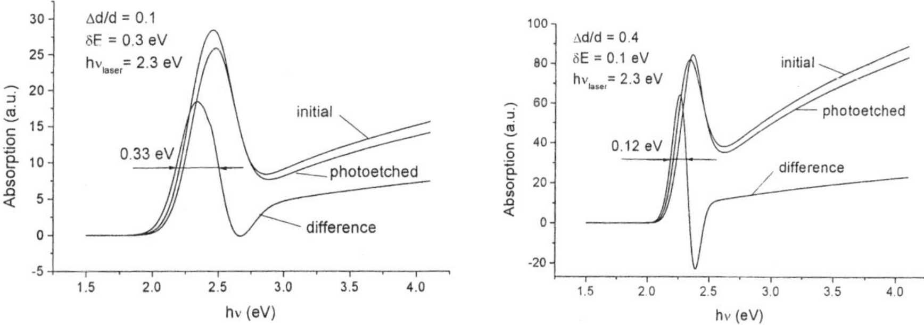

Figure 10. Simulation of differential absorption spectra for the particles with different homogeneous broadening δE.

8. Acknowledgments

The authors are thankful to Prof. T.Goto for providing his laboratory for optical measurements, to Prof. O.Terasaki and Dr. Z.Liu for high-resolution electron microscopy of our samples. Presenting author appreciates support provided by Organizing Committee, which made it possible to attend the workshop and present this report.

9. References

1. M.G.Bawendi, W.L.Wilson, L.Rolhberg, P.J.Carroll, T.M.Jedju, M.L.leigerwald, and L.E.Brus (1990) Electronic Structure and Photoexcited-Carrier Dynamics in Nanometer-Size CdSe Clusters. *Phys. Rev. Lett.* **65**, 1623-1626.
2. O.Yamamoto, T.Sasamoto, and M. Inagaki (1998) Preparation of crystalline CdSe particles by chemical bath deposition, *J. Mater. Res.* **13**, 3394-3398.
3. M.A.Hines and P.Guyot-Sionnest (1996) Synthesis and Characterization of Strongly Luminescing ZnS-Capped CdSe Nanocrystals, *J. Phys. Chem.* **100**, 468-471.
4. M.Danek, K.F.Jensen, C.B.Murray, M.G.Bawendi (1996) Synthesis of Luminescent Thin-Film CdSe/ZnSe Quantum Dot Composites Using CdSe Quantum Dots Passivated with an Overlayer of ZnSe, *Chem. Mater.* **8**, 173-180.
5. A.Henglein, (1982) Photo-Degradation and Fluorescence of Colloidal-Cadmium Sulfide in Aqueos Solution, *Ber. Bunsenges. Phys. Chem.* **86**, 301-305.
6. A.Fojtik, H.Weller, U.Koch, and A.Henglein, (1984) Photo-Chemistry of Colloidal Metal Sulfides 8.Photo-Physics of Extremely Small CdS Particles: Q-State CdS and Magic Agglomeration Numbers, *Ber. Bunsenges. Phys. Chem.* **88**, 969-977.
7. A. van Dijken, D.Vanmaekelbergh, A.Meijerink (1997) Size selective photoetching of nanocrystalline CdS particles, *Chem. Phys. Letters* **269**, 494-499.
8. A. van Dijken, A.H.Janssen, M.H.P. Smitsmans, D.Vanmaekelbergh, and A.Meijerink (1998) Size-Selective Photoetching of Nanocrystalline Semiconductor Particles, *Chem. Mater.* **10**, 3513-3522.
9. A.I.Ekimov, Al.L.Efros, A.A.Onushchenko (1985) Quantum size effect in semiconductor microcrystals, *Sol. St. Comm.* **56**, 921-924.
10. S.V.Gaponenko, U.Woggon, A.Uhrig, W.Langbein, C.Klingshirn (1994) Narrow-band spectral hole burning in quantum dots, *Journal of Luminescence*, **60&61**, 302-307.
11. Murray C.B., Norris D.J. and Bawendi M.G. (1993) Synthesis and Characterization of Nearly Monodisperse CdE (E=S, Se, Te) Semiconductor Nanocrystals, *J.Am.Chem.Soc.* **115** ,8706-8715.

GROWTH AND CHARACTERIZATIONS OF WELL-ALIGNED CARBON NANOTUBES

Z. F. REN,[a] Z. P. HUANG,[b] Y. TU,[a] D. Z. WANG,[a] W. Z. LI,[a] J. G. WEN,[a] M. SENNETT,[c] M. GIERSIG,[d] and K. KEMPA[a]

[a]*Boston College, Department of Physics, Chestnut Hill, MA 02467*

[b]*NanoLab, Inc., Brighton, MA 02135*

[c]*U.S. Army Soldier and Biological Chemical Command, Natick Soldier Center, Materials Science Team, Natick, MA 01760*

[d]*Hahn-Meitner-Institute, Berlin, Germany*

1. Introduction

Large arrays of well-aligned carbon nanotubes are first made possible on substrates in 1998 by plasma enhanced chemical deposition [1,2] in which the diameter and length of each carbon nanotube are under control, but not the growth angle, location, nor the spacing between them. Soon after, the titled growth has been achieved by controlling the plasma direction using the same growth technique [3]. Almost at the same time, the control of location and spacing of the nanotubes have been accomplished using electron beam (e-beam) lithography to pattern the nickel dots first at where they are needed and then to grow the carbon nanotubes using the same growth technique [4,5]. However, e-beam is not possible to be commercialized for large scale. Therefore, alternative cheap and scalable technique is sought. Fortunately, the catalytic dots have been fabricated by electrochemistry and excellent aligned carbon nanotubes arrays have been grown [6]. Due to the nature of electrochemistry, the control on location of each nanotube is lacking. For applications that do not require the pre-determined location of each nanotube such as regular electron source, the arrays grown using the dots by electrochemistry is good enough. However, for applications that do require the pre-determined location of each nanotube such as microscopic probing tips, nanophotonics, etc., the control of location of each nanotube is crucial. Recently, we have been successful to grow large arrays of carbon nanotubes with diameter, length, location, and spacing under control by a simple and scalable technique, nanosphere lithography [7]. Since the very first report on large arrays of well-aligned carbon nanotubes, numerous papers have used the same or a slightly modified technique to grow aligned carbon nanotube arrays by either DC or microvave plasma CVD [8-18].

2. Experimental

The catalytic dots have been prepared by four techniques: magnetron sputtering, e-beam lithography, pulse current electrochemistry, and nanosphere lithography. The growth was accomplished by plasma enhanced chemical vapor deposition. The gases used are acetylene and ammonia that provides the carbon source and catalytic effect respectively. Scanning electron microscope (SEM) and transmission electron microscope (TEM) were used to characterize the arrays.

L.M. Liz-Marzán and M. Giersig (eds.),
Low-Dimensional Systems: Theory, Preparation, and Some Applications, 133–140.

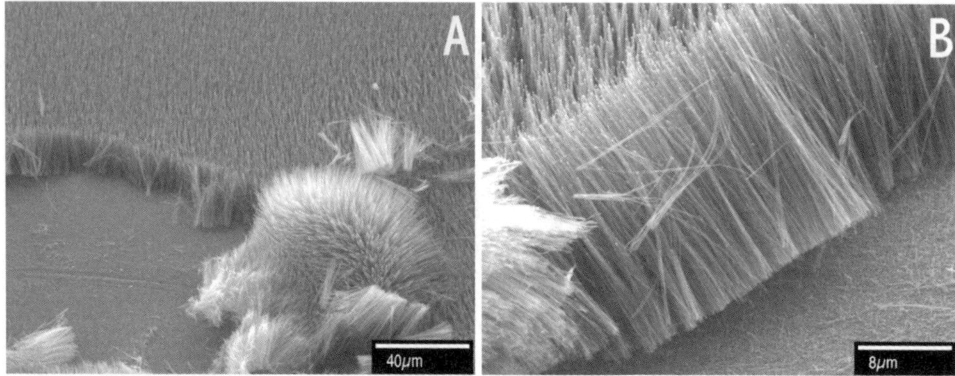

Figure 1. SEM images of aligned carbon nanotubes. A) low magnification to show the alignment over large area, B) medium magnification to show the length of the nanotubes.

3. Results

3.1. ALIGNED GROWTH OF CARBON NANOTUBES ON SPUTTERED NI FILMS

Figure 1 shows the SEM images of the arrays of carbon nanotubes grown on glass substrates by plasma enhanced chemical vapor deposition [1]. They clearly show the excellent alignment with the glass substrate. In order to measure the alignment, diameter, and length of the carbon nanotubes, part of the sample was scratched as shown. For this particular sample, the tubes are about 20 μm long. For diameter estimate, higher magnification SEM exam was carried out.

Figure 2 shows SEM images of the aligned carbon nanotube arrays in higher magnifications. Clearly, nanotubes with different diameters have been produced. The diameters are controlled by the thickness of the catalytic Ni thickness: the thinner the Ni, the smaller the nanotubes.

Figure 2. SEM images of carbon nanotube arrays grown with different thickness of Ni layers. A) 15 nm thick Ni, B) 40 nm thick Ni.

For a thickness of 15 nm, nanotubes are about 50 nm in diameter as shown in Figure 2A, whereas a 40 nm thick Ni yielded nanotubes of about 250 nm in diameter as shown in Figure 2B. Further reduction of Ni thickness has yielded nanotubes with diameters in the range of 10 – 25 nm, but without excellent alignment if the length is more than a few μm.

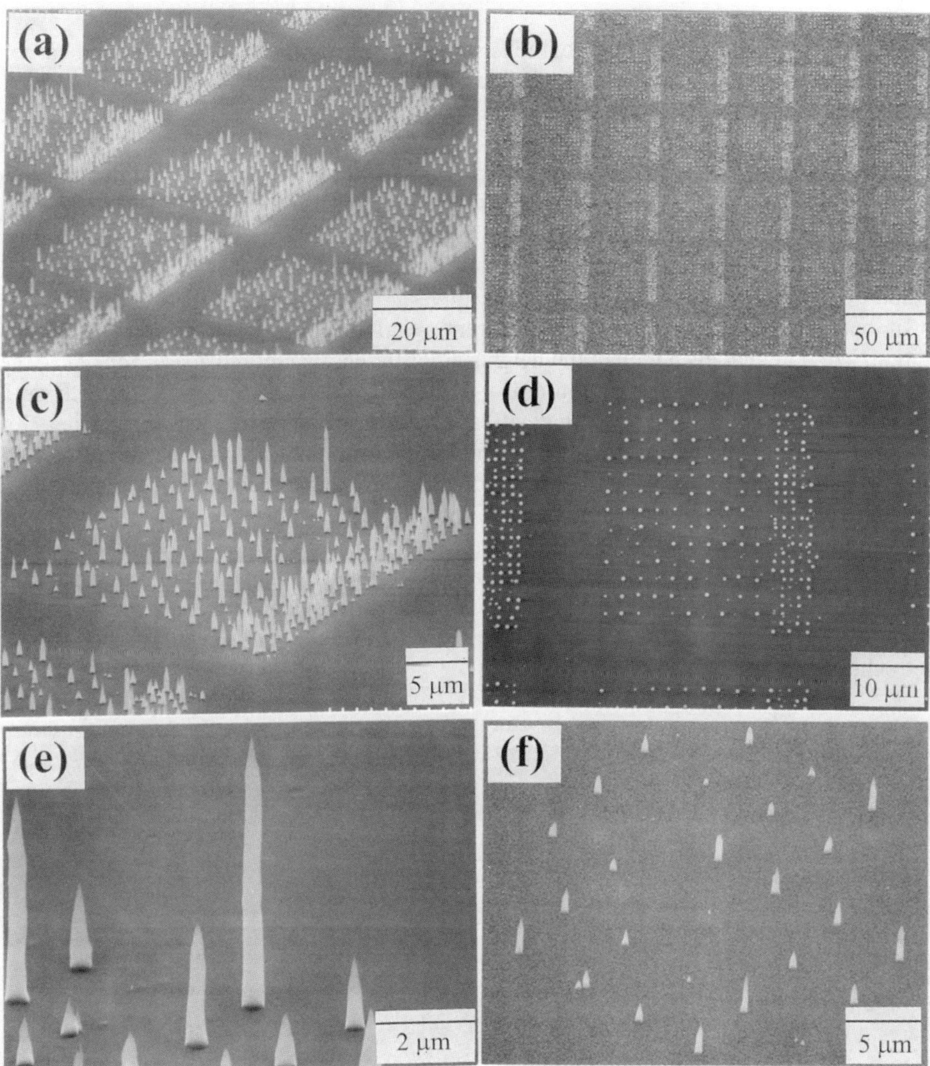

Figure 3. A series of SEM micrographs from different viewing angles showing growth of carbon nanotube obelisks on an array of submicron nickel dots. (a) An inclined view of a repeated array pattern. (b) A top (normal) view of a repeated array pattern. (c) An inclined view of one array pattern. (d) A top (normal) view of one array pattern. The initial Ni dots (and subsequently the grown carbon structures) are spaced either 2 μm apart (left) or 1 μm apart (right). (e) A magnified view along the edge of one pattern. A sharp, tapered tip is evident. (f) An inclined view of carbon obelisks grown on nickel dots separated by 5 μm.

3.2. ALIGNED GROWTH OF CARBON NANOTUBES ON NI DOTS MADE BY E-BEAM LITHOGRAPHY

For e-beam lithography, thin film nickel (Ni) patterns were fabricated on a p-type boron doped (100) silicon substrate. Ni layer of 150 Å was deposited by thermal evaporation. The patterned substrate was loaded into a PE-HF-CVD system. The growth was carried out at the similar conditions as above.

Figure 3 is a series of SEM micrographs showing the growth of single multiwall carbon nanotubes on each dot of an array of ~100 nm nickel dots. Figures 2a, 2c, 2e, and 2f were taken at an inclined angle, and Figures 2b and 2d are top views taken normal to the substrate.

Figures 2a and 2b demonstrate selective growth of the carbon structures on the multiply repeated array patterns. The grown structures accurately reflect the spacing and periodicity of the lithographically patterned Ni dots. Figures 2c and 2d were taken at a higher magnification and show the repeated array pattern where the nanotubes are spaced either 2 μm apart (left) or 1 μm apart (right). Significant variation in the height (0.1 to 5 μm) of the grown bundles is observed, with no apparent relationship between height and spatial position. We note that even though the heights are different by more than a factor of 10, the base diameters are approximately uniform (~ 150 nm). Figure 2f shows the growth on a grid of Ni dots spaced 5 μm apart, indicating little dependence of growth on spacing (for spacings > 1 μm). The reason for such aligned growth of single carbon nanotubes in our system is due to the plasma, which is also used widely by others since our initial report [1].

Soon after, we were successful on obtaining patterns with more uniform length as shown in Figure 4. Each nanotube has a sharp tip with radius of about 2-3 nm that may be very useful for high current field emission applications. Unfortunately, the nice patterns grown using e-beam lithography is only good for concept proving. In the next, we present a cheap way for fabricating arrays with spacing, but not location, controlled: electrochemistry.

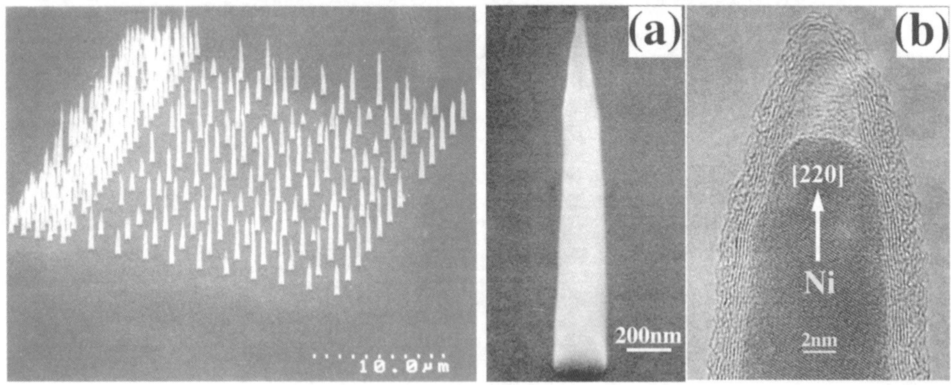

Figure 4. SEM and TEM images of carbon nanotube arrays.

3.3. GROWTH OF ALIGNED CARBON NANOTUBES WITH SPACING CONTROLLED BY ELECTROCHEMISTRY

Figure 5 shows the different nucleation site densities of the Ni nanoparticles from about 7.5×10^5 cm^{-2} to 3×10^8 cm^{-2} by electrochemistry. The white dots shown in the pictures were the Ni nanoparticles that had been confirmed by Energy Dispersive X-ray Spectroscopy (EDX). Most particles had diameter from 100 nm to 200 nm with some nanoparticles smaller than 50 nm. The Ni nanoparticles were randomly located on the surface of the substrate.

Figure 6 (a) to (e) shows the different site densities of CNTs grown from the electrodeposited Ni nanoparticles. The CNTs site densities of the samples were about 7.5×10^5, 2.0×10^6, 6.0×10^6, 2.0×10^7, and 3×10^8 cm^{-2}, respectively. Figure 6 (f) provides a closer look at one of the well-aligned CNTs. Figure 5 shows the typical TEM image of the CNTs. It shows that CNTs have bamboo structure, which is similar with that of the CNTs grown from the Ni thin films prepared by sputtering using the same PE-HF-CVD method. Clearly, the microstructure does not depend on the way the Ni catalyst was prepared.

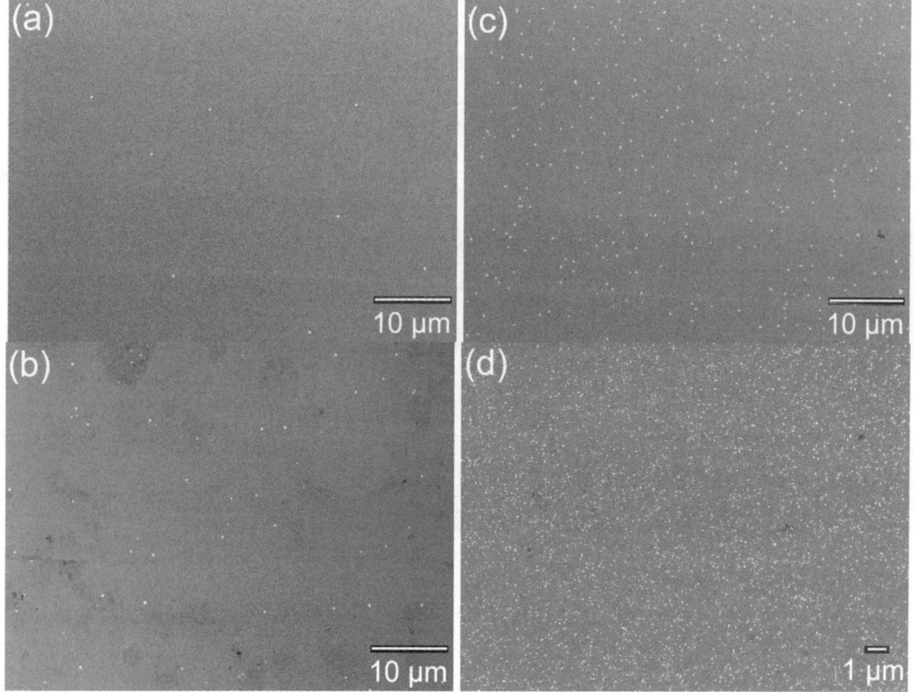

Figure 5. SEM images of Ni nanoparticles deposited electrochemically with site density of (a) 7.5×10^5 cm^{-2}, (b) 2×10^6 cm^{-2}, (c) 2×10^7 cm^{-2}, and (d) 3×10^8 cm^{-2}.

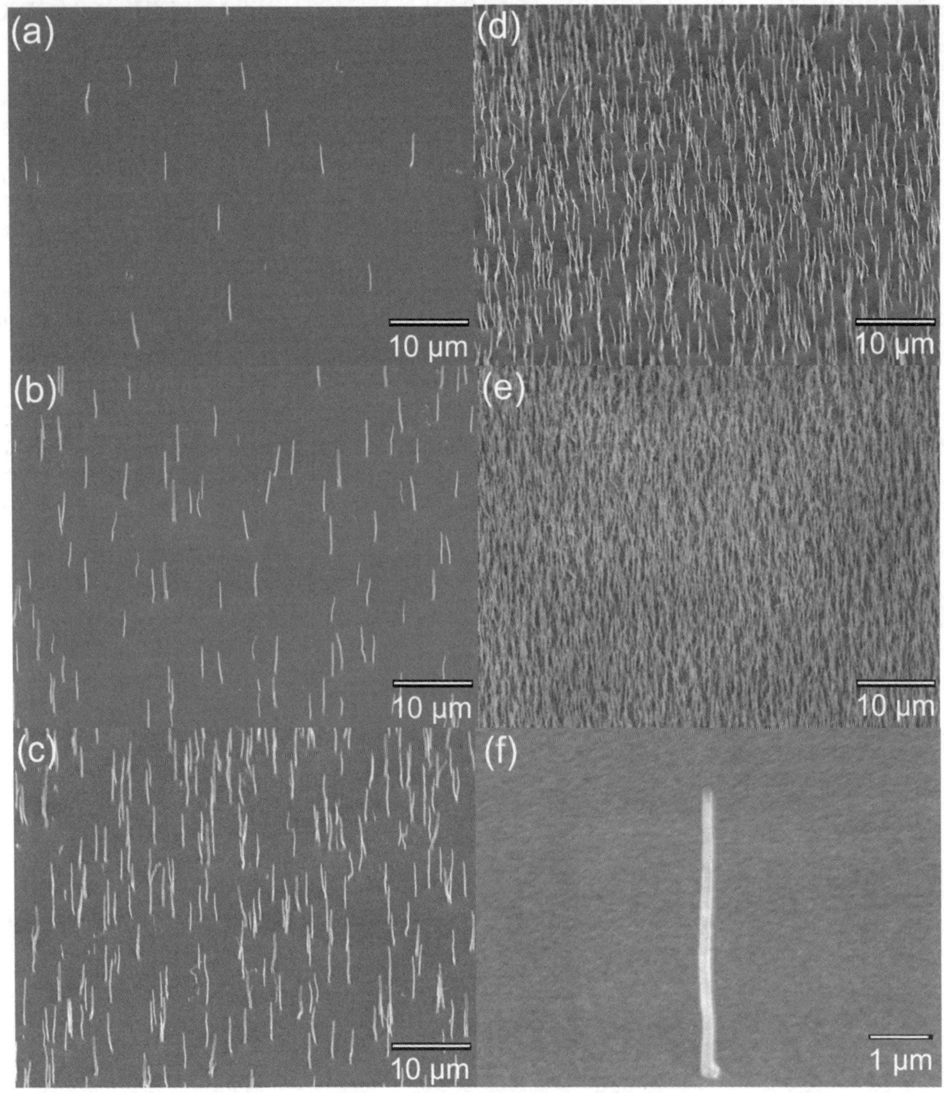

Figure 6. SEM images of aligned CNTs with site density of (a) 7.5×10^5 cm^{-2}, (b) 2×10^6 cm^{-2}, (c) 6×10^6 cm^{-2}, (d) 2×10^7 cm^{-2}, (e) 3×10^8 cm^{-2}, and (f) a single standing CNT.

3.4. GROWTH OF ALIGNED CARBON NANOTUBE ARRAYS WITH PERIODICITY

Even though electrochemistry provided excellent control on site density, but could not produce Ni dots at the pre-determined locations. Fortunately, we have developed another technique, nanosphere lithography, to fulfill the goal. Figure 7 shows the SEM images of the Ni dots made by nanolithography. Figure 8 shows the nanotubes made from the dots shown in Figure 7.

Figure 7. SEM images of Ni dots made by nanolithography in low (left) and medium (right) magnifications.

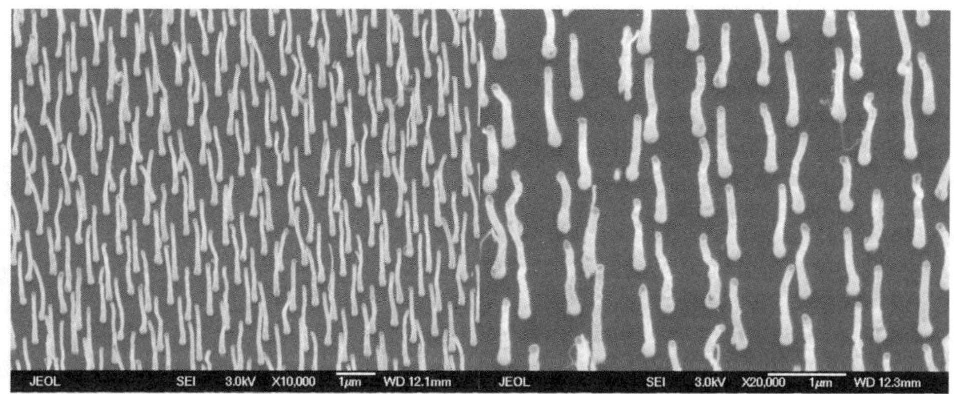

Figure 8. SEM images of low (left) and medium (right) magnifications of carbon nanotube arrays.

4. Conclusions

Well-aligned carbon nanotube arrays without periodicity have been achieved on either Ni films made by magnetron sputtering or Ni dots made by electrochemical deposition, whereas the arrays with periodicity have been realized on either Ni dots made by e-beam lithography or on Ni dots made by nanosphere lithography. Obviously, the electrochemical deposition is much better than magnetron sputtering, and the nanosphere lithography is much better than e-beam lithography in the sense of both the cost and scalability.

5. Acknowledgements

This work is partially funded by US. Army Natick Soldier Systems Center under grant DAAD16-00-C-9227, Army Research Office under grant DAAD19-00-1-0001, DOE under grant DE-FG02-00ER45805, and National Science Foundation under grants DMR-9996289 and ECS-0103012.

6. References

1. Ren, Z. F., Huang, Z. P., Xu, J. W., Wang, J. H., Bush, P., Siegal, M. P., and Provencio, P. N., (1998) Synthesis of large arrays of well-aligned carbon nanotubes on glass, *Science* **282**, 1105 – 1107.
2. Huang, Z. P., Xu, J.W., Ren, Z. F., Wang, J. H., Siegal, M. P., and Provencio, P. N. (1998) Growth of large-scale well-aligned carbon nanotubes by plasma enhanced hot filament chemical vapor deposition, *Appl. Phys. Lett.* **73**, 3845-3847.
3. Ren, Z. F., Huang, Z. P., Xu, J. W., Wang, D. Z., Wang, J. H., Calvet, L., Chen, J., Klemic, J. F., and Reed, M. A. (1999) Large arrays of well-aligned carbon nanotubes, AIP Conf. Proc. **486** (Electronic Properties of Novel Materials—Science and Technology of Molecular Nanostructures), 263 - 267.
4. Ren, Z. F., Huang, Z. P., Xu, J. W., Wang, D. Z., Wen, J. G., Wang, J. H., Calvet, L., Chen, J., Klemic, J. F., and Reed, M. A. (1999) Growth of a single freestanding multiwall carbon nanotube on each nanonickel dot, *Appl. Phys. Lett.* **75**, 1086 – 1088.
5. Wen, J. G., Huang, Z. P., Wang, D. Z., Chen, J. H., Yang, S. X., Ren, Z. F. Wang, J. H., Calvet, L. E., Chen, J., Klemic, J. F., Reed, M. A. (2001) Growth and characterization of aligned carbon nanotubes from patterned nickel nanodots and uniform thin films, *J. Mater. Res.* **16**, 3246-3253.
6. Tu, Y., Huang, Z. P., Wang, D. Z., Wen, J. G., Ren, Z. F. (2002) Growth of aligned carbon nanotubes with controlled site density, *Appl. Phys. Lett.* **80**, 4018 - 4020.
7. Kempa, K., Rybczynski, J., Huang, Z. P., Wu, P. F., Wang, D. Z., Giersig, M., Kimball, B., Sennett, M., Rao, D. V. G. L. N., Steeves, D., Carnahan, D. L., and Ren, Z. F. (in preparation).
8. Bower C, Zhu W, Jin SH, Zhou O. (2000) Plasma-induced alignment of carbon nanotubes, *Appl. Phys. Lett.*, **77,** 830-832.
9. Cui H, Zhou O, Stoner BR (2000) Deposition of aligned bamboo-like carbon nanotubes via microwave plasma enhanced chemical vapor deposition, *J. Appl. Phys.* **88**, 6072-6074.
10. Baylor LR, Merkulov VI, Ellis ED, Guillorn MA, Lowndes DH, Melechko AV, Simpson ML, Whealton, JH (2002) Field emission from isolated individual vertically aligned carbon nanocones, *J. Appl. Phys.* **91**, 4602-4606.
11. Lee CJ, Lyu SC, Kim HW, Park CY, Yang CW (2002) Large-scale production of aligned carbon nanotubes by the vapor phase growth method, *Chem. Phys. Lett.* **359,** 109-114.
12. Merkulov VI, Melechko AV, Guillorn MA, Simpson ML, Lowndes DH, Whealton JH, Raridon RJ (2002) Controlled alignment of carbon nanofibers in a large-scale synthesis process, *Appl. Phys. Lett.* **80**, 4816-4818.
13. Wang SG, Wang JH, Qin Y (2002) Synthesis of carbon nanotubes by microwave plasma chemical vapor deposition at low temperature, *Acta Chim. Sinica* **60,** 957-960.
14. Kim J, No K (2002) Growth of carbon nanotubes on the glass substrate for flat panel display applications, *Int. J. Mod. Phys. B* **16**, 979-982.
15. Zhang WD, Wen Y, Tjiu WC, Xu GQ, Gan LM (2002) Growth of vertically aligned carbon-nanotube array on large area of quartz plates by chemical vapor deposition, *Appl. Phys. A-Mater. Sci. & Proc.* **74**, 419-422.
16. Huczko A (2002) Synthesis of aligned carbon nanotubes, *Appl. Phys. A-Mater. Sci. & Proc.* **74**, 617-638.
17. Teo KBK, Chhowalla M, Amaratunga GAJ, Milne WI, Pirio G, Legagneux P, Wyczisk F, Pribat D, Hasko DG (2002) Field emission from dense, sparse, and patterned arrays of carbon nanofibers, *Appl. Phys. Lett.* **80**, 2011-2013.
18. Chhowalla M, Teo KBK, Ducati C, Rupesinghe NL, Amaratunga GAJ, Ferrari AC, Roy D, Robertson J, Milne WI (2001) Growth process conditions of vertically aligned carbon nanotubes using plasma enhanced chemical vapor deposition, *J. Appl. Phys.* **90**, 5308-5317.

ENCAPSULATION OF CO NANOCLUSTERS IN A-C:H

V. I. IVANOV-OMSKII

Ioffe Physical-Technical Institute RAS
St. Petersburg, 194021 Russia

Abstract. Films with various cobalt concentrations were grown by co- sputtering of graphite and cobalt targets with DC magnetron in Ar-H_2 plasma. A set of diagnostics such as high resolution electron microscopy, X-ray absorption, infrared and Raman spectroscopy are presented to show the mechanism of cobalt-carbon interaction in hydrogenated amorphous carbon films modified with cobalt A strong amplification of absorption by the C-H bond deformation modes and appearance of some C-C bond stretching modes induced by cobalt are reported.

1. Introduction

Nanocluster engineering has a clear relation to the modern tendency in solid-state physics and technology and is thereby motivated. As to Co nanoclusters the special interest is to develop unique magnetic materials with high density of insulated magnetic domains. The importance of this problem is well illustrated in the case of nanostructures based on magnetic metals [1], because, in this case, the encapsulation is able to play a dual role in the protection of nanoclusters from the aggressive action of its environment and the weakening of the exchange interaction between neighboring particles. This opens up attractive prospects for using such nanocomposites as media for ultrahigh-density magnetic recording. In fact, a-C:H can adequately protect a metal surface from degradation in aggressive media, as was shown by the example of Ag, which is extremely sensitive to such action [2].

So, why is carbon amorphous? The atomic arrangement in amorphous carbon films shows several types of bonding configurations due to the unique ability of carbon to form differently hybridized valence states. This property endows the material structure with a very high degree of adaptability to embedded foreign elements. It offers a distinct opportunity for the production of various carbon-based nanocomposites with intentionally introduced nanoclusters of foreign substances. An interest to this new basket of possibilities of the material (which is traditionally considered simply as universal coating due to the mentioned adaptability to the properties of materials to be covered) emerged comparatively recently.

A special goal of this work is to contribute a share to the piggy bank of our modern knowledge of the problem of metal nanoclusters encapsulation. Our main interest relates to Co nanoclusters and the first of whole to their interaction with carbon matrix, and we pay special attention to the diagnostics of carbon matrix subjected to invasion of metal

L.M. Liz-Marzán and M. Giersig (eds.),
Low-Dimensional Systems: Theory, Preparation, and Some Applications, 141–150.

nanoclusters. The problem of encapsulating nanoclusters of metals with large magnetic moments into a carbon host was considered in a number of studies (see, for example, [3,4]). However, the range of methods for the control and characterization of the obtained composites was restricted to electron microscopy, X-ray diffraction, and to measurements of macroscopic magnetic characteristics. At the same time, for optimizing the synthesis and magnetic properties of nanoclusters, it was found to be important to know the nature of the interaction between the nanoclusters and the carbon host. Methods of vibration spectroscopy, such as infrared (IR) absorption and the Raman scattering, together with X-ray absorption spectroscopy (EXAFS) are useful sources of information about the properties of the composites.

2. A-C:H(Co) Films

The DC magnetron co-sputtering of graphite and cobalt targets technique in argon–hydrogen plasma (80%Ar + 20%H_2) was used for deposition of a-C:H(Co) films and resulting encapsulation of cobalt nanoclusters [6]. Si(100) was used as substrate and the film thickness was some hundreds of nm. The relative atomic concentrations of carbon and cobalt were controlled by adjustment of the area ratio for cobalt and graphite surfaces subjected to ion bombardment. The resulting cobalt concentration was determined using proton Rutherford backscattering and the method of nuclear reactions [6].

3. Vibration Spectroscopy

Although our topic is the encapsulation of Co nanoclusters into a-C:H, it is instructive to remind some results of modification of a-C:H with Cu, which will be useful for understanding of Co-C interaction in a-C:H(Co). In Figure 1 the Raman and the IR absorption spectra of a-C:H(Cu) are shown together with IR spectrum for undoped a-C :H.

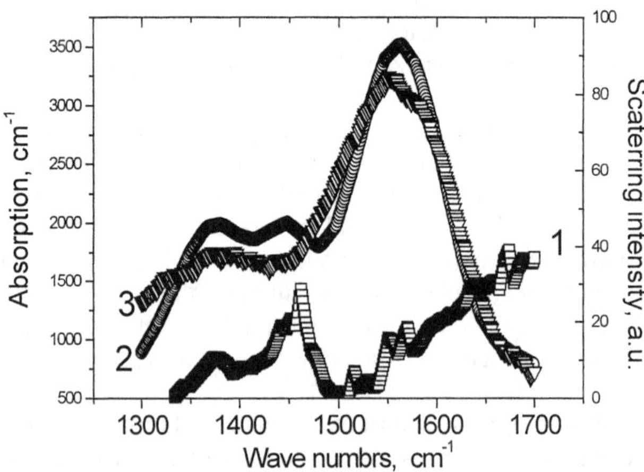

Figure 1. IR Absorption spectra: 1-a-C:H, 2-a-C:H(Cu), and 3-a-C:H(Cu) Raman spectrum.

In the last case only the absorption by bending modes of the C–H (1375 cm^{-1} and 1450 cm^{-1}) and O-H (~1700 cm^{-1}) bonds vibrations dominates in the spectra, while any noticeable absorption in the region of stretching vibrations of the C–C-bond (1500-1600 cm^{-1}) practically is absent. The IR absorption spectrum of a-C:H(Cu) demonstrates a dramatic difference in this relation. Along with an essential absorption enlargement, an intense absorption band emerges at the frequency coinciding with the frequency of G-line in Raman spectra shown alongside. This phenomenon was early shown [7] to arise from the interaction of copper with graphen-rings built in the graphite-like nanoclusters which are known to be important components of amorphous carbon structure. The interaction activates in the IR optical absorption the graphene-rings vibration modes which were otherwise silent. We call this line pseudo Raman bearing in mind that its appearance evidences that at least part of copper is atomically dispersed.

In Figure 2, we show the fragments of absorption spectra for the a-C:H and a-C:H(Co) films in the frequency region of stretching vibrations of the C–C bonds and the bending vibrations of the C–H bonds for various cobalt content. It can be seen that the introduction of cobalt into the carbon host initially increases considerably the absorption coefficient in the spectral range under consideration. When the ratio [Co]/[C] between the atomic concentrations attains a value of ~0.9, the absorption growth gives way to its decrease. In order to consider the influence of cobalt on the IR absorption by the carbon–hydrogen host more carefully, we analyze the structure of the absorption bands in Figure 2, by deconvoluting them into Gaussian contours.

As an example of the deconvolution, in Figure 3 we show together the spectra of the Raman scattering and of the IR absorption for the a-C:H(Co) sample with [Co]/[C] = 0.5. In the Raman spectrum, we observe two pronounced bands at the frequencies of ~1350 and ~1580 cm^{-1} characteristic of fine-crystalline graphite rather than amorphous carbon in whose spectrum such good resolution of similar bands (G- and D-bands) is routinely observed only in annealed samples.

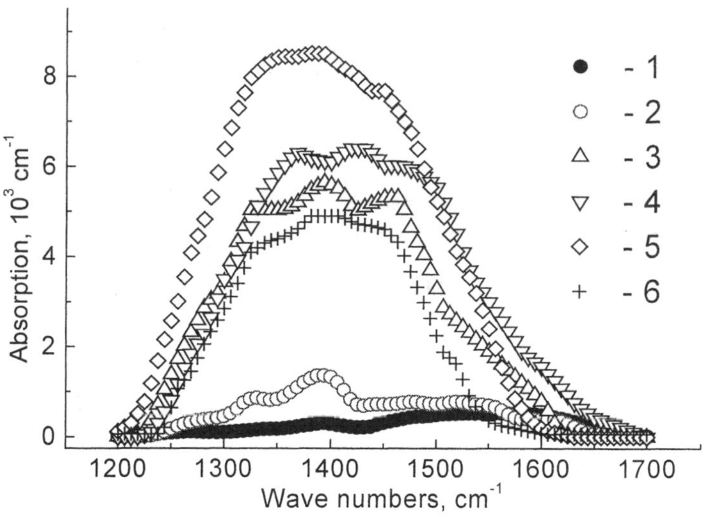

Figure 2. Absorption spectra for (1) a-C:H film and a-C:H(Co) films with [Co]/[C] = 0.08 (2), 0.5 (3), 0.73 (4), 0.88 (5), and 1.4 (6).

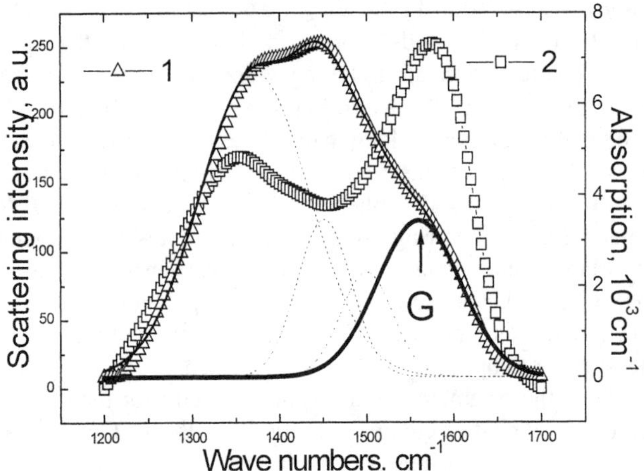

Figure 3. IR absorption (1) and Raman scattering (2) spectra for the a-C:H(Co) sample with [Co]/[C] = 0.5.

It can be seen that the position of a special feature in the IR spectrum of absorption in the region of 1500–1600 cm^{-1} correlates well with the position of the G-band in the Raman spectra, which can be associated with the activation of the quasi-Raman G-band in the optical absorption spectrum. It justifies consideration of this band in the analysis of the shape of the absorption-spectrum and it means that Co is partially atomically dispersed in the carbon matrix. The good resolution of G- and D-bands confirms, at the atomic level, that the modification by cobalt induces graphitization of the carbon host [3, 4].

A similar deconvolution of spectra for the a-C:H(Co) films with various Co concentrations was performed for the same set of frequencies with only the amplitude and band width varied. We list in Table 1 frequencies of the Gaussian bands of deconvolution for the spectra shown in Figure 2. These sets provide the best agreement with the experiment.

Table 1. Deconvolution of experimental spectra into Gaussian contours

Content [Co]/[C]	Gaussian contours frequency, cm^{-1}						
0.085	1288	1372	1436	1500	1560	1600	
0.5	1288	1372	1444	1508	1567		1652
0.73	1290	1368	1450	1500	1560		1650
0.88	1290	1373	1440	1500	1560	1600	
1.4	1290	1372	1440	1500	1560	1598	1650
Attribution:	sp^3CH	sp^3CH_3 sym	sp^3CH_2 sp^2CH	Benzene ring	G-line	C=C Aromat.	C=C

A striking feature is the fact that (as follows from Table 1) the introduction of cobalt leaves the frequencies of vibration modes of the hydrogen–carbon host virtually intact, but a drastic intensification of absorption takes place at some of these frequencies, as the comparison of spectra for various concentrations in Figure 2 shows. It can be seen from Figure 2 that the intensity first increases with concentration and then starts to decrease for high concentrations ([Co]/[C] > 0.9). It seems to be natural to associate this fact with the relative decrease in the fraction of carbon in the film as the cobalt concentration increases. In this case, we observe the highest intensification of the band associated with the symmetric vibration mode sp^3 CH_3 sym, which can likely be ascribed to the polarizing action of the chemical bond of this group with a cobalt atom belonging to a metal nanocluster or with a cobalt-carbide molecule. The former effect seems to be more probable with allowance made for the fact that the highest intensification corresponds to an approximately equiatomic ratio between carbon and cobalt atoms. The behavior of the carbon–carbon vibration modes near the metal surface was considered previously, and it was shown that these are only symmetric modes that prove to be optically active [16], which corresponds to our interpretation of the band at 1372 cm^{-1} (see, Table 1). Similar behavior is observed for the band at 1500 cm^{-1} corresponding in frequency to the vibrations of a benzene ring presumably connected with cobalt. The appearance of vibration bands of the benzene ring in the IR spectrum of a-C:H(Co) points to a strong interaction between Co atoms and the carbon host, which decomposes the graphene planes of the a-C: H(Co) graphitic fragments into isolated benzene rings. A supplementary argument in favor of forming benzene derivatives of cobalt is the appearance of a band at 1600 cm^{-1} in the a-C:H(Co) spectrum. In these cases, this band is an unavoidable satellite of the band at 1500 cm^{-1} as a result of the resonance splitting of the absorption band by the vibrations of three double bonds in the benzene ring [9].

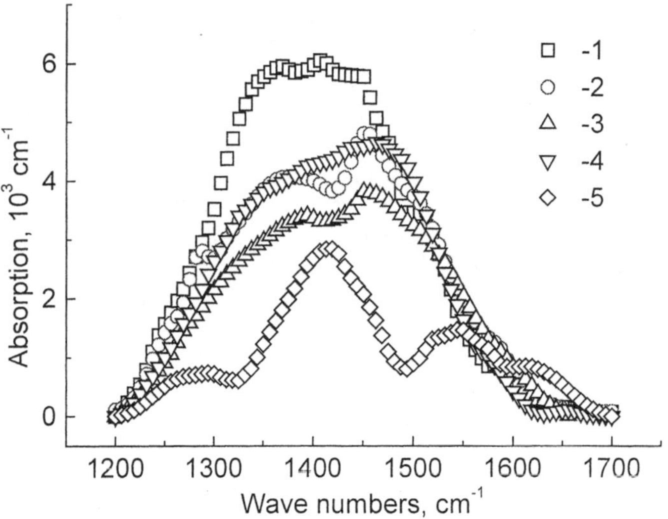

Figure 4. Absorption spectra of a-C:H(Co) for [Co]/[C] = 0.35 after annealing: as deposited (1), at 190 °C (2), at 260 °C (3), at 320 °C (4), and at 380 °C.

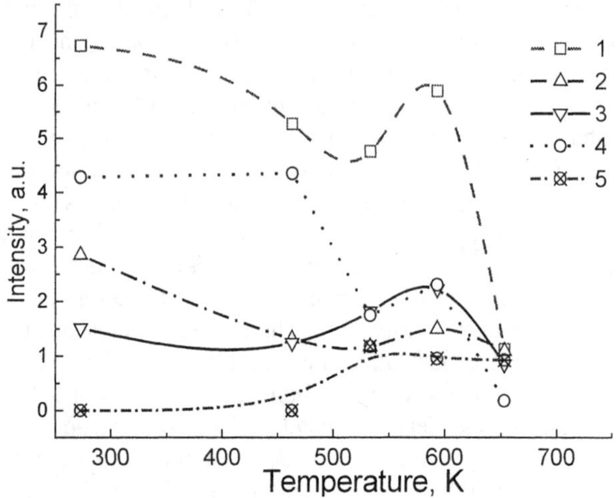

Figure 5. Areas under the deconvolution contours as a function of annealing temperature for the [Co]/[C] = 0.35 film, for the peaks at 1372 (1), 1450 (2), 1500 (3), 1552 (4), and 1600 (5) cm^{-1}.

In the intermediate case of weaker interaction of the intercalation of the graphene planes of Cu [10], the quasi-Raman band near 1560 cm^{-1} is activated in the IR spectrum. The simultaneous appearance of these three bands indicates that the investigated films both contain Co atoms interacting with the graphen rings and Co atoms combining with carbon and forming benzene derivatives of cobalt carbide. As the results of electron-microscope analysis [3,4] show, the former type of interaction leads to the formation of graphitic shells around the metal nanoclusters, which increase their stability, whereas the latter type retards the formation of shells and must be suppressed by an annealing procedure. In Figure 4, we show the absorption spectra for a-C:H(Co) films with [Co]/[C] = 0.35 annealed in vacuum at various temperatures. In the range of frequencies of 1400–1600 cm^{-1} assigned to the vibration modes of the carbon–carbon bonds, the non-monotonous dependence of the absorption on the annealing temperature caught our attention. In Figure 5, with the aim of gaining deeper insight into this feature, we display the dependence of areas under the Gaussian contours of decomposition of the absorption spectra for this sample on the annealing temperature. It can be seen that there are a number of characteristic features in the modification of the spectrum under the effect of thermal annealing. First, we should note a very substantial change in the absorption by the modes which were initially intensified anomalously due to the interaction of cobalt atoms with the carbon host. A temperature of ~350 °C, at which the anomalous intensification of the 1372 cm^{-1} mode almost completely disappears, coincides with the temperature of the breaking of the Co–C bond in the metastable carbide Co_2C [5]. This circumstance corroborates the mechanism proposed above for the intensification of absorption by the bending modes of vibrations of the CH_3 group due to formation of a bond between this group and a cobalt atom. The disappearance of the 1552 cm^{-1} quasi-Raman band as the annealing temperature is increased also corresponds to weakening of the interaction of cobalt with graphitic fragments of the carbon host, which facilitates the coalescence of the cobalt nanoclusters observed at these temperatures. To some degree, the bond between cobalt and carbon atoms in the benzene rings (bands at 1500 and 1600 cm^{-1}) behaves in a similar manner.

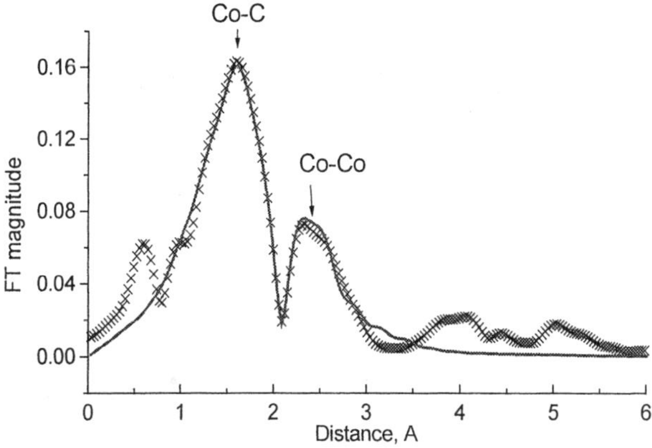

Figure 6. FT EXAFS spectrum for as-deposited a-C:H(38% Co) film. The crosses are experimental, the solid line is a fit.

The inadequate reliability of the procedure of decomposition of the spectrum makes it senseless to discuss the less pronounced features in Figure 6; however, an abrupt fall in the absorption coefficient at ~350 °C (~650 K), which can be attributed to the dissociation of cobalt carbide and its derivatives, can apparently be determined with reasonable reliability

4. X-ray Spectroscopy

The well known kind of X-ray absorption spectroscopy known as extended X-ray absorption fine structure (EXAFS) has been applied to the study of the local structure of a-C:H(Co) films under discussion. The measurements were performed at room temperatures at beam line 13B of Photon Factory of National Institute for Advanced Interdisciplinary Research in Tsukuba, Japan [11]. The EXAFS Fourier transformed spectra were back transformed using the R-range of 1 to 5 Å and the fitting was performed in R-space. In the fitting, we used Co-Co and Co-C correlations. A typical fitting result for as-deposited sample is shown in Figure 6. The results of the curve-fitting in Figure 6 are presented also in Table 2 and provide direct evidence that the majority of Co atoms are coordinated by C atoms.

Table 2. Structural parameters obtained from curve-fitting for As grown films
The notations are: N – coordination number, BL – bond length

Co content	$N_{Co\text{-}C}$	$BL_{Co\text{-}C}$, Å	$N_{Co\text{-}Co}$	$BL_{Co\text{-}Co}$, Å
27%	5.6±1.2	2.04±0.02	1.6±0.8	2.45±0.02
32%	4.9±1.1	2.03±0.02	5.1±1.5	2.48±0.02
38%	4.1±1.1	2.03±0.02	6.0±1.6	2.47±0.02

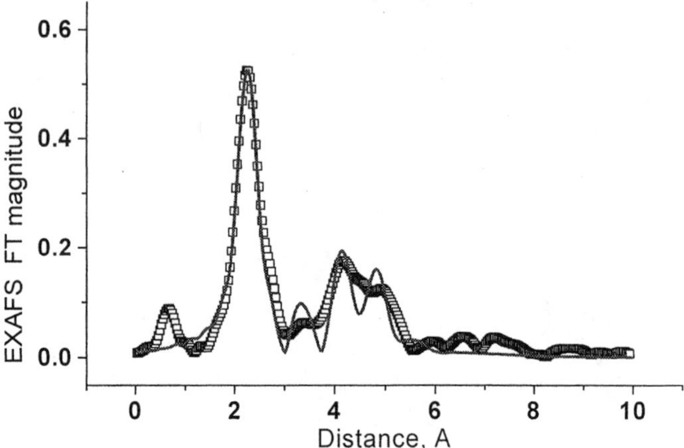

Figure 7. FT EXAFS spectrum for a-C:H(38% Co) film after annealing: □ experiment, — fit for Co-Co correlation.

Besides, Co-Co correlations are additionally found which indicates that a certain number of Co atoms have Co as the first-nearest neighbor. The determined Co-Co coordination number of 1.6 to 6 and the fact that the Co-Co bond length in our samples (~2.46 Å) is shorter than that in the bulk (2.50 Å) is indicative of a small-size metallic cluster. The results of the FT EXAFS curve-fitting for a film annealed at ~500 °C are exhibited in Figure 7. The corresponding structural parameters as obtained from the fitting procedure are tabulated in Table 3. The increase of BL_{Co-Co} and N_{Co-Co} values resulting from the anneal clearly indicate the formation of Co-Co bonds. The FT spectra of the annealed samples are very similar among themselves indicating that the local structure of Co (almost) does not depend on the concentration of the metal. The higher shells in the FT spectra are clearly observed, which is an evidence for high crystal quality as well as for the rather large size of the nanocrystals. Figure 8 shows a TEM image of the a-C:H(38%Co) film after annealing at ~500 °C. It is actually seen that Co nanocrystals are formed, their size being ~ 20 nm. From data for BL_{Co-Co} in Table 3 it follows that the local structure of the Co phase is the same as in the bulk Co. Nevertheless, the obtained value for the Co-Co first-nearest coordination number of ca. 8.5 is smaller than 12 in bulk Co.

Table 3. Structural parameters for annealed films

Sample	N_{Co-C}	N_{Co-Co}	BL_{Co-Co}, Å
27 %	---	8.5 ± 0.9	2.50 ± 0.01
32 %	---	8.9 ± 1.1	2.50 ± 0.01
38 %	---	8.1 ± 0.9	2.50 ± 0.01

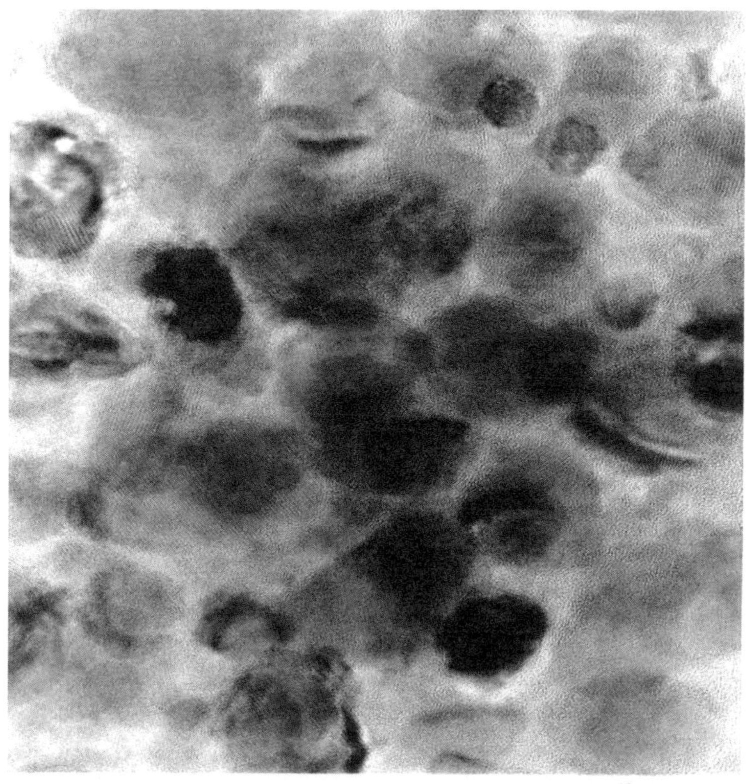

Fig 8. Cobalt nanoclusters TEM image

An explanation would be that only a fraction of Co atoms (~66 %) form the nanocrystalline phase, the rest being dissipated in the matrix. In such a case, the Co atoms located within the matrix will not contribute to the Co-Co coordination number determined from the fit which assumes only the presence of Co phase Since the EXAFS from Co-C pairs is considerably weaker than that from Co-Co pairs, the contribution of the Co-C pairs into the raw EXAFS and FT spectra is small and may be mistaken for the background.

5. Conclusion

IR spectroscopy in the region of vibration frequencies of the carbon–carbon bonds in a-C:H(Co), together with Raman spectroscopy, makes it possible to clarify the main mechanisms of interaction between cobalt and the carbon host in relation to the carbon concentration and the annealing temperature. At least three types ofinteraction are revealed: (a) a strong interaction associated with the formation of a metastable chemical bond in methyl derivatives of cobalt carbide; (b) an interaction destroying the graphitic fragments with the formation of benzene derivatives of carbon; and (c) a weak interaction characteristic of metals that do not form carbides, e.g., copper. However, thermal annealing at ~350 °C suppresses these interactions favoring, as might be expected, the formation of cobalt nanoclusters in the amorphous carbon host. X-ray spectroscopy analysis is in accord with the above conclusions.

6. References

1. Hayashi, T., Hirono, S., Tomita, M., and Umemura, S. (1996) Magnetic thin films of cobalt nanocrystals encapsulated in graphite-like carbon, *Nature* **381**, 772-774
2. Dyuzhev, G. A., Ivanov-Omskii, V. I., Kuznetsova, E. K., Rumyantsev, V.D., Yastrebov, S.G., Zvonareva, T.K., and Abaev, M.I. (1996) Diamond-like carbon and fullerene as a protective coating for Ag-based reflective optics, *Mol. Mater.* **8**, 103-106
3. Jiao, J., and Seraphin, S. (1998) Carbon encapsulated nanoparticles of Ni, Co, Cu, and Ti, *J. Appl. Phys.* **83** (5), 2442-2448
4. Delaunay J.-J., Hayashi T., Tomita M., and Hirono S. (1997) Formation and microstructural analysis of co-sputtered thin films consisting of cobalt nanograins embedded in carbon, *J. Appl. Phys.* **82** (5), 2200-2210
5. Zvonareva, T.K., Ivanova, E.I., Frolova, G.S., Lebedev, V.M., and Ivanov-Omskii, V.I.. (2002) Vibratonal Spectroscopy of a-C:H(Co), *Semiconductors* **36**(6), 695-700
6. Ivanov-Omski, V. I., and Smorgonskaya, E. A. (1999) Charge displacement induced by intercalation of graphite-like nanoclusters in amorphous carbon with copper, *Phys.Solid State* **41**, 786-788
7. Ferrari, C.A., and Robertson, J. (2001) Resonant Raman spectroscopy of disordered, amorphous, and diamondlike carbon, *Phys. Rev.* B **64**, 075414
8. Bellamy, L. J. (1954) *The Infrared Spectra of Complex Molecules*, Methuen, London, p. 59.
9. Ivanov-Omski V. I., Zvonareva T. K., and Frolova G. S. (2000) Vibration Modes of Carbon in Hydrogenated Amorphous Carbon Modified with Copper, *Semiconductors* **34**, 1391-1397
10. Kolobov, A.V., Oyanagi, H., Ivanov-Omskii, V.I., and Tanaka, K. (2002) Local structure of Co nanocrystals embedded in hydrogenated amorphous carbon: an x-ray absorption study (*in press*)

SYNTHESIS OF COLLOIDAL MAGNETIC NANOPARTICLES:

Properties and Applications

M. HILGENDORFF and M. GIERSIG
Hahn-Meitner-Institut Berlin
Glienicker Str. 100, 14109 Berlin, Germany

1. Introduction

Ferrofluids have received much attention in the past several decades for their interesting properties which can be exploited to develop new technologies [1] such as new refrigerators which employ the magneto caloric effect [2], new inks for inkjet printers [3], novel spin valves [4], or for new cancer therapies, such as hyperthermia [5], and apherese [6]. Therefore, the colloidal chemical synthesis of ferrofluids is of great interest. Many well documented references and patents have described successful syntheses of magnetic nanoparticles and the investigation of their properties and applications. A nearly complete list can be found in the magnetic fluids bibliography [7].

Theoretical calculations on the structures and properties of clusters consisting of some hundred atoms have been done in principle. The properties of nanoparticles consisting of several thousand atoms (e.g. a spherical close packed Co nanoparticle, 10 nm in diameter, consists of about 47000 atoms) can not be theoretically calculated because of the unavailability of sufficient computing power. This can be solved in part by the development of "bits" of reduced size. Let us assume that a magnetic particle of 10 nm in diameter can act as a spin up/spin down 1/0 binary code element. A close packed monolayer of these crystals will then have a data storage capacity in the Tbit/inch^2 range, which is several orders of magnitude higher than what is currently possible. Thus, and in accordance with Moore's law, it is no surprise that colloidal chemical preparations of magnetic particles (ferrofluids) have enjoyed increased attention over the past few years [8].

Most of the possible applications of ferrofluids require high remanent magnetization or high anisotropy at room temperature. High remanent magnetization is a characteristic feature of soft ferromagnetic materials with diameters above a critical minimum dependent on the material (> 10 nm for Co). On the other hand, hard ferromagnetic materials, which often exist as alloys, exhibit high anisotropies. Soft ferromagnetic materials embedded in a noble metallic matrix show a technologically important effect called the giant magneto resistance (GMR) [9].The ability to prepare air-stable ferromagnetic/noble metal bimetallic particles is therefore a challenge. Well-known ferromagnetic alloys consist of ferromagnetic 3d- and noble metal elements of the platinum group. They are known to have large magnetic moments and large magnetic anisotropies [10]. Accordingly, the synthesis of monodisperse ferromagnetic/noble metal bimetallic particle alloys (especially Fe/Pt) or core-shell particles and their assembly into PPA has been investigated intensely in recent years [11].

L.M. Liz-Marzán and M. Giersig (eds.),
Low-Dimensional Systems: Theory, Preparation, and Some Applications, 151–161.

To facilitate the preparation of nanocrystalline particles appropriate for technological applications, easy methods must exist for the preparation of non-aggregated, monodisperse colloids, as well as highly ordered layers on different substrates. Naturally, it is preferable to have inexpensive materials and efficient syntheses, as well as state of the art industrial-scale preparation techniques. One must strive to lower to costs associated with scaling up new laboratory techniques for industrial requirements.

Unfortunately, most requirements are limited by the possibilities offered by chemical preparation techniques in the field of colloid chemistry. For example, different applications require suspensions of colloidal particles in different solvents (e.g. aqueous solutions for biocompatability or non-polar solvents for development of well ordered self assembled monolayers), just as various materials may require preparation methods tailored to their chemical behavior.

Although noble metals are often used as model systems in many chemical fields they are unfortunately not suitable for non-noble materials like Fe_3O_4, Fe, Co, or Ni, which are sensitive to oxidation and complexation reactions in many cases. This is particularly the case in the field of colloid chemistry, where the enhanced dispersivity of the surface-to-volume ratio of nanoparticles means an enhancement of reactivity compared to bulk materials. Many successful methods developed for noble metal nanoparticles could not be transferred to ferrofluids. Thus, many researchers dealt with Fe_3O_4 and Co as model systems for ferrofluids and their assembly into large PPA.

Fe_3O_4 ferrofluids can easily be prepared in water and have been therefore investigated as a model system for applications in bioscience [5,6] and for assembly methods using electrostatic forces such as layer-by-layer (LbL) assembly [12]. Co ferrofluids which are usually prepared in organic solvents because of the necessity to use bulky (i.e. hydrophobic) stabilizers to prevent oxidation and to overcome the magnetic dipole-dipole interaction, have been investigated as a model system for self assembly [13].

Many successful synthesis have been developed to get ferrofluids which are suitable as building blocks for different applications. New applications require usually a further optimization of well established preparation techniques or the development of new synthesis.

The colloid chemical synthesis of stable ferrofluids of ternary magnetic particles such as garnets or hexaferrites has not been published to our knowledge. Therefore, we will give in the following some examples about the optimization of the preparation of magnetite in aqueous solutions, as well as new advances of the colloid chemical preparation of magnetic metal particles in organic solutions, further developed in accordance to the requirements of new applications.

2. Wet Chemical Synthesis of Magnetic Nanoparticles

Iron oxide ferrofluids are usually prepared in aqueous solutions by hydrolysis and condensation [14], well-known in the field of sol-gel chemistry [15].Due to the enhanced reactivity of non-noble 3-d transition elements and the large magnetic dipole-dipole interaction of Fe, Co, and Ni ferrofluids with those elements, they are usually synthesized using the water-in-oil "reverse micelles" emulsion technique [13a),b)]. In addition one can perform the reduction or thermal decomposition of metal precursors in organic solvents in the presence of bulky stabilizers such as fatty acids [13c)-i)]. As a result, Fe- [16], Co- [13], Ni- [17], and FeN_x [18] ferrofluids, as well as their alloys with noble metal particles of the

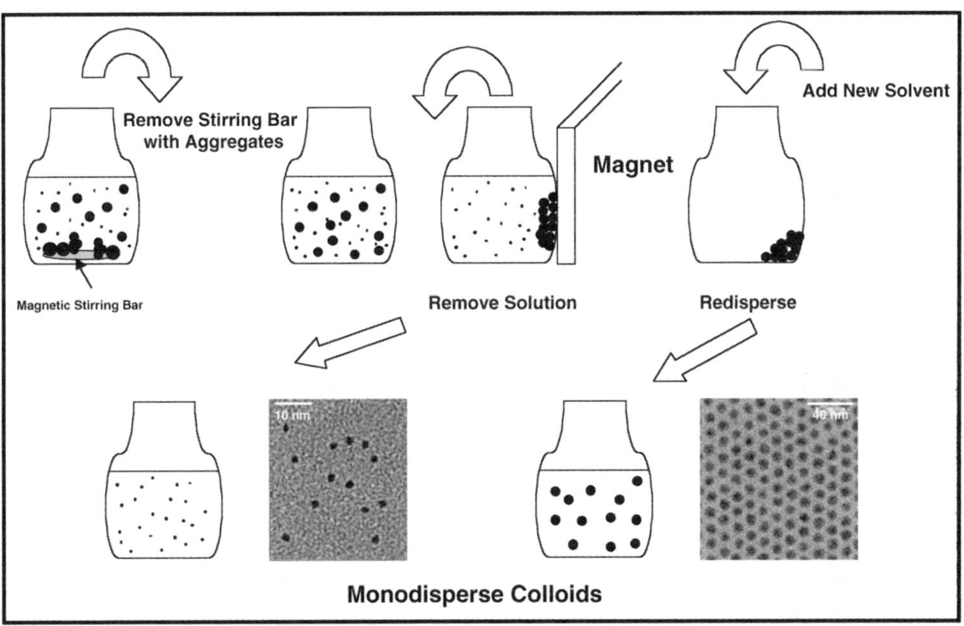

Figure 1. Sketch of the size selective precipitation of magnetic nanoparticles to obtain a standard deviation of the size distribution << 5 %. The TEM images show Co particles having a diameter of 3 nm and 11 nm respectively.

platinum group, e.g. Fe/Pt [11b)], Co/Pt [11c)], Co/Rh [11f)], are formed with enhanced anisotropy and oxidation stability. To further increase the stability against oxidation and to use ferrofluids as building blocks for GMR devices, the preparation of ferromagnetic particles covered with a noble metal shell, e.g. Ag@Co [11a)], Pt@Co [11c)], or Au@Co [11e)], and vice versa, e.g. Co@Ag [11g)], or Co@Fe [11d)] have been published.

A narrow particle size distribution is required for the self assembly of colloids dried on a substrate. Two principle methods have shown, after much investigation, to produce the best results. The first method involves a size selective precipitation which exploits the dependency of nanoparticle stability on size to separate out monodisperse systems of particles. Upon adding a solvent having a larger Hamaker constant, larger particles aggregate and precipitate faster than smaller particles. The precipitate can then be separated from the solution by centrifugation or, in the case of magnetic particles, by application of an external magnetic field [19]. A sketch of this method is shown in Figure 1.

The second method is based on the optimization of nucleation and crystal growth during the synthesis. The latter method is of course preferred as it would allow for the direct *in situ* control of nanoparticle formation, but it is usually more expensive as it often requires non-commercial precursors and an increased expense. Moreover, the narrowest obtainable size distributions from this latter method have been experimentally, as well as theoretically, limited to ~ 5 % [20]. So, there is little wonder why many scientists, especially industrial scientists, deal with size selective precipitation, which reliably allows for preparation of colloids having a standard size distribution << 5 %.

Nevertheless, there are application requirements that may favor the latter method's use of non-commercial or highly expensive compounds as precursors. These requirements are concerned with large anisotropies often found in alloys or rod-like particles.

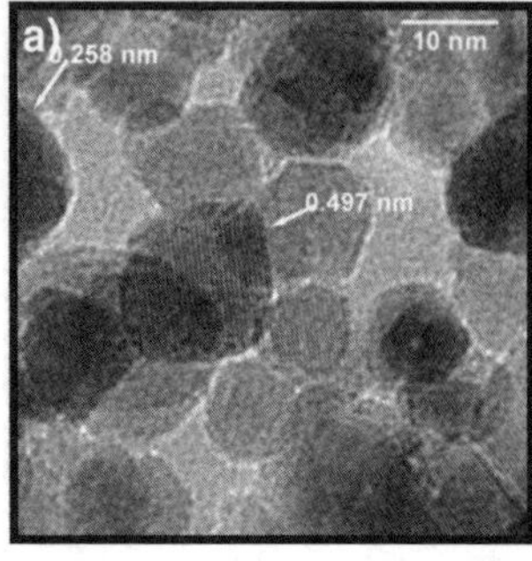

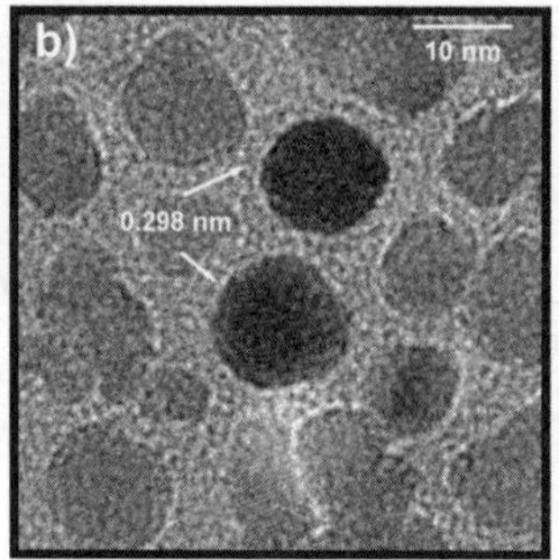

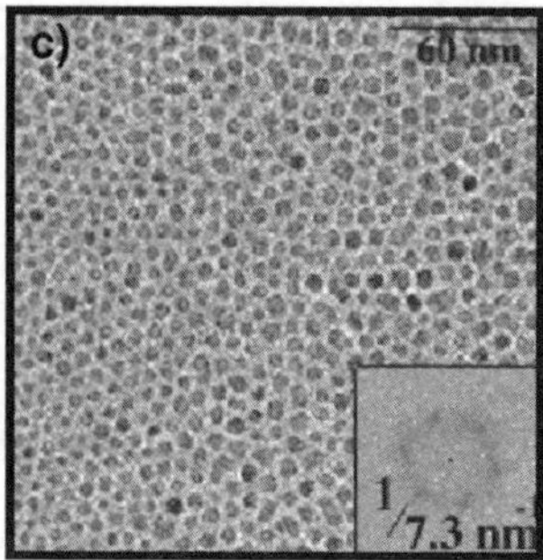

Figure 2. Typical TEM images of iron oxide particles; a) prepared in aqueous solution via Massart and b) the same particles dried from a toluene solution after adsorbing oleic acid on their surface. The distances shown in the images are the result of fast Fourier analysis, indicating typical lattice planes of magnetite. c) A self assembled layer of particles shown in b) after size selective precipitation applying the method shown in Figure 1. Figure 2c reprinted with permission from [13d)], copyright 2000, World Scientific.

Rod-like Ni particles have already been prepared successfully using nickel dicyclo-1,5-octadiene as a precursor [17b)]. Indeed, the preparation of alloys may even be favored through use of "exotic" stabilizers.

2.1 IRON OXIDE

The most common preparation of magnetite (Fe_3O_4) ferrofluids was developed by Massart about 20 years ago and described in detail in 1987 [14]. The synthesis was based on the coprecipitation of Fe(II) and Fe(III) salts in aqueous solutions stabilized by repulsive electrostatic forces. This can be either done in acidic media, giving positively charged particles, or in alkaline media, giving negatively charged particles. Stable ferrofluids are available if the pH of the final solution is far from the pH_{ZPC} which is the case if a ζ–potential large enough to overcome attractive van der Waals forces is present, as pointed out in section 2.1.3. As a typical result, particles with a standard size deviation of ~25 % are formed. TEM investigations show typically aggregated particles, as presented in Figure 7a, consisting usually of a mixture of ferrimagnetic magnetite, and maghemite (γ-Fe_2O_3), and paramagnetic hematite (α-Fe_2O_3). The ratio of the different iron oxide phases is directly compared to the expense associated with performing the synthesis under oxygen free conditions, because γ-Fe_2O_3 and α-Fe_2O_3 are oxidation products of Fe_3O_4.

Since these reports, several optimizations have been discovered to produce ferrofluids useful for new applications. One of them was concerned with the possibility of changing the magnetic properties of the inverse Fe_3O_4 spinell by replacing Fe(II) with Co(II), Ni(II), Mn(II), or Zn(II) ions [21].

Additionally, many reports dealt with solving the classic problem of nanoparticle aggregation in aqueous solutions. To overcome such aggregation, new preparations have involved micellar solutions [13b),22] as well as the transfer of iron oxide particles from aqueous- to non-polar solvents by hydrophobizing the surface by adsorbing bulky stabilizers such as fatty acids [23]. Figure 2b shows impressively the difference in iron oxide ferrofluid prepared in aqueous solution (Figure 2a) after its transfer to toluene using oleic acid molecules adsorbed on the surface of the particles as a hydrophobizing agents.

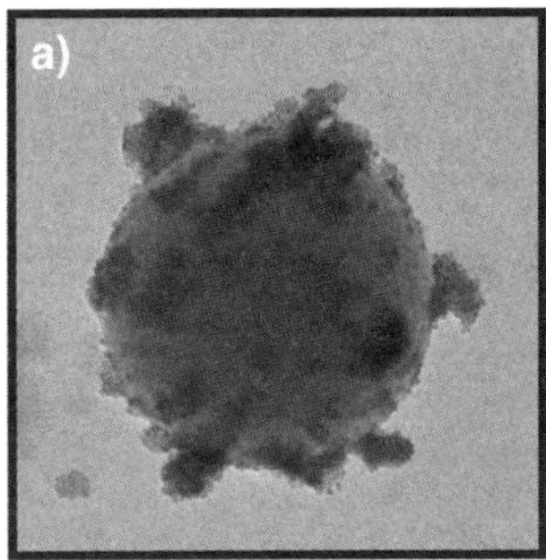

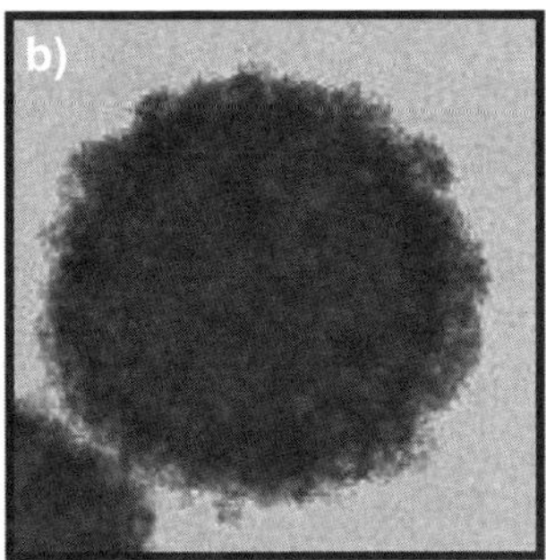

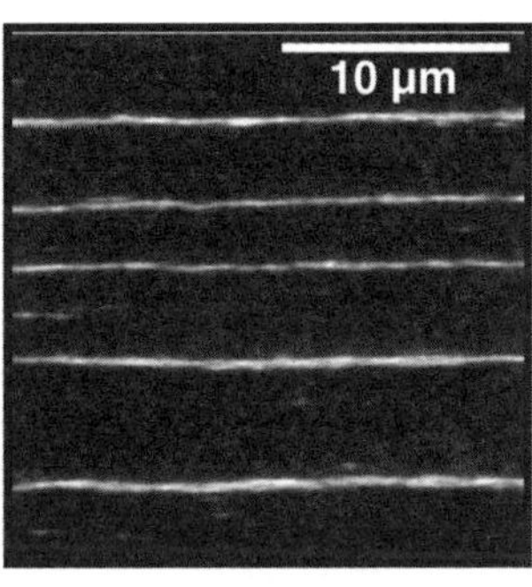

Figure 3. PS-Latex (diameter: 640 nm) covered with magnetite nanoparticles applying LbL assembly before a) and after the further optimization of the magnetite synthesis b). c) Well aligned stripes of magnetite coated PS-Latex obtained by magnetophoretic deposition. Figure 3b) and 3c) reprinted with permission from [12b)], copyright 2002, Elsevier Science.

Different methods have been applied to decrease particle size distribution. One method that has been successfully applied involves synthesis of iron oxides in reverse micelles. A second method dealt with a size selective precipitation (Figure 1) after transfer of iron oxide particles to non-polar organic solvents [23b)]. Particle size distributions of < 10 % have been available for magnetite ferrofluids. Figure 2c presents a monolayer of magnetite particles self assembled from an organic solution after size selective precipitation.

Finally, some reports dealt with the problem of reducing the size distribution of aggregates in aqueous solutions. Besides obvious application potential in the biosciences, aqueous solutions of magnetic particles are of interest for new techniques dealing with charge assembly (such as LbL). Since charged particles are only available in polar solvents, mainly water, a further optimization of Massart's method has been developed by Caruso et al. [12a)] to cover polystyrene latex spheres (PS-Latex) with a shell of magnetic nanoparticles.

Figure 3a demonstrates the less successful coating of PS-Latex with magnetic particles prepared according to the Massart method. The uneven coating can be understood as the consequence of a broad size distribution of the aggregates. Smaller aggregates with a narrower size distribution could be obtained by first hydrolyzing Fe(II) and Fe(III) salts in separate flasks, and by then using the more steric stabilizer tetrabutylammonium hydroxide, instead of the tetramethylammonium hydroxide used by Massart. Lastly, the hydrolyzed iron salts are mixed and heated to complete the condensation. As a result, the PS-Latex particles are homogenously coated with magnetic particles as shown in Figure 3b. Figure 3c shows magnetite coated PS-Latex particles arranged in well aligned parallel stripes as the result of magnetophoretic deposition experiments [12b)].

A second solution of the "aggregate size problem" was achieved through preparation of magnetite via the Massart method but in ethanol instead of water, followed by a solvent exchange from ethanol to water. As a result, very small particles less than 4nm in size have been synthesized for use in LbL experiments [12a)].

Most recently, Sun et al. [24] have developed a very successful method to achieve monodisperse Fe_3O_4 nanoparticles. The method is based on the reduction of iron(III)-acetylacetonate by 1,2-hexadecanediol and further thermal decomposition at high temperatures (solvent: diphenylether, bp. 265°C) in the presence of oleic acid and oleylamine as stabilizers.

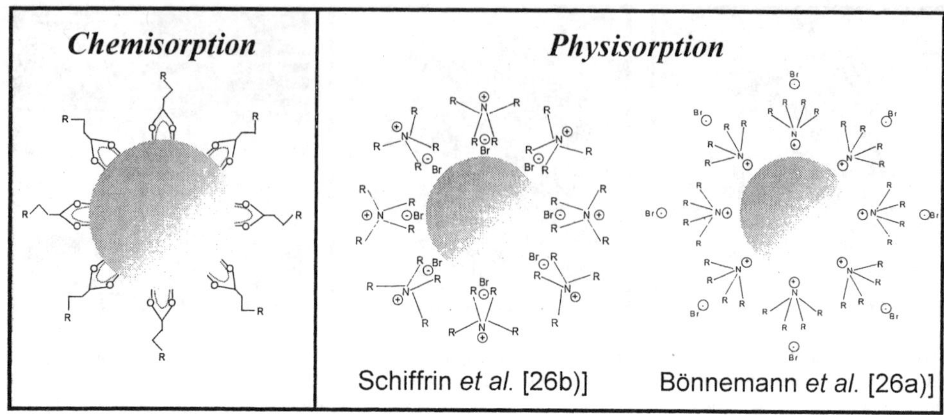

Figure 4. Sketches of different models describing the particle surfactant interface.

As a result, small Fe_3O_4 nanoparticles (diameter ~4 nm) having a particle size distribution of ~5% are formed and can be further grown by the seed-mediated growth method [25]. The authors have mentioned that the mechanism leading to Fe_3O_4 formation under the current conditions is not yet understood.

In the following section, the size selective precipitation scheme for production of extremely narrow size distributions with a standard deviation of << 5 % for Co particles in a ferrofluid will be presented. The reason for this difference in the obtained standard deviation of the particle size distribution between Co- and FeO_x particles is mainly the coexistence of different iron oxide phases.

2.2 METAL PARTICLES

The synthesis of colloidal solutions of non-noble metal nanoparticles has been the topic of intense research in the past several decades. During this time, methods have been designed and optimized to obtain colloids suitable as building blocks for new applications. Various synthetic routes have been investigated to develop smart ferromagnetic particles with novel and application specific properties. As a partial consequence of this research, stable fluids of non-noble metallic nanoparticles are now available in hydrophobic solvents. Models have argued that either electrostatic forces between particles and their stabilizers (tetraoctylammonium (TOA) salts were used for Au or Ag metal particles) [26] or strong chemisorption, as demanded by Buske [27], are responsible for the physical nature of the surfactant-particle interface in organic solvents.

Colloidal particles enjoyed a warm reception in the field of catalysis. Since, catalysis requires extremely small particles possessing a large surface-to-volume ratio and an amorphous, active shell of metal compound, it came as no surprise that they would find increased use as highly reactive and efficient catalysts.

Bönnemann et al. developed a method for the preparation of noble metal, as well as non-noble metal, particles via the reduction of metal salts by a non-commercial, tetraoctylammonium modified, super hydride [26a)]. The authors explained that the binding between particles and their stabilizers is electrostatic in nature.

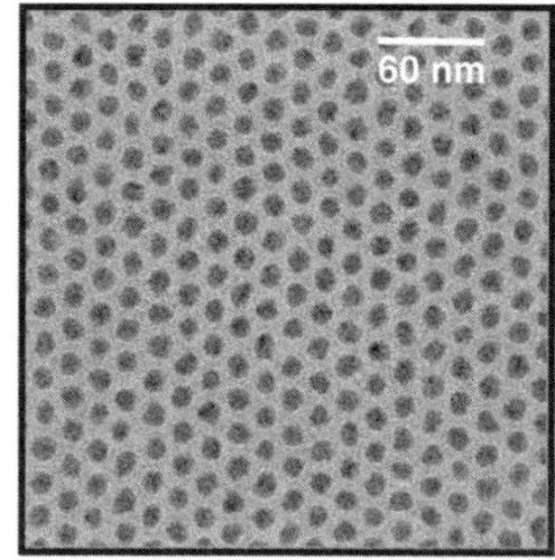

Figure 5. TEM images at different magnifications of monodisperse 11 nm Co particles stabilized by oleic acid, obtained after size selective precipitation, and self assembled on a carbon coated Cu grid.

It should therefore be possible to transfer these particles to aqueous solutions using the method developed by Gittins et al. While a summary of Bönnemann's work [11e)] has recently been published, the authors did not discuss the varying chemical reactivities, during redox and complexation reactions, of transitition elements towards stabilizers.

Completely different particle-surfactant models have been published by Bönnemann and Schiffrin [26]. Bönnemann assumed positively charged metal particles surrounded by a negatively charged electron cloud, where the positive TOA ion is located at the particles surface. In contrast, Schiffrin assumed positively charged noble metal particles surrounded by negatively charged ions. Moreover, Schiffrin pointed out that it is important to note that the charge of the surfactant ions points away from the particles. The different models are shown in Figure 4.

However, both ionic models describe the physisorption of stabilizers. Physisorption, as opposed to chemisorption, allows for exchange of stabilizers which enables the transfer of particles from non-aqueous to aqueous systems. This method has been recently studied by Gittins et al. [28].

As mentioned earlier, chemisorption, more than physisorption, is required to obtain stable non-noble metal particles of larger size. Figure 5 presents TEM images of self assembled layers of monodisperse 11 nm Co particles, which were stabilized by a chemisorbed shell of oleic acid molecules obtained after size selective precipitation. Due to chemisorption, no further ligand exchange is possible meaning that the particles can no longer be transferred to polar solvents. As a consequence, such colloids would not be suitable for biological applications and LbL experiments.

The compounds used for a preparation determine not only the shape but also the stability of the final particles. Small changes in preparation conditions can give completely different crystalline phases. Moreover, it is well known in the field of colloid chemistry that crystal structures different from bulk materials are available [29]. Figure 6 presents Co nanoparticles with different crystal structures. While the hcp phase of the particle in Figure 6a [13f),30] is a typical phase of bulk Co, the bcc- (Figure 6b) [13e),31] and ε–phase (Figure 6c) [13g),32,] have been only found in nanoparticles.

Different crystal structures possess unique properties and anisotropies, which must be considered before the application of such particles for memory devices. Consider a monolayer of magnetic crystals.

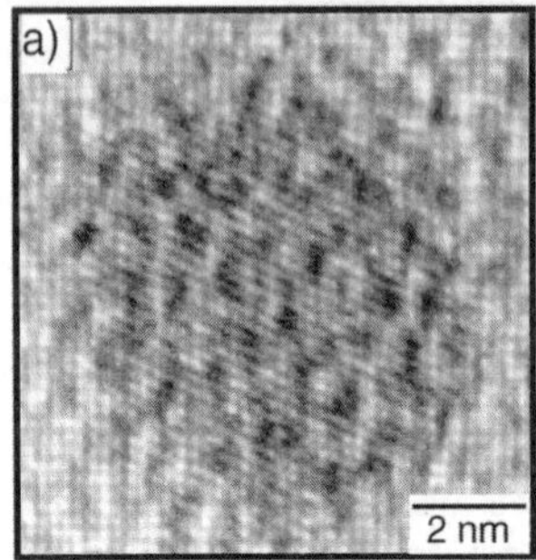

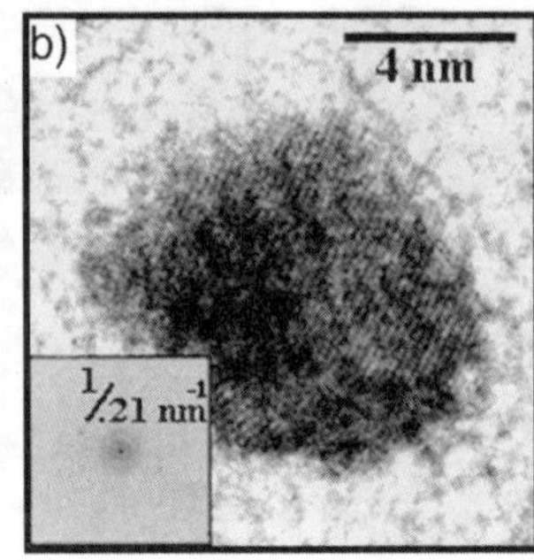

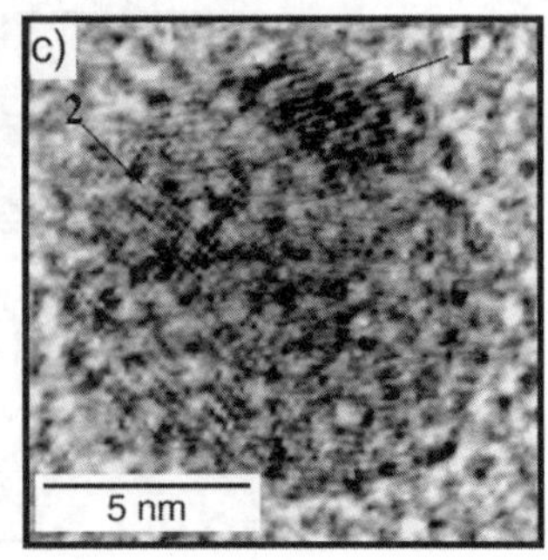

Figure 6. TEM images of Co nanoparticles with different crystal structures. a) hcp, reproduced with permission from [13f)], copyright 2001, MRS Publishing; b) bcc, reproduced with permission from [13e)], copyright 1999, IOP Publishing; and c) ε-fcc, reproduced with permission from [13g)], copyright 2001, CSIRO Publishing.

It has been simulated that if 10 % of this array were to consist of crystals of a different structure (a lower anisotropy will cause the percolation of distorted spins), then this fact alone would be the main reason for media noise and an unknown loss of stored data [33].

3. Summary

Colloid chemistry is rather complex. Small changes in preparation conditions lead to completely different results, a fact which is a consequence of enhanced surface to-volume ratio of nanoparticles. Therefore, the properties of nanoparticles are mainly determined by the interaction of surfaces and interfaces. In particular, enhancement of the surface-to-volume ratio in non-noble metals results in an increased reactivity that makes these materials difficult to handle. Moreover, this increased reactivity complicates the scalability of ground state science processes, often requiring expensive setups, to an industrial level development that demands low cost production methods.

The development of materials for new applications is a dynamic process. Careful investigation of advances in colloid chemistry often gives rise to novel ideas and solutions which drive the development of applications in response to technological demands.

4. Acknowledgment

The authors gratefully acknowledge the financial support of the German Science Foundation (DFG), AZ.: II C10-322 10 72 and T. Kempa, Boston College, USA, for carefully reading the manuscript.

5. References

1. Rosensweig, R.E. (1998) *Ferrohydrodynamics,* Dover Publishing, New York.
2. a) Shao, Y.Z., Lai, J.K.L., and Shek, C.H. (1996) Preparation of nanocomposite working substances for room-temperature magnetic refrigeration, *J. Magn. Magn. Mater.* **163**, 103-108; b) Pecharsky, V.K. and Gschneidner Jr., K.A. (1999) Magnetocaloric effect and magnetic refrigeration, *J. Magn. Magn. Mater.* **200**, 44-56; c) Yamamoto, T.A., Tanaka, M., Misaka, Y., Nakagawa, T., Nakayama, T., Niihara, K., and

Numazawa, T. (2002) Dependence of the magnetocaloric effect in superparamagnetic nanocomposites on the distribution of magnetic moment size, *Scr. Mater.* **46**, 89-94.

3. a) Ronay, M. (1976) Preparation of magnetic particles and magnetic fluids by chemical reaction in a magnetic field, *IBM Technol. Discl. Bull.* **19**, 2753-2763; b) Kormann, C., Schwab, E., Raulfs, F.-W., and Beck, K. H. (1996) Magnetic ink concentrate, U.S. Patent 5,500,141.
4. Lodder, J.C., Monsma, D.J., Vlutters, R., and Shimatsu, T. (1999) The spin-valve transistor: technologies and progress, *J. Magn. Magn. Mater.* **198-199**, 119-124.
5. Jordan, A., Scholz, R., Wust, P., Schirra, H., Schiestel, T., Schmidt, H., and Felix, R. (1999) Endocytosis of dextran and silan-coated magnetite nanoparticles and the effect of intracellular hyperthermia on human mammary carcinoma cells in vitro, *J. Magn. Magn. Mater.* **194**, 185-196.
6. Mapra, M.Y., Körner, I.J., Hildebrandt, M., Bargou, R., Krahl, D., Reichardt, P., and Dörken, B. (1997) Monitoring of tumor cell purging after highly efficient immunomagnetic selection of CD34 cells from leukapheresis products in breast cancer patients: comparison of immunocytochemical tumor cell staining and reverse transcriptase-polymerase chain reaction, *Blood* **89**, 337-344.
7. a) Charles, S.W. and Rosensweig, R.E. (1983) Magnetic fluids bibliography, *J. Magn. Magn. Mater.* **39**, 192-220; b) Kamiyama, S. and Rosensweig, R.E. (1987) Magnetic fluids bibliography, *J. Magn. Magn. Mater.* **65**, 403-439; c) Blum, E., Osols, R., and Rosensweig, R.E. (1990) Magnetic fluids bibliography, *J. Magn. Magn. Mater.* **85**, 305-378; d) Cabuil, V., Neveu, S., and Rosensweig, R.E. (1993) Magnetic fluids bibliography, *J. Magn. Magn. Mater.* **122**, 439-482; e) Bhatnagar, S.P. and Rosensweig, R.E. (1995) Magnetic fluids bibliography, *J. Magn. Magn. Mater.* **149**, 199-232; f) Vékás, L., Sofonea, V., and Balau, O. (1999) Magnetic fluids bibliography, *J. Magn. Magn. Mater.* **201**, 454-489.
8. a) Weller, D. and Moser, A. (1999) Thermal effect limits in ultra-high density magnetic recording, *IEEE Trans. Magn.* **35**, 4423-4439; b) Sellmyer, D.J., Yu, M., and Kirby, M.D. (1999) Nanostructured magnetic films for extremely high density recording, *Nanostruct. Mater.* **12**, 1021-1026.
9. Baibich, M.N., Broto, J.M., Fert, A., Van Dau, F.N., Petroff, F., Eitenne, P., Creuzet, G., Friederich, A., and Chazelas, J. (1988) Giant magnetoresistance of (100) Fe/(001) Cr magnetic superlattices, *Phys. Rev Lett.* **61**, 2472-2475.
10. a) Shull, R.D. and Bennet, L.H. (1992) Nanocomposite magnetic materials, *Nanostruct. Mater.* **1**, 83-88; b) Harp, G.R., Parkin, S.S.P., O'Brian, W.L., and Tonner, B.P. (1995) Induced Rh magnetic moments in Fe-Rh and Co-Rh alloys using x-ray magnetic circular dicroism, *Phys. Rev. B* **51**, 12037-12040; c) Moraïtis, G., Dreyssé, H., and Khan, M.A. (1996) Band theory of induced magnetic moments in CoM (M = Rh, Ru) alloys, *Phys. Rev. B* **54**, 7140-7142.
11. a) Rivas, J., Sánchez, R.D., Fondado, A., Izco, C., García-Bastida, A.J., García-Otero, J., Mira, J., Baldomir, D., Gonzáles, A., Lado, I., López-Quintela, M.A., Oseroff, S.B. (1994) Structural and magnetic characterization of Co particles coated with Ag, *J. Appl. Phys.* **76**, 6564-6566; b) Sun, S., Murray, C.D., Weller, D., Folks, L., and Moser, A. (2000) Monodisperse FePt nanoparticles and ferromagnetic FePt nanocrystal superlattices, *Science* **287**, 1989-1992; c) Park, J.-I. and Cheon, J. (2001) Synthesis of "solid solution" and "core-shell" type cobalt-platinum magnetic nanoparticles via transmetallation reaction, *J. Am. Chem. Soc.* **123**, 5743-5746; d) O'Conner, C.J., Kolesnichenko, V., Carpenter, E., Sangregorio, C., Zhou, W., Kumbhar, A., Sims, J., and Agnoli, F. (2001) Fabrication and properties of magnetic particles with nanometer dimensions, *Synth. Met.* **122**, 547-557; e) Bönnemann, H. (2001) Nanostructured metal colloids – chemistry and potential applications, in H.S. Nalwa (ed.), *Handbook of Surfaces and Interfaces of Materials* **Volume 3**, Nanostructured Materials, Micelles and colloids, Academic Press, San Diego, pp. 41-64; f) Fromen, M.C., Serres, A., Zitoun, D., Respaud, M., Amiens, C., Chaudret, B., Lecante, P., and Casanove, M.J. (2001) Structural and magnetic study of bimetallic $Co_{1-x}Rh_x$ particles, *J. Magn. Magn. Mater.* **242-245**, 610-612; g) Sobal, N.S. Hilgendorff, M., Möhwald, H. Giersig, M., Spasova, M., Radetic, T., and Farle, M. (2002) Synthesis and structure of colloidal bimetallic nanocrystals: the non-alloying system Ag/Co, *Nano Lett.* **2**, 621-624.
12. a) Caruso, F., Spasova, M., Susha, A., Giersig, M., and Caruso, R.A. (2001) Magnetic nanocomposite particles and hollow spheres constructed by a sequential layering approach, *Chem. Mater.* **13**, 109-116; b) Bizdoaca, E.L., Spasova, M., Farle, M., Hilgendorff, M., and Caruso, F. (2002) Magnetically directed self-assembly of submicron spheres with a Fe_3O_4 nanoparticle shell, *J. Magn. Magn. Mater.* **240**, 44-46.
13. a) Chen, J.P., Sørensen, C.M., and Klabunde, K.J. (1995) Enhanced magnetization of nanoscale colloidal nanoparticles, *Phys. Rev. B* **51**, 11527-11532; b) Pileni, M.-P. (2001) Magnetic fluids: fabrication, magnetic properties, and organization of nanocrystals, *Adv. Funct. Mater.* **11**, 323-336; c) Sun, S. and Murray, C.B. (1999) Synthesis of monodisperse cobalt nanocrystals and their assembly into magnetic superlattices, *J. Appl. Phys.* **85**, 4325-4330; d) Giersig, M. Hilgendorff, M. (2000) Ordered colloidal magnetic particles by magnetophoretic deposition, in P. Jena, S.N. Khanna, and B.K. Rao (eds.), *Cluster and Nanostructure Interfaces*, World Scientific, Singapore, pp. 203-208; e) Giersig, M., Hilgendorff, M. (1999) The preparation

of ordered colloidal magnetic particles by magnetophoretic deposition, *J. Phys. D: Appl. Phys.* **32**, L111-L113; f) Murray, C.B., Sun, S., Doyle, H., and Betley, T. (2001) Monodisperse 3d transition-metal (Co, Ni, Fe) nanoparticles, *Mater. Res. Soc. Bull.* **26**, 985-991; g) Hilgendorff, M., Tesche, B., Giersig, M. (2001) Creation of 3-d crystals from single cobalt nanoparticles in external magnetic fields, *Aust. J. Chem.* **54**, 497-501. h) Papirer, E, Horny, P., Balard, H., Anthore, R., Petipas, C, and Martinet, A. (1983) The preparation of a ferrofluid by decomposition of dicobalt octacarbonyl, *J. Colloid Intrface Sci.* **94**, 207-219; i) Osuna, J., de Caro, D., Amiens, C., Chaudret, B., Snoeck, E., Respaud, M., Broto, J.-M., and Frert, A. (1996) Synthesis, characterization, and magnetic properties of cobalt nanoparticles from an organometallic precursor, *J. Phys. Chem.* **100**, 14571-14574.

14. Massart, R. and Cabuil, V. (1987) Synthèse en milieu alcalin de magnétite colloïdale: contrôle du rendement et de la taille des particules, *J. Chim. Phys.* **84**, 967-973.
15. Brinker, C.J. and Scherer, G.W. (1990) *Sol-Gel Science,* Academic Press Inc., San Diego.
16. a) Smith, T.W. and Wychick, D. (1980) Colloidal iron dispersions prepared via the polymer-catalyzed decomposition of iron pentacarbonyl, *J. Phys. Chem.* **84**, 1621-1629; b) Suslick, K.S., Fang, M., and Hyeon, T. (1996) Sonochemical Synthesis of iron colloids, *J. Am. Chem. Soc.* **118**, 11960-11961.
17. a) Hoon, S.R., Kilner, M., Russel, G.J., and Tanner, B.K. (1983) Preparation and properties of nickel ferrofluids, *J. Magn. Magn. Mater.* **39**, 107-110; b) Cordente, N., Respaud, M., Senocq, F., Casanove, M.-J., Amiens, C., and Chaudret, B. (2001) Synthesis and magnetic properties of nickel rods, *Nano Lett.* **1**, 565-568.
18. Nakatani, I., Hijikata, M., and Ozawa, K. (1993) Iron-nitride magnetic fluids prepared by vapour-liquid reaction and their magnetic properties, *J. Magn. Magn. Mater.* **122**, 10-14.
19. Giersig, M. and Hilgendorff, M. (2002) On the road from single, nanosized, magnetic clusters to multi-dimensional nanostructures, *Colloids and Surfaces A* **202**, 207-213.
20. Talapin, D.V., Rogach, A.L., Haase, M., and Weller, H. (2001) Evolution of an ensemble of nanoparticles in a colloidal solution: Theoretical study, *J. Phys. Chem. B* **105**, 12278-12285.
21. a) Upadhyay, R.V., Davies, K.J., Wells, S., and Charles, S.W. (1995) Preparation and characterization of ultra-fine $MnFe_2O_4$ and $Mn_xFe_{1-x}O_4$ spinel systems: II. Magnetic fluids, *J. Magn. Magn. Mater.* **139**, 249-254; b) Davies, K.J. , Wells, S., Upadhyay, R.V., Charles, S.W., O'Grady, K., El Hilo, M., Meaz, T., and Mørup, S. (1995) The observation of multi-axial anisotropy in ultrafine cobalt ferrite particles used in magnetic fluids, *J. Magn. Magn. Mater.* **149**, 14-18; c) Fannin, P.C., Charles, S.W., and Dormann, J.L. (1999) Field dependence of the dynamic properties of colloidale suspensions of $Mn_{0.66}Zn_{0.34}Fe_2O_4$ and $Ni_{0.5}Zn_{0.5}Fe_2O_4$ particles, *J. Magn Magn. Mater.* **201**, 98-101.
22. a) Liz, L., López Quintela, M.A., Mira, J., and Rivas, J. (1994) Preparation of colloidale Fe3O4 ultrafine particles in microemulsions, *J. Mater. Sci.* **29**, 3797-3801; b) López Pérez, J.A., López Quintela, M.A., Mira, J., Rivas, J., and Charles, S.W. (1997) Advances in the preparation of magnetic nanoparticles by the microemulsion method, *J. Phys. Chem. B* **101**, 8045-8047.
23. a) Davies, K.J., Wells, S., and Charles, S.W. (1993) The effect of temperature and oleate adsorption on the growth of maghemite particles, *J. Magn. Magn. Mater.* **122**, 24-28; b) Mălăescu, I., Gabor, L., Claici, F., and Ştefu, N. (2000) Study of some magnetic properties of ferrofluids filtered in magnetic field gradient, *J. Magn. Magn. Mater.* **222**, 8-12; c) Fried, T., Shemer, G., and Markovich, G. (2001) Ordered two-dimensional arrays of ferrite nanoparticles, *Adv. Mater.* **13**, 1158-1161.
24. S. Sun and H. Zeng, Size-controlled synthesis of magnetite nanoparticles, *J. Am. Chem. Soc* **124**(28), 8204-8205 (2002).
25. a) Brown, K.R. and Natan, M.J. (1998) Hydroxylamine seeding of colloidal Au nanoparticles in solution and on surfaces, *Langmuir* **14**, 726-728; b) Jana, N.R., Gearheart, L., and Murphy, C.J. (2001) Evidence for seed-mediated nucleation in the chemical reduction of gold salts to gold nanoparticles, *Chem. Mater.* **13**, 2313-2322; c) Yu, H., Gibbons, P.C., Kelton, K.F., and Buhro, W.E. (2001) Heterogeneous seeded growth: a potentially general synthesis of monodisperse metallic nanoparticles, *J. Am. Chem. Soc.* **123**, 9198-9199.
26. a) Bönnemann, H., Brijoux, W., Brinkmann, R., Dijius, E., Joußen, T., and Korall, B. (1991) Erzeugung von kolloidalen Übergangsmetallen in organischer Phase und ihre Anwendung in der Katalyse, *Angew. Chem.* **103**, 1344-1346, *Angew. Chem. Int. Ed.* **30**, 1344-1346; b) Fink, J., Kiely, C.J., Bethell, D., and Schiffrin, D.J. (1998) Self-organization of nanosized god particles, *Chem. Mater.* **10**, 922-926.
27. Buske, N., Sonntag, H., and Götze, T. (1984) Magnetic fluids- their preparation, stabilization and applications in colloid science, *Colloids and Surfaces* **12**, 195-202.
28. Gittins, D.I. and Caruso, F. (2001) Spontaneous phase transfer of nanoparticulate metals from organic to aqueous media, *Angew. Chem. Int. Ed.* **40**, 3001-3004.
29. a) Katsikas, L., Eichmüller, A., Giersig, M., and Weller, H. (1990) Discrete exitonic transitions in quantum-sized CdS particles, *Chem. Phys. Lett.* **172**, 201-204; b) Dance, I.G., Garbutt, R.G. and Bailey, T.D. (1990) Aggregated structures of the compounds $Cd(SC_6H_4X\text{-}4)_2$ in DMF solution, *Inorg. Chem.* **29**(4), 603-608.

30. Murray, C.B., Sun, S., Gaschler, W., Doyle, H., Betley, T.A., and Kagan, C.R. (2001) Colloidal synthesis of nanocrystals and nanocrystal superlattices, *IBM J. Res. Dev.* **45**, 47-56.
31. Respaud, M., Broto, J.M., Rakoto, H., Fert, A.R., Thomas, L., Barbara, B., Verelst, M., Snoeck, E., Lecante, P., Mosset, A., Osuna, J., Ould Ely, T., Amiens, C., and Chaudret, B. (1998) Surface effects on the magnetic properties of ultrafine cobalt particles, *Phys. Rev. B* **57**, 2925-2935.
32. a) Dinega, D.P. and Bawendi, M.G. (1999) Eine aus der Lösung zugängliche neue Kristallstruktur von Cobalt, *Angew. Chem.* **111**, 1906-1909; *Angew. Chem. Int. Ed.* **38**, 1788-1791; b) Puntes, V.F., Krishnan, K.M., and Alivisatos, P. (2001) Synthesis, self-assembly, and magnetic behavior of a two-dimensional superlattice of single-crystal ε-Co nanoparticles, *Appl. Phys. Lett.* **78**, 2187-2189.
33. Laidler, H. and O'Grady, K. (1998) Crystallographic effects in Co alloy media, *www.datatech-online.com* **1**, 93-97.

NANOSPHERE LITHOGRAPHY – FABRICATION OF VARIOUS PERIODIC MAGNETIC PARTICLE ARRAYS USING VERSATILE NANOSPHERE MASKS

J. RYBCZYNSKI, M. HILGENDORFF AND M. GIERSIG
Hahn-Meitner Institut Berlin GmbH,
Glienicker Str. 100, 14109 Berlin, Germany
and
Poznan University of Technology,
ul. Nieszawska 13a, 60-965, Poland

1. Introduction

The formation of monolayers of self-assembled colloid particles (mainly polystyrene submicron-sized, monodisperse spheres) is well-established and widely used in various fields of research [1]. Well-ordered latex particle films allow production of regularly arranged triangular-shaped structures on almost any substrate. By the evaporation of different metals through the mask it is possible to prepare nanosized particles with diverse optical [1,2] or magnetic properties [2]. These can be used in many potential applications in optics [2,3], data storage or e.g. nanotubes growth [4,5].

There are many different fabrication methods based on electrophoresis [6,7], electrostatic deposition [8,9], the Langmuir-Blodgett technique, spin-coating [10] the controlled evaporation of a solvent from the solution containing latex particles on a hydrophilic substrate [11], or non-photolithographic methods [12,13].

Many authors use Micheletto's method [11] or other drying-based methods and they point out that a hydrophilic surface is crucial for monolayer deposition [14]. To deposit PS-latex particles onto hydrophobic substrates it is necessary to use an alternative method, such as assembly at a liquid-gas or a liquid-liquid interface. To overcome this one can also use the monolayer transfer technique showed e.g. by F. Burmeister et al. [14]. Unfortunately, this causes the generation of new structural defects. Because of that we looked for a new solution to cover hydrophobic silicon substrates with latex monolayers quickly. To achieve this, we developed another preparation technique – deposition of PS particles similar to a Langmuir-Blodgett film on water. Our preparation method, based on self-assembly at a liquid-gas interface, allows one to apply large (a few cm^2) monolayers directly onto any kind of surfaces. The main advantages of this new technique are a short preparation time (in comparison to drying-based methods) and a high level of hexagonal structure orientation as will be shown further.

The main problem with formation of 2D latex monolayers is there always a high number of different structural defects, such as: point defects (vacancies), line defects (dislocations) and also many disordered areas or grain boundaries. In our method we lowered the volume of structural defects by increasing the average grain size and consolidation of monolayers. Using this simple fabrication technique it was possible to prepare monolayers as large as a few square cm with grain size of 5×5 mm^2 and larger (using 1040 and 496 nm latex particles) and areas of even 50 μm^2 containing no structural

L.M. Liz-Marzán and M. Giersig (eds.),
Low-Dimensional Systems: Theory, Preparation, and Some Applications, 163–172.

defects (Figure 5). The quality of the NSL mask is outstanding when using bigger 1040, 925 and 496 nm PS-Latex particles. However, due to stronger interactions between smaller and lighter particles (217 or 127 nm) it is much more complicated to obtain uniform monolayers with these smaller particles.

2. Experimental Section

The monolayers were prepared using monodisperse polystyrene particles with a diameter of 1040, 925, 496, 217 and 127 nm. All particles were purchased from Microparticles GmbH (Germany) as a 10 wt.% water solution and diluted by mixing with an equal amount of ethanol. Solution were then kept for 10 sec. in an ultrasonic bath to improve mixing.

Oriented (111) p-silicon substrates, boron doped, with resistivity 10-20 Ωcm and thickness 525±25 μm, came from SilChem and were cut into small 10x10 mm or 5x5 mm squares. The surface roughness parameter R_A was approx. 1 nm. Each square was thoroughly cleaned with acetone and additionally in ethanol, in an ultrasonic bath for 20 sec.

For the PS-latex particles self-assembly distilled, ultrafiltrated, UV sterilized and deionized Milli-Q water from Millipore systems was used. The pH value was exactly 7,00 and the measured resistance: 18,2 MΩcm.

About 4-6 μl of prepared solutions was applied onto the surface of a clean, approx. 4x2 cm large, silicon wafer (which was kept in 10% dodecylsodiumsulfate solution for 24 hours previously) and distributed using an Eppendorf pipette. The wafer was then slowly immersed in the ∅15 cm glass vessel filled with 150 ml of Milli-Q water, and PS particles started to form an unordered monolayer on the water surface (Figure 1a). To consolidate the particles about 4 μl of 2% dodecylsodiumsulfate solution was added (Figure 1b) and a large monolayer with highly ordered areas was obtained. Such monolayers were then lifted off from the water surface using previously mentioned clean silicon wafers (Figure 1c).

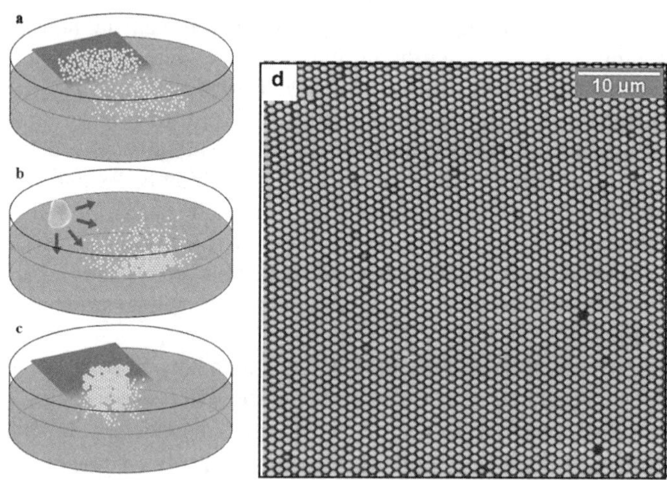

Figure 1. The preparation process of monolayered masks consisted of PS-latex particles: application of latex particles onto a water surface (a), consolidation of particles (b), liftoff of ordered monolayer (c). Typical structure of ordered monolayer (50x50 μm^2 area) consisted of 1040 nm PS-latex particles (d).

We used 1040 and 496 nm PS-latex particles to prepare double-layered masks. Such covered silicon substrates were produced by immersing dry, once-covered substrates, into water and depositing a second layer.

The electron beam evaporation of metal particles was performed in two different vacuum chambers: one constructed in HMI and the other by BesTech (Germany). Each of them was working with similar parameters. The evaporated metal was initially melted in a water-cooled crucible using an electron beam bent in a magnetic field. By varying the magnetic field the electron beam oscillated to heat the whole metal surface. Electrons were emitted from a tungsten cathode with 10kV high tension applied. Each part of this process was performed in $2 \cdot 10^{-8}$ mbar vacuum. Evaporation rate depended on the type of material and for nickel was about 0,8 nm/min. The thickness of the growing metal layer was estimated using an oscillating quartz plate (whose oscillation frequency depends on the thickness) placed on the same distance from the crucible as the sample (25 cm).

Nickel in the range of 15-40 nm, gold or Ni-Au-Ni and Au-Ni-Au multilayers were evaporated. In the last case sandwich-like structures of e.g. 10 nm Au, 5 nm Ni and 10 nm Au were prepared.

After the evaporation process, liftoff of latex layers was performed. All samples were immersed in tetrahydrofuran (THF), purchased from Merck, for 15 min. to dissolve PS-latex particles. The rest of the latex layer was removed by keeping the samples in an ultrasonic bath for 20 sec. The substrates were then washed with ethanol and dried.

For some applications it would be suitable to prepare spherical magnetic particles instead of triangular-shaped ones. It would facilitate also theoretical calculations of possible properties. To overcome the restriction of triangular structures a part of the samples was annealed in an argon atmosphere under 500 mbar pressure. A quartz glass pipe, with the sample inside, was previously washed 3 times with argon 4.8 gas. Samples containing nickel were annealed for 40-50 min. at 900-950 °C to obtain round, quasi-spherical particles.

Because deposited PS-latex particle masks are very versatile, they can be used for many various applications. We also chose a different approach for fabrication of periodic particle arrays. Following the wet chemistry methods we also prepared solutions of colloidal magnetite particles [15,16] and developed spraying thereof as a possible method for their application on different substrates. Aqueous solutions were prepared using optimized Massart synthesis [17]. Fe_3O_4, $Mn_{0,25}Fe_{2,75}O_4$ and other various combinations of Mn, Co and Ni with Fe_3O_4 were prepared. The average diameter of these nanoparticles was about 12 nm. Tetrabutylammoniumhydroxide (TBAOH) as a stabilizing agent (surfactant) was used. The first step of preparation was the hydrolysis of two solutions, one containing $FeCl_2 \cdot 4H_2O$ and the other with $FeCl_3 \cdot 6H_2O$, in 12 mmol of 0.5% ammonium and TBAOH solution each, saturated with nitrogen for 30 min. After that both solutions were mixed together and boiled for 20 min. By mixing the solutions we obtained black suspension of magnetite nanoparticles. The final mixture was then cooled down and washed three times with nitrogen-saturated Milli-Q water to remove the rest of the chlorides from the solution. Finally, since the fact that during washing a part of the surfactant was removed also, 0.2 ml of TBAOH was added.

Figure 2 shows the spray pyrolysis apparatus we used for application of the magnetite nanoparticles. Silicon wafer covered with PS-latex monolayer was placed in the chamber filled with nitrogen-oxygen mixture. Magnetite solution was transported by the pump to the ultrasonic nozzle and then was sprayed onto the sample. To accelerate drying of the sprayed droplets the sample was heated.

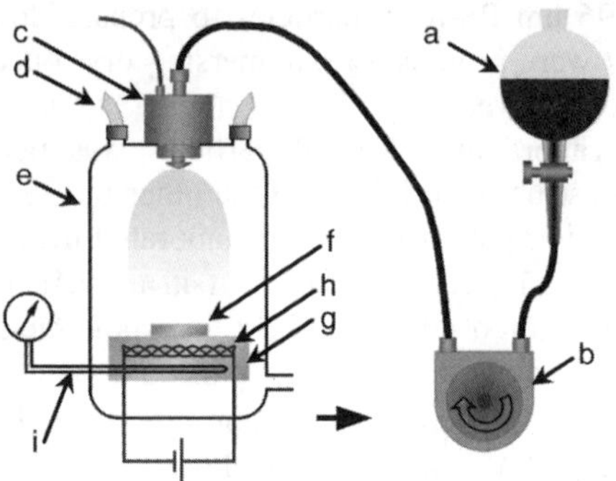

Figure 2. Magnetite spraying equipment. Colloid particles as a water solution from glass vessel (a) are transported using the pump (b) to the ultrasonic nozzle (c) and sprayed in the gas atmosphere. N_2+O_2 gas mixture is loaded through two openings (d) into the chamber (e), where the sample is located (f). To accelerate drying process the sample lies on the heated table (g). Thermocouple (i) enables measuring of temperature of the heating element (h).

The entire spraying process was controlled by varying such parameters as: flux rate (by controlling the rotation speed of the pump), spraying force (by controlling the nozzle output), heating temperature, spraying time and amount of gas mixture in the chamber. By preparation of over 20 samples we determined the best process parameters. Thus we managed to obtain the periodic array surfaces with satisfactory reproducibility.

During the fabrication of monolayers and further – periodic particle arrays, we used various techniques to investigate the structure. Each step of the preparation process was followed by optical microscopy investigations, AFM or SEM measurements. Monolayers built from large PS-latex particles with a diameter of 1040, 925 and 496 nm were investigated firstly using optical microscopy, which enables us to see the results quickly. We used the Carl-Zeiss Axiophot2 optical microscope equipped with the AxioCam CCD camera to take the pictures. Magnification of all pictures was 1000x and 500x. All monolayers and evaporated metal particle structures were also imaged using atomic force microscopy, which was our basic investigation method. We used a Digital Instruments Nanoscope IIIa AFM microscope working in contact as well as tapping mode. Pointprobe Si_3N_4 (for contact mode) and n^+-silicon (for tapping mode) cantilevers were used.

3. Results and discussion

Presented pictures (Figure 3a and 3b) show initial results of the preparation of monolayer structure – arranged hexagonal latex particle layers containing also a certain amount of structural defects. To lower the number of these we managed to increase the mean grain size by careful selection of the experimental latex particle concentration, volume of deposited particles and deposition speed. However some additional particles creating a second layer were always present. Figure 3b shows the morphology of such defects. According to SEM images these could be separated or agglomerated latex particles. The number of these spots depends on the quality of the PS-latex solution and on the share of the solution volume that takes part in the formation of the monolayer. Because of the fact that not all of the deposited particles are creating a monolayer it was necessary to find experimentally the right volume of the deposited solution to lower the number of such defects.

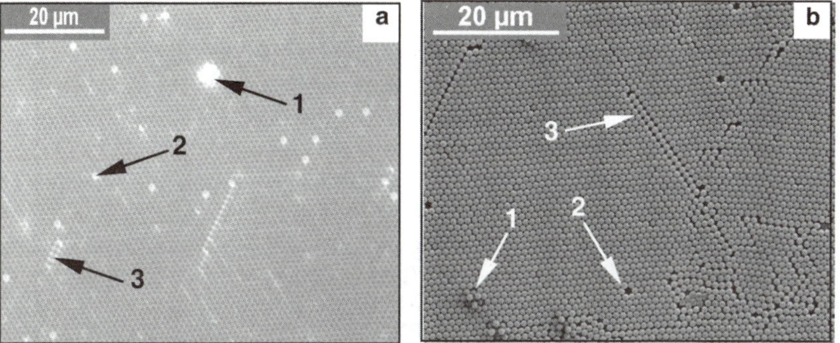

Figure 3. Optical microscopy image (a) and SEM image (b) showing a typical structure defects of initial monolayers built from 1040 nm PS-latex particles: additional particles creating a second layer (1), vacancies (2) and dislocations (3).

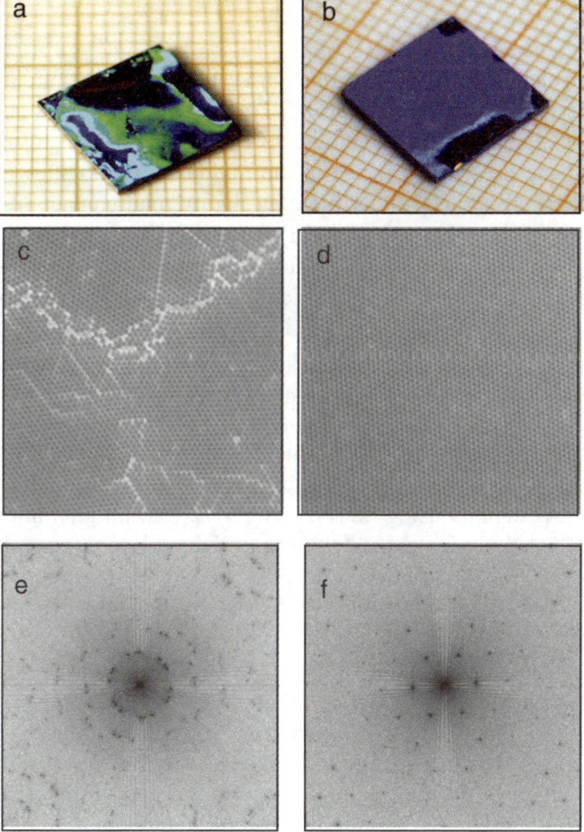

Figure 4. The 1×1 cm^2 silicon wafers covered with 496 nm PS-latex particles deposited as a monolayer (a, b). In contradistinction to the left sample (a) most of the surface of the right one does not contain any grain boundaries, which is represented as a monochrome light interference color of the surface (b). Fast Fourier Transformation made from above presented pictures shows many additional reflexes in the left picture (e) because of the disordering, in contradistinction to the right one (f), showing FFT of perfectly ordered latex particles.

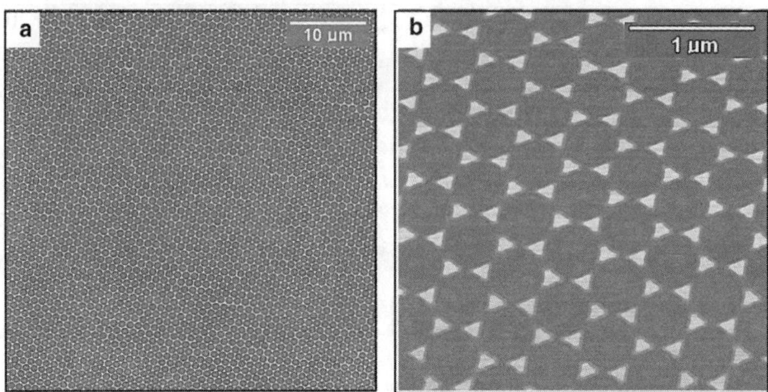

Figure 5. AFM image of 15 nm high Ni particles evaporated onto Si substrate through the 1040 nm PS-latex particle mask, showing a very well-ordered structure (a). The 35 nm high Ni particles evaporated through the 496 nm PS-latex particle mask (b).

A very good correlation between previously investigated hexagonal latex particle arrays and evaporated metal structures was observed. The presented pictures show well-ordered, triangular-shaped nickel particles dictated by dense hexagonal packing of latex spheres.

We also noticed the presence of rings in the middle of round sites left after liftoff of latex spheres. These small structures surrounding the original contact area of the particles were observed e.g. by J. Boneberg et al. [11d] According to them, these rings came from the rest of the dissolved latex particles which fully corresponds to our investigations. Due to high adhesion of evaporated nickel to the silicon substrate, it was possible to remove the rings quickly by an additional ultrasonication in THF without damaging the evaporated particle array.

To prove the presence of nickel and gold and measure the volume of these elements in three-layered samples we performed additionally EDX investigations. Samples for such analysis were prepared by applying silicon-nickel powder, obtained from diamond knife scratched silicon surfaces, onto a copper grid covered with carbon film. Such prepared samples were then investigated using the Philips CM12 transmission microscope, working with 120 kV, and EDAX equipment. In the case of three-layered samples, qualitative as well as quantitative analysis was done. The presented EDX spectra show unequivocally high gold and nickel content. The quantitative measurement results exhibit good conformity with calculated theoretical values. By comparison between L-lines for nickel and gold we estimated the percentage of weight content of both elements. Although the measured values: 11.5 wt.% for Ni and 88.5 wt.% for Au (and their ratio $M_{Au}/M_{Ni} = 7.7$), seem at first to be different to the theoretically calculated gold-to-nickel ratio, which is 4:1 that represents 20% Ni and 80% Au in the 10Au5Ni10Au sample, one must also take into account the weight share of each element. Thus, the real gold-to-nickel ratio is about 8.6 what corresponds to 10.34 wt.% Ni and 89.66 wt.% Au in the sample.

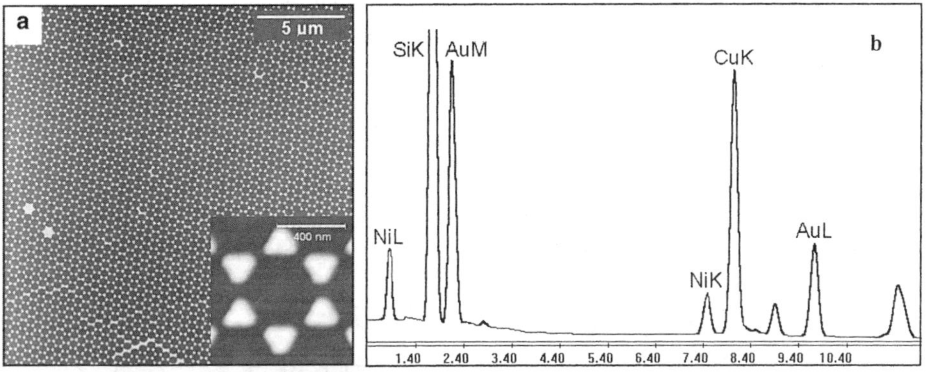

Figure 6. The structure of sandwich-like 10nm Au - 5nm Ni - 10nm Au particles made by evaporation of metals onto 496 nm PS-latex spheres (a). A typical EDX spectra of Au-Ni-Au structure. The strongest peaks correspond to Si substrate and copper microscope grid (b).

After the annealing process of the nickel and gold structures, all the evaporated particles exhibited, as expected, a quasi-spherical shape. The height of these structures increased slightly while a diameter has been reduced due to volume conservation. According to our observations the right annealing temperature should be at least 65-70% of melting point temperature. Samples prepared in lower temperatures don't exhibit any morphological changes.

Figure 7 shows the result of annealing 10 nm nickel particles evaporated through a 496 nm PS-latex particle mask. The obtained structure exhibits a very high homogeneity of nickel particles. Their mean size is about 65 nm. Thus we can be sure that annealing allows fabrication of perfectly round and highly ordered magnetic particle arrays. In the presented case, the particle density is, on the average, 11 part./μm^2, which corresponds to 7.1 billion per in^2. That means theoretically 7.1 Gbit/in^2 (1 particle = 1 bit). However, when using smaller 217 nm PS-latex particles this value increases to 10.2 Gbit/in^2.

Figure 7. The SEM image showing 10 nm nickel particles, evaporated through 496 nm PS-latex particle mask, after annealing at 900°C for 40 min.

To obtain a periodic array with an increased distance between each evaporated particle, double-layered masks were prepared. Evaporation of nickel onto such fabricated surfaces results theoretically in the formation of round nickel structures with √3 times enlarged distance between them. Because of the fact that in the case of double layer masks, the probability of appearance of various structural defects is twice as high than when using monolayers, such prepared samples most often exhibit a different structure than the theory indicates. During our experiments we obtained two interesting structures (Figure 8a and 9a). According to AFM measurements these particles are considerably smaller than triangle-shaped ones. The average distance between them is also larger. Such a structure was observed in every part of the sample.

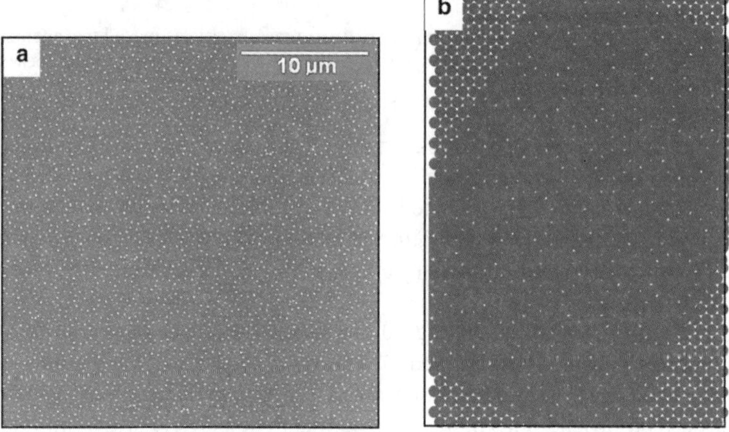

Figure 8. The result of evaporation of 20 nm nickel through double layer 496 nm PS-latex particles mask. The angle between both latex layers is about 30° (a). Simulation performed to prove theoretical assumptions (b).

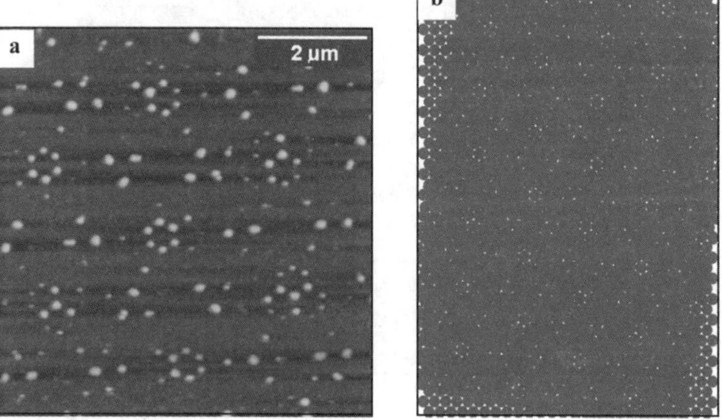

Figure 9. The result of evaporation of 15 nm nickel through double layer 496 nm PS-latex particles mask. The angle between both latex layers is about 10° (a). Simulation performed to prove theoretical assumptions (b).

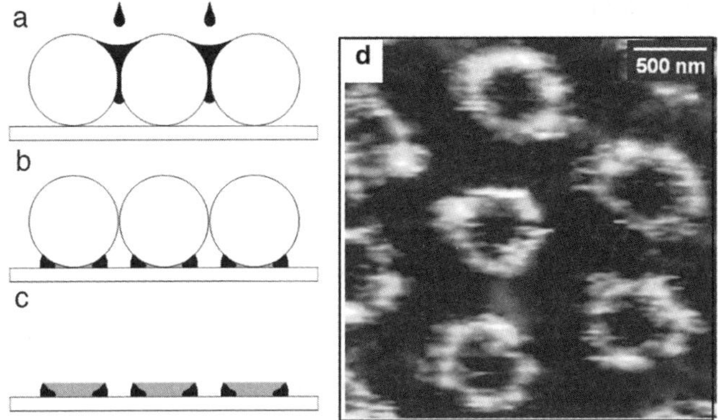

Figure 10. An explanation for the formation process of magnetite rings. Sprayed magnetite droplets are injected through the triangle-shaped spacings between the PS particles (a). Because of e.g. capillary forces these droplets dry as rings surrounding the PS particles (b). After liftoff of PS-latex particles ring-like strucutre can be observed (c). The AFM image showing such structure, formed by spraying of $Mn_{0.25}Fe_{2.75}O_4$ through 1040 nm PS-latex particle mask (d).

Since the order of these small structures does not correspond to the theory, it can be explained as a different orientation between both layers than the 3D-hexagonal close packing. To investigate this orientation we performed a simple computer simulation, which indicates that the angle between both layers is about 30° (Figure 8b) or 10° (Figure 9b).

Figure 10b shows the initial result of spraying magnetite nanoparticles at the flux rate set to 0.30 ml/min. during 40s. Nitrogen-to-oxygen ratio was 3:1. Sample was heated to 80°C. By reason of a movement of small magnetite particles when scanning the surface with AFM cantilever in the contact mode, the blurred edges of these rings can be observed.

4. Conclusions

In this paper we have shown that the presented fabrication method – the deposition of PS-latex particles as a Langmuir-Blodgett film on water surface, is another way to prepare periodically arranged, evaporated particles. This allows the application of large monolayered masks directly onto arbitrary surfaces. Moreover, other advantages of this technique are: a short preparation time and a high level of hexagonal orientation. Further improvement of evaporated structures could be carried out through annealing which enables a high unification of particle size and considerably increases their homogeneity. Fabrication of sandwich-like multilayer particles by the evaporation of three layers of metal was also presented as a possible method of magnetic properties increase of evaporated metal particles. Furthermore, we showed that versatile PS-latex masks can be used also for other applications such as e.g. preparation of different structures by spraying of aqueous colloid particles solutions.

5. Acknowledgment

We thank S. Kubala and S. Fichter from Hahn-Meitner Institut in Berlin for help in evaporation and annealing. We also gratefully acknowledge stimulating discussions with

M. Hilgendorff and we thank U. Bloeck for help in SEM and EDX measurements. This work was financially supported by the European Union under contract No. HPRN-CT-1999-00150 and German chemical industry.

6. References

1. Park, S. H., Gates, B., and Xia, Y. (1999) A three-dimensional photonic crystal operating in the visible region, *Adv. Mat.* **11**, 462-466.
2. Gates, B. and Xia, Y. (2001) Photonic crystals that can be addressed with an external magnetic field, *Adv Mat.* **13**, 1605-1608.
3. Xia, Y., Gates, B., and Yin, Y. (2001) Current chemistry. Building complex structures from monodisperse spherical colloids, *Aust J. Chem.* **54**, 287-290.
4. Wen, J. G., Huang, Z. P., Wang, D. Z., Chen, J. H., Yang, S. X., Ren, Z. F., Wang, J. H., Calvet, L. E., Chen, J., Klemic, J. F., and Reed, M. A. (2001) Growth and characterization of aligned carbon nanotubes from patterned nickel nanodots and uniform thin films, *J. Mater. Res.* **16**, 3246-3253.
5. Read, M. E., Schwarz, W. G., Kremer, M. J., Lennhoff, J. D., Carnahan, D. L., Kempa, K., and Ren, Z. F. (2001) Carbon nanotube-based cathodes for microwave tubes, *Proc 2001 Part. Accelerator Conf., Chicago* 1026-1028.
6. Giersig, M. and Mulvaney, P. (1993) Formation of ordered two-dimensional gold colloid lattices by electrophoretic deposition, J. Phys. *Chem.* **97**, 6334-6336.
7. Giersig, M. and Mulvaney, P. (1993) Preparation of ordered colloid monolayers by electrophoretic deposition, *Langmuir* **9**, 3408-3413.
8. Rogach, A., Susha, A., Caruso, F., Sukhorukov, G., Kornowski, A., Kershaw, S., Möhwald, H., Eychmüller, A., and Weller, H. (2000) Nano- and microengineering: 3-D colloidal photonic crystals prepared from sub-μm-sized polystyrene latex spheres pre-coated with luminiscent polyelectrolyte/nanocrystal shells, *Adv. Mat.* **12**, 333-337.
9. Deckman, H. W. and Dunsmuir, J. H. (1982) Natural lithography, *Appl. Phys. Lett.* **41**, 377-379.
10. (a) Hulteen, J. C. and Van Duyne, P. R.(1995) Nanosphere lithography: A materials general fabrication process for periodic particle array surfaces, *J. Vac. Sci. Techn. A* **13**, 1553-1558; (b) Hulteen, J. C., Treichel, D. A., Smith, M. T., Duval, M. L., Jensen, T. R. and Van Duyne, R. P. (1999) Nanosphere lithography: size-tunable silver nanoparticle and surface cluster arrays, *J. Phys. Chem. B* **103**, 3854-3863; (c) Winzer, M., Kleiber, M., Dix, N., and Wiesendanger, R. (1996) Rapid communication fabrication of nano-dot and nano-ring-arrays by nanosphere lithography, *Appl. Phys. A* **63**, 617-619.
11. (a) Micheletto, R., Fukuda, H., and Ohtsu, M. (1995) A simple method for the production of a two-dimensional, ordered array of small particles, *Langmuir* **11**, 3333-3336; (b) Boneberg, J., Burmeister, F., Schäfle, C., and Leiderer, P. (1997) The formation of nano-dot and nano-ring structures in colloidal monolayer lithography, *Langmuir* **13**, 7080-7084; (c) Burmeister, F., Badowsky, W., Braun, T., Wieprich, S., Boneberg, J., and Leiderer, P. (1999) Colloid monolayer lithography – a flexible approach for nanostructuring of surfaces, *App. Surf. Sci.* **144-145**, 461-466; (d) Burmeister, F., Schäfle, C., Keilhofer, B., Bechinger, K. M., Boneberg, J., and Leiderer, P. (1998) From mesoscopic to nanoscopic surface structures: Lithography with colloid monolayers, *Adv. Mat.* **10**, 495-497; (e) Denkov, N. D., Velev, O. D., Kralchevsky, P. A., Ivanov, I. B., Yoshimura, H., and Nagayama, K. (1992) Mechanism of formation of two-dimensional crystals from latex particles on substrates, *Langmuir* **8**, 3183-3190; (f) Rakers, S., Chi, L. F., and Fuchs, H. (1997) Influence of the evaporation rate on the packing order of polydisperse latex monofilms, *Langmuir* **13**, 7121-7124; (g) Adachi, E., Dimitrov, A. S., and Nagayama, K. (1995) Stripe patterns formed on a glass surface during droplet evaporation, *Langmuir* **11**, 1057-1060.
12. Lu, Y., Yin, Y., Gates, B., and Xia, Y. (2001) Growth of large crystals of monodispersed spherical colloids in fluidic cells fabricated using non-photolithographic method, *Langmuir* **17**, 6344-6350.
13. Park, S. H. and Xia, Y. (1999) Assembly of mesoscale particles over large areas and its application in fabricating tunable optical filters, *Langmuir* **15**, 266-273.
14. Burmeister, F., Schäfle, C., Matthes, T., Böhmisch, M., Boneberg, J., and Leiderer, P. (1997) Colloid monolayer as versative lithographic masks, *Langmuir* **13**, 2983-2987.
15. Massart, R., Dubois, E., Cabuil, V., and Hasmonay, E. (1995) Preparation and properties of monodisperse magnetic fluids, *J. Magn. Magn. Mat.* **149**, 1-5.
16. Liz, L., Lopez-Quintela, M. A., Mira, J., and Rivas, J. (1994) Preparation of colloidal Fe_3O_4 ultrafine particles in microemulsions, *J. Mat. Sc.* **29**, 3797-3801.
17. Massart, R. and Cabuil, V. (1987) Synthèse en milieu alcalin de magnetite colloïdale: contrôle du rendement et de la taille des particules, *J. Chim. Phys.* **84**, 967-969.

MAGNETISM OF MONODISPERSE CORE/SHELL PARTICLES

M. SPASOVA and M. FARLE
Institut für Physik,
Gerhard-Mercator-Universität Duisburg,
Lotharstr. 1, 47048 Duisburg, Germany

1. Introduction

The magnetism of small particles or colloids has been investigated over many decades, and several reviews have dealt with the magnetic response of such particles ensembles [1,2]. The sizes which can be prepared and are interesting for nanostructured materials range from a few atoms ("clusters") to nanoparticles and colloids of a few micrometer diameter. All efforts on a detailed understanding of the magnetism of individual particles are hindered by the fact that particles prepared by either chemical synthesis, cluster beam techniques or Oswald ripening on single crystal surfaces are never perfectly alike. Hence, all magnetic measurements average over the properties of particles with different sizes , surfaces and shapes.

Nevertheless some general conclusion on the magnetic properties can be drawn which are based on the increasing importance of finite-size effects in smaller particles. Surface effects and temperature become a major player in the determination of the magnetic response. The importance of the surface of the nanoparticle is easily understood, if one considers that 10 mg of a powder of 3 nm Co particles contain roughly 5 mg of Co surface atoms. The importance of temperature enters as follows: when thermal energy k_BT becomes comparable to the magnetic anisotropy energy $K_{eff}V$ of the particle – K_{eff} being the effective anisotropy energy density and V the particle volume, the magnetization starts to fluctuate in time, and the particle acts like a giant spin, the "superparamagnet". For typical bulk values of the anisotropy of Co for example this critical volume is reached for a diameter of about 10 nm. Obviously, the time scale of fluctuations varies with temperature. On the other hand by increasing the magnetic anisotropy energy by either increasing the particle volume, by increasing the microscopic (intrinsic) magnetic anisotropy or by changing the shape of the particles (shape anisotropy) one can decrease the fluctuations of the particle. For a given particle shape and volume the "blocking temperature", that is the cross-over temperature from the rapidly fluctuating state of the magnetization to the static case aligned along an easy direction, can be controlled by modifying the intrinsic anisotropy of the nanocrystal. Based on thin film magnetism results one may conclude that surface anisotropy can be a major handle in tuning the magnetic response of the particle. An additional tuning parameter is given by the strength of the particle – particle interaction which in general is of dipolar character. Only if particles come so close that orbitals of surface atoms of neighboring particles overlap, the much larger exchange interaction takes over. Based on these simple deductions one immediately comes to the conclusion that the synthesis of magnetic particles with controlled surface parameters

L.M. Liz-Marzán and M. Giersig (eds.),
Low-Dimensional Systems: Theory, Preparation, and Some Applications, 173–192.

will offer the greatest flexibility in designing the magnetic properties of particles for different applications. One way to achieve the control of surface properties is to select the surface atoms, that is to create a shell around a core, the core-shell particle, which is the topic of this article.

Such types of monodisperse colloids consisting of a ferromagnetic core and a diamagnetic shell (or vice versa) are interesting for applications in ultrahigh-density magnetic storage [3] and medicine as well as for basic research in terms of model systems for the investigation of phase transitions in bulk matter (e.g. melting). Core/shell particles can be synthesized ranging in diameter from few nanometer to several microns [4] with a shell thickness ranging between 1 nm and several tens of nanometers, providing unique control over their magnetic properties. Here we restrict our discussion to the properties of a prototype system, namely CoO@Co (a CoO shell around a Co core), which is easy to prepare and consists of the ferromagnet Co (hcp and fcc, T_C = 1404 K) and an antiferomagnet CoO (fcc, T_N = 293 K). Other examples like Ag@Co nanoparticles can be found in the recent literature [5].

2. Magnetic Anisotropy in Nanoparticles

Magnetic anisotropy energy MAE is the energy difference between two directions of the magnetization with respect to a crystallographic axis. It is present in any bulk ferromagnet. There are only two microscopic interactions responsible for the energetic difference between the so-called easy and hard axes of magnetization: a) the long-range dipole-dipole coupling of magnetic moments is responsible for shape anisotropy and b) spin-orbit interaction is responsible for the intrinsic (magnetocrystalline) anisotropy. All phenomenological contributions to the MAE like strain induced MAE and surface MAE are due to spin-orbit interaction, which is the only interaction which couples the lattice degrees of freedom (x, y, z) to spin degrees of freedom (S_x, S_y, S_z). The anisotropic part of the free energy density for a thin film with a tetragonal crystal structure is given by:

$$E = 2\pi\,(N_\perp - N_{||})\,M^2 \cos^2\theta \;-\; K_2 \cos^2\theta \;-\; \tfrac{1}{2} K_{4\perp} \cos^4\theta \;-\tfrac{1}{2} K_{4||}\, \tfrac{1}{4}\,(3+\cos 4\varphi)\, \sin^4\theta \qquad (1)$$

The first term represents the shape anisotropy of a thin film with demagnetization factors $N_{||}$ and $N_\perp$. The phenomenological constants K_2, $K_{4\perp}$, $K_{4||}$ are the second and fourth order anisotropy constants in units of energy per volume. For a cubic system K_2 is zero, and for a hexagonal system like Co $K_{4||}$ is zero. It is worth mentioning that the $K_{4\perp}$ in Co is almost as large as K_2 – a fact which is often neglected. All the anisotropy constants are temperature dependent.

In thin films the surface effects become more important as the film thickness d decreases. This effect has been phenomenologically introduced by Néel by describing the total intrinsic anisotropy as a sum of volume anisotropy K_i^V and surface anisotropy K_i^S (in units of energy per unit surface) according to

$$K_i = K_i^V + 2\,K_i^S/d \qquad (2)$$

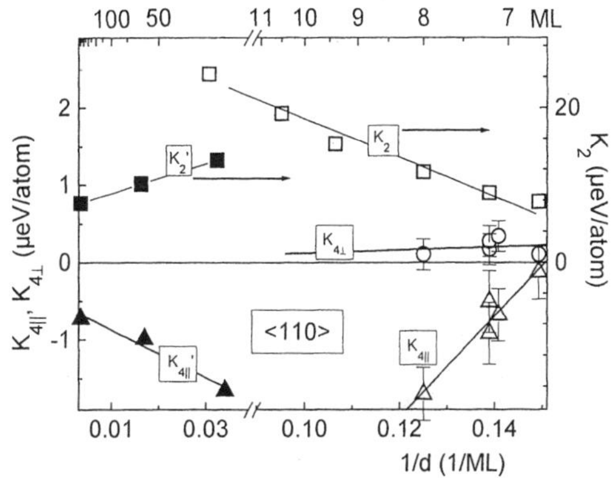

Figure 1. Magnetic anisotropy constants of Ni/Cu(001) determined by ferromagnetic resonance. Fourth- and second order anisotropy constants are shown.

The indices i refer to the second and fourth order anisotropy constants. The sign of both the volume and surface contribution can be either positive or negative, which means that either in-plane or out-of-plane direction is the easy axis of magnetization. As a consequence one can find a situation like in Ni films on Cu(001) where the thinner film is in-plane magnetized while a thicker film is out-of-plane magnetized ($d > 8$ monolayers (ML)) or vice versa as in the cases of Co films on Cu substrates.

In Figure 1 an illustrative example of the complexity of surface and volume effects is shown for the case of Ni on Cu(001) [6]. The second and fourth order anisotropy constants are shown as a function of reciprocal film thickness. One observes two regimes with either positive or negative slopes (surface anisotropy) and different volume anisotropy constants (for $1/d \rightarrow 0$). Since the magnitude and sign of the surface anisotropy and the bulk anisotropy should not depend on film thickness, the interpretation is clear and has been confirmed by structural analysis. Below a critical thickness of about 15 monolayers the film growths pseudomorphic with a negative K_2^S and a positive $K_{4\|}$.

Above 15 ML the slopes change sign revealing the relaxation of the crystal structure. In other words the "surface" anisotropy for $d > 15$ ML (app. 3 nm) is an effective quantity which contains $1/d$ contributions related to the strain relaxation. Any real surface or interface contains steps and defect sites. This causes an additional reduction of the symmetry and of nearest neighbor coordination. Following the rule of thumb that lower symmetry corresponds to larger MAE one expects a large contribution to the surface and interface anisotropy which was also confirmed in the cases of Ni on Cu(001) [7] and Fe on Ag(001) [8]. Preferentially oriented step edges lead to an uniaxial in-plane anisotropy on the order of 1 meV/atom ($\approx 5 \times 10^{-8}$ erg/cm) which increases quadratically with the step density. In principle two different step induced contributions arise at the substrate and the vacuum interface. The substrate step density does not change with film thickness. The step density at the vacuum interface changes with thickness resulting in an unpredictable thickness dependent contribution to $(K_i^{surf} + K_i^{int})/d$ [9]. Gradmann et al. [10] showed that step anisotropies at Fe(110) and Fe/Pd interfaces are on the order of -0.39 and -0.16 meV/atom favoring perpendicular magnetization. Their finding contradicts the theoretical prediction by the Néel model [11] that roughness always reduces perpendicular anisotropy. Note that dipolar and spin-orbit induced roughness give contributions of opposite sign to MAE.

Values of K_2^S and K_2^V have been tabulated for the 3d elements in several reviews (for example [12-14]). The thickness independent part K_2^V is in general one order of magnitude larger than the bulk value. K_2^V of fcc Co films on various substrates has the same magnitude as K_2 of bulk hcp Co. This is no contradiction to the previous statement, since bulk fcc Co has a much lower cubic K_4. In general, the averaged K_2^S of Co is positive (favoring perpendicular magnetization) and K_2^V is negative. For the hcp Co(0001) and for the fcc Co(100) surface/vacuum interface one finds a negative K_2^S. It must be concluded that strong hybridization with the substrate and defects/steps at the interfaces cause the positive K_2^S leading to a strong perpendicular anisotropy in most ultrathin Co layers. In some systems good quantitative agreement between measured and calculated Néel-type surface anisotropies is obtained, in most cases however, the sign and magnitude differ.

Usually the anisotropy is given as energy per volume, e.g. J/m^3. The surface and step anisotropy have the dimension of energy per area (J/m^2) and energy per line (J/m) which makes numbers hard to compare. A better way is to give the anisotropy in energy per atom, which means that the atomic volume in a sample consisting of N atoms must be estimated. The different faces ((111) versus (100)) contain a different number of atoms per unit area. This gives a different surface anisotropy in eV/atom, when K^S in J/m^2 is the same for both. A dimension of energy/atom allows a convenient comparison of volume-, surface-, and step-type anisotropy. Also the correlation to calculated values is much more straight forward.

At a first thought, it is tempting to assume that similar to thin film magnetism the surface anisotropy will dominate the magnetic anisotropy below a certain particle diameter. However, for large enough spherical particles the effects of a surface anisotropy normal to the surface would average out. This is evident from symmetry considerations. However, particles on the nanometer scale cannot be considered spherical. Facets form on the surface which are usually the closest packed orientation for a given crystal structure (e.g. (111) for fcc and (110) for bcc). The smaller the particle the more atoms are located at the boundaries of different facets and all the modifications due to step induced anisotropies become important. Hence, it is surprising that the phenomenological approach for the description of the MAE of a nanoparticle $K = K^V + 6\,K^S/D$ which takes the surface-to-volume fraction as a function of diameter D into account seems to work reasonable well.

3. Single-Domain Particles

For the application in magnetic recording media, it is desirable that the particles are single-domain magnets with all of the spins aligned in a single direction. The inclusion of a domain wall would lower the coercivity and presents a source of noise. Magnetic nanoparticles are in a single domain state, if it costs more energy to create a domain wall than to support the external magnetostatic energy (stray field) of the single-domain state, that is to say if the energy $E_{dw} = \sigma_{dw}\pi r^2 = 4\pi r^2(A\,K)^{1/2}$ needed to create a domain wall with a width of the radius r *of* a spherical particle, (σ_{dw} is a the domain wall energy density; A is the exchange stiffness constant $A \approx 10^{-11}$ J/m; K_2 is the uniaxial anisotropy energy density) exceeds the magnetostatic energy ΔE_{MS} saved by reducing the single domain state to a multidomain state, $\Delta E_{MS} \approx \frac{1}{3}\mu_0 M_s^2 V = \frac{4}{9}\mu_0 M_s^2 \pi r^3$ (M_S is the saturation magnetization, V is the particle volume) [15]. For the materials with a strong anisotropy, the critical diameter of the sphere d_C is reached when $E_{DW} = \Delta E_{MS}$:

$$d_c \approx 18 \frac{(AK_2)^{1/2}}{\mu_0 M_s^2} \tag{3}$$

For hcp Co assuming bulk values of $\mu_0 M_s^2 = 24*10^5$ J/m^3 and $K_2 = 4.1*10^5$ J/m^3, one calculates $d_c \approx 15$ nm at room temperature. For fcc Co the cubic magnetocrystalline anisotropy constant is $K_4 = 0.85*10^5$ J/m^3 and the critical diameter for a single-domain state is about 7 nm. It should be noted that the effective magnetic anisotropy energy for small particles can differ much from the bulk value due to a large contribution of the surface anisotropy which can be several orders of magnitude greater than that of the bulk magnetocrystalline anisotropy.

Single-domain particles can show a broad range of coercivities from zero to $H_C = 2K_2/M_S$. The lower limit applies for very small particles and small anisotropies and the upper limit is approached when the particle size is close to the critical diameter d_c.

4. Superparamagnetism

The magnetization process of a single-domain particle involves the rotation of the magnetic moment. The anisotropy energy which has to be overcome is usually larger than the one required for domain wall motion, resulting in a higher coercivity. A magnetization reversal in a magnetic nanoparticle was first investigated theoretically by Stoner and Wohlfarth (SW) [16]. The rotation of magnetization direction in a single domain particle with uniaxial anisotropy is described by a double well potential

$$E(\theta) = VK_{eff} \sin^2 \theta \tag{4}$$

where θ is the angle between the easy axis of magnetization and the magnetization vector, K_{eff} is the effective magnetic anisotropy constant (including shape anisotropy) and V is the particle volume (Figure 2). The energy barrier $E_B = VK_{eff}$ separates the two minima at $\theta=0$ and $\theta=\pi$ when the particle magnetization is parallel (antiparallel) to the easy axis. In small particles E_B becomes comparable to the thermal energy ($k_B T$) at 300 K. This results in superparamagnetic relaxation, i.e. the magnetization fluctuates between the two energy minima. The temperature dependence of the superparamagnetic relaxation time, τ, is described by the Néel-Brown expression [17]

$$\tau = \tau_0 \exp\left(\frac{K_{eff} V}{k_B T}\right) \tag{5}$$

where k_B is Boltzmann's constant, $\tau_0 \sim 10^{-12}$-10^{-9} s.

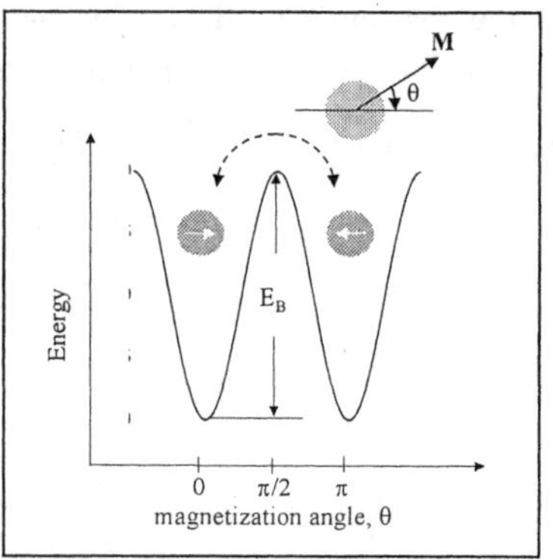

Figure 2. Schematic illustration of the energy of a single-domain particle with uniaxial anisotropy in dependence on the magnetization orientation. E_B is the energy barrier for the rotation of the magnetization. The left and right arrows represent spin directions corresponding to two energy minima when the magnetization is parallel or antiparallel to the easy axis. θ is the angle between **M** and the easy axis.

According to Eq. (5) the probability that the magnetization remains in the original position during time t can be written as $P(t) = \exp(-t/\tau)$, i.e. for the remanence we have $M_r(t)/M_0 = \exp(-t/\tau)$. If $\tau >> t_{\exp}$, no change of the magnetization can be observed during the observation time t_{exp}. This is the region of stable ferromagnetism. For SQUID magnetometry, the measuring time is usually 100 s and the magnetic transition occurs during this time, if the condition $25k_BT \geq K_{eff}V$ is fulfilled. For a given particle size the equation $25k_BT_B = K_{eff}V$ defines the blocking temperature T_B. For example, due to the different anisotropy 8 nm hcp Co of 300 K and 14 nm fcc Co [18] have a blocking temperature of 300 K. Below T_B the particles exhibit ferromagnetic behavior. Above the blocking temperature and below the Curie temperature of the individual particle, the particles have a spontaneous magnetization (given by $N\mu_m/V$, where N is the number of magnetic moments μ_m in the particle). In zero applied field the time averaged magnetization will be zero, since many flips of the magnetization occur during the time of experiment. In an applied magnetic field, the magnetization of such superparamagnetic or thermally demagnetized particles follows a Langevin behavior

$$M(H) = M_s L\left(\frac{\mu H}{k_B T}\right) \tag{6}$$

M_s is the saturation magnetization of the particle, μ=Nμ_m is the magnetic moment of the particle and $L\left(\frac{\mu H}{k_B T}\right)$ is the Langevin function defined by

$$L\left(\frac{\mu H}{k_B T}\right) = \coth\left(\frac{\mu H}{k_B T}\right) - \frac{k_B T}{\mu H} \tag{7}$$

The field and temperature dependence of the magnetic response of the superparamagnetic particles without interaction is the same as for a paramagnet, if one uses the particle magnetic moment μ instead of the atomic magnetic moment μ_m of the paramagnet. The susceptibility of a superparamagnet is increased N^2-fold. The saturation magnetization is defined when magnetic moments of all particles are aligned in an applied field.

The results presented above are strictly valid for individual particles, or arrays of non-interacting particles with the same diameter, anisotropy and with parallel easy axes. Realistic particles have some size distribution. Such particles would give rise to a superposition of Langevin functions with different values of $\mu = M_S V$. The magnetization of superparamagnetic particles in a magnetic field can be described by a weighted sum of Langevin functions and Eq. (6) should be modified

$$M(H) = M_s \int L\left(\frac{\mu(V)H}{k_B T}\right) f(V)dV \tag{8}$$

where $f(V)$ is a distribution function of the particle volume.

The distribution of particle sizes results in a blocking temperature distribution. Magnetic interactions between nanoparticles have a strong influence on the superparamagnetic relaxation [19]. Dormann et al. [20] considered the dipole-dipole interaction between the particles and predicted that the energy barrier E_B increases when the interaction E_{int} increases. The relaxation time for an ensemble of interacting particles is then given by

$$\tau = \tau_0 \exp\left(\frac{K_{eff} V + E_{\text{int}}}{k_B T}\right) \tag{9}$$

where E_{int} is the interaction induced anisotropy energy of the particle which can be found by summing up the contributions from all neighbouring particles.

The increase of relaxation time with increasing interaction strength is in accordance with the results on a large number of samples, but in contradiction with Mössbauer studies of γ-Fe_2O_3 particles with weak inter-particle interaction [21]. These experimental observations can be explained by an interaction-dependent variation of the phenomenological damping parameter, η, which enters in the theoretical expression for τ_0 [22,23]. The damping parameter, η, increases with increasing interaction strength due to two contributions: the first contribution is due to disordered surface spins which makes the coherent spin rotation more difficult. The authors claimed that the surface disorder decreases with increasing inter-particle interaction that results in a decreasing damping parameter. The second contribution, which was claimed to be dominant, is due to spatial variation of the dipole field from neighbouring particles within the particle or interaction induced increase of the surface contribution of the magnetic anisotropy constant, Hansen and Mørup [24] found this explanation to be unrealistic since the interaction fields are much smaller than exchange fields which are responsible for the surface spin structure and the interaction field cannot effect the surface spin structure and the surface anisotropy significantly. Another model [21] considered the particle in an effective dipolar field due to surrounding particles. τ_0 was assumed independent on interactions. For the weak interactions ($\mu B/2K_{eff}V \ll 1$, B is an effective dipolar field arising from the surrounding particles) and in the limit $k_BT < K_{eff}V$ (which is the case for the blocking temperature determined by Mössbauer spectroscopy with its small time frame), this model has predicted that the interactions lead to a decrease of the relaxation time.

Three-dimensional arrays of magnetic dipoles, which interact only via dipolar coupling, become long-range ordered below a critical temperature [25] depending on the coupling strength. For a simple cubic lattice the ground state is antiferromagnetic, for fcc and bcc it is ferromagnetic. In chain-like structures the ordering is also ferromagnetic [26]. In a sample with a random distribution of magnetic dipoles the ordered state will be the one of a spin glass [27].

5. Experimental

5.1. PREPARATION OF Co NANOPARTICLES

Monodisperse chemically stable Co nanoparticles were prepared by thermal decomposition of dicobaltoctacarbonyl [$Co_2(CO)_8$] in an Ar atmosphere [28]. A combination of oleic acid [$CH_3(CH_2)_7CH{=}CH(CH_2)_7COOH$] and oleyl amine [$CH_3(CH_2)_7CH = CH(CH_2)_8NH_2$] was used as a surfactant, which covers the growing particle, controls its size and prevents the agglomeration of particles. A typical synthetic procedure is as follows: Under argon atmosphere, a mixture of oleyl amine (1 mmol), deoxygenated decane (100 ml), and oleic acid (1 mmol) is heated to 320 K. 1 mol of dicobaltocta-carbonyl is added and stirred for 15 min. Finally, the solution is allowed to simmer at 170 °C for 60 minutes. As-prepared particles have size distribution of about 15 %. The magnetic field separation [29] and size-selective precipitation [30] were used to narrow the size distribution and to produce a colloidal suspension of highly monodisperse Co nanoparticles. After cooling to room temperature, the Co particles were separated from the solvent (decane) by a magnet.

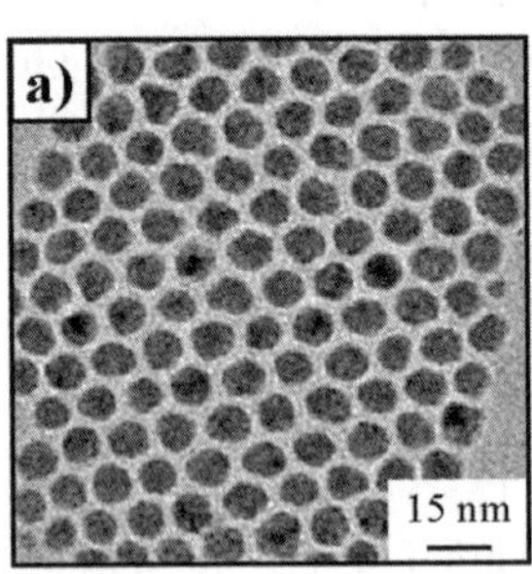

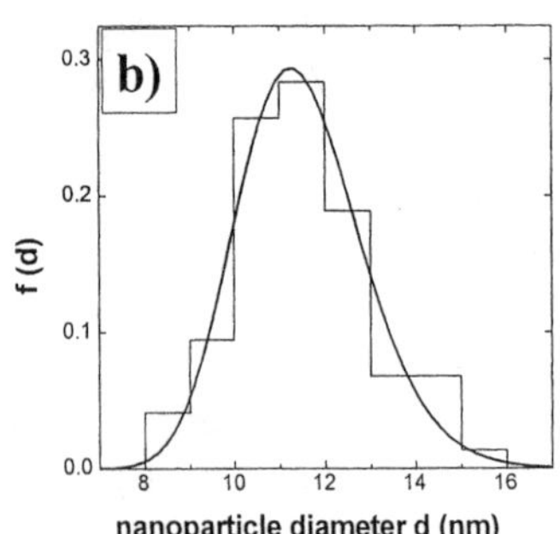

Figure 3. (a) Low magnification TEM micrograph of Co nanoparticles deposited on a carbon-coated copper grid. (b) The size distribution of the Co nanoparticles determined by TEM and its simulation according to the log-normal law as described in the text.

Ultrasmall particles remain in the solvent due to their smaller effective magnetic moment and can be removed from the larger particles together with the solvent. The separated wet powder of larger particles is redispersed in 100 ml toluene. The stability of the particles in toluene is increased. For size-selective precipitation, a small amount of ethanol was added in the solution. Ethanol reduces the barrier to aggregation and destabilizes the nanoparticle dispersion. Since larger nanoparticles experience the greatest attractive forces, they aggregate first and can be separated from the dispersion by filtering, centrifugation or applying a magnetic field. This procedure is repeated till the desired monodispersity is achieved.

Figure 3a shows a TEM image of the resultant particles deposited on a copper grid coated with 5 nm thick amorphous carbon film. The size distribution determined from the TEM image (Figure 3b) follows a log-normal law [31]

$$f(d) = \frac{1}{\sqrt{2\pi}\,\sigma d} \exp\left(-\frac{1}{2\sigma^2} \ln^2\left(\frac{d}{d_{mp}} \right) \right) \tag{10}$$

where $f(d)$ is the probability density of finding a particle with diameter d. σ and d_{mp} are the standard deviation and the most probable diameter, respectively. A fit according to Eq. (10) yields the mean diameter d_{mp} = 11.4 nm and the standard deviation σ = 0.18 (about 2 %).

5.2. STRUCTURE AND COMPOSITION

High resolution TEM studies [32] showed that the particles were polycrystalline consisting of a core-shell structure (Figure 4). It was established that the core was made of Co with fcc structure, often containing twins. Digital Fourier transforms from segments of high resolution images were utilized to confirmed the fcc crystal structure for the Co core (Figure 4c) and for the polycrystalline CoO shell (Figure 4b).

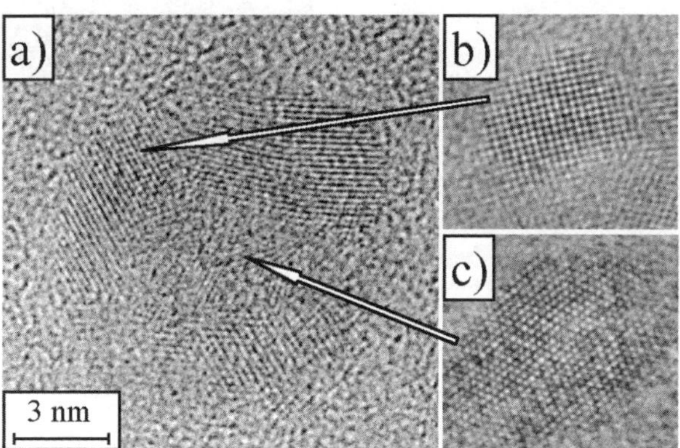

Figure 4. A high resolution TEM image of a Co nanoparticle on an amorphous C-coated copper grid. Magnified high resolution micrographs of the CoO crystallite in the shell in [001] orienttation (b) and twinned fcc Co core in [110] orientation (c) recorded at the locations indicated by the arrows. Images (b) and (c) are 1.3 times magnified compared to image (a). A twin boundary parallel to the common {111} plane is visible in the core (c).

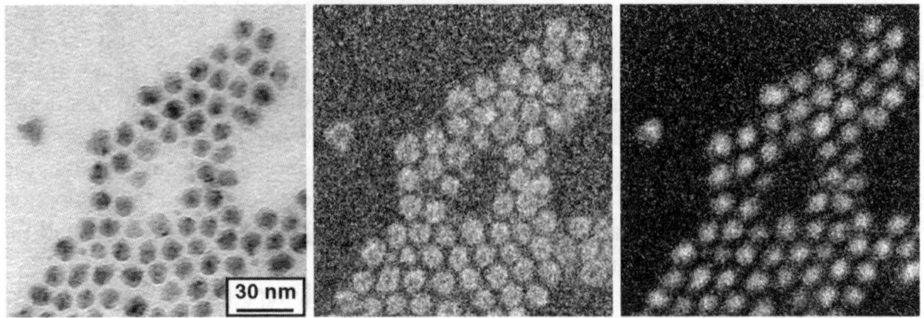

Figure 5. TEM image of the Co nanoparticles (a) and corresponding EELS element mapping of the Co-component (b) and oxygen-component (c). Energy filtered images are recorded at the Co L_3-edge (778 eV) and at the oxygen K-edge (~ 543 eV).

Figure 4c shows a typical atomic resolution image of a particle core. Detailed analysis of the images using Fourier diffractograms indicates that the core structure is fcc Co, containing twins parallel to the common {111} plane. In agreement with these HRTEM observations, selected area electron diffraction (SAED) patterns (not shown) of the Co nanoparticles exhibit reflection rings corresponding to fcc Co and CoO phases with lattice parameters consistent with those of bulk fcc Co and CoO structures. Spatially resolved compositional analysis of the nanoparticles was performed by spectral imaging of the sample using electron energy-loss signal. Figure 5 shows a low magnification TEM image of an array of the nanoparticles (Figure 5a) and energy filtered images using the cobalt L_3-edge at ~ 778 eV loss (Figure 5 b) and the oxygen K-edge at ~ 543 eV loss (Figure 5c).

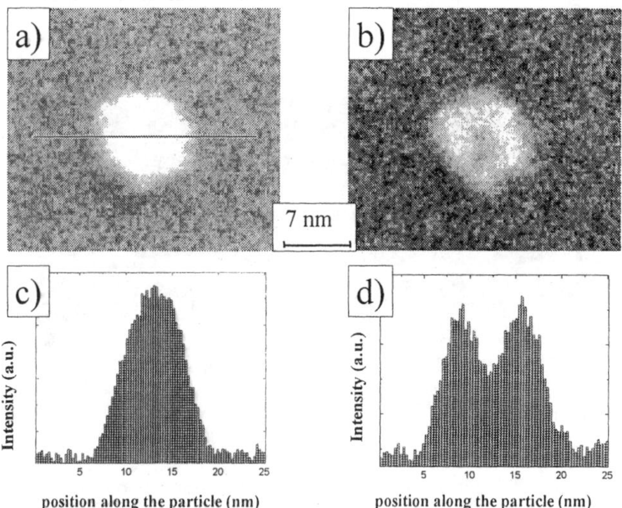

Figure 6. High resolution EELS element mapping of the Co- (a) and the oxygen-component (b) in the Co/CoO nanoparticle. Intensity profiles across the particle show the distribution of Co (c) and oxygen (d).

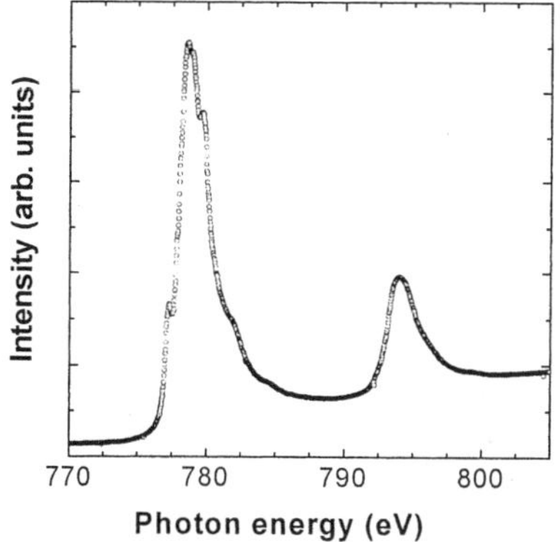

Figure 7. Co $L_{2,3}$-edge photoabsorption spectrum of the Co nanoparticles deposited on a carbon coated grid. Multiplet structure of the spectra indicates partial oxidation of the Co nanoparticles.

In the TEM image, contrast changes within particles indicate their polycrystalline structure. Contrast changes in the energy-filtered images indicates non-uniform distribution of Co and oxygen across the particles. Figure 6 shows high resolution energy filtered images of the Co nanoparticle. The cobalt content (Figure 6a) is larger in the center of the particles (higher brightness) while the oxygen content is obviously larger at the particle edge (Figure 6b). Histograms of the Co (Figure 6c) and oxygen (Figure 6d) intensities yield a Co core diameter of 7-8 nm and an oxide shell thickness of 2-2.5 nm that is in a full agreement with the HRTEM observations.

The X-ray absorption spectrum (Figure 7) confirms the presence of CoO showing a multiplett structure in the Co L_3 edge that is due to a superposition of metallic Co and Co in an oxidic environment.

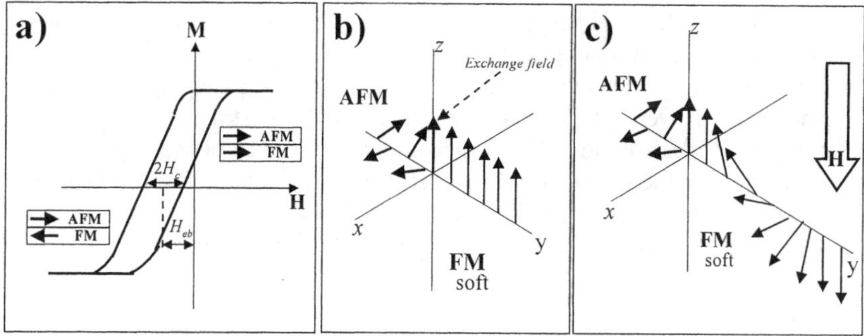

Figure 8. (a) Schematic representation of effect of exchange coupling on M(H) hysteresis loop for the material with AFM/FM interface. The arrow in AFM layer indicates the direction of the exchange field exerted by AFM layer on FM layer. (b) Direction of the spin near the AFM/FM interface after cooling through Néel temperature of AFM material. (c) Effect on spins in soft FM material of a field applied opposite to the exchange field. (the Figure is adapted from Ref. [15]).

6. Magnetization and Exchange Bias

CoO is antiferromagnetic (AFM) with a Néel temperature of T_N = 293 K. Above this temperature CoO is paramagnetic, and below T_N the Co moments are ferromagnetically coupled in {111} planes with the moments in adjacent {111} planes antiparrallel. The exchange coupling between the ferromagnetic (FM) Co core and AFM CoO shell in the nanoparticles leads to highly interesting phenomena at the interface. Exchange bias or unidirectional exchange anisotropy (UEA) is a direct consequence of this interaction.

When the oxidized Co particles are cooled through the Néel temperature in the presence of an external field, the transition from the paramagnetic to the AFM state in CoO occurs in the presence of the magnetized Co. As a result, the magnetic moment of the CoO are aligned to minimize their energy of interaction with the Co moments across the interface. For $T < T_N$, the Co magnetization is biased to the direction of the cooling field. This bias field is shown by an arrow in the AFM layer in Figure 8a. A stronger negative field is required to demagnetize the sample than if it is cooled in zero field. Majklejohn and Bean [33] explained the field-displaced hysteresis loop as exhibiting exchange anisotropy which is due to an exchange coupling between the moments of Co and CoO. Figure 8a shows the definitions of coercivity and exchange bias field in systems exhibiting exchange anisotropy. When the field is applied in the same direction used during cooling, the hysteresis loop is shifted toward the negative field direction by H_{eb}. The coercivity H_c is half the width of the $M(H)$ loop at M =0.

The exchange coupling across the AFM/FM interface is such that the moments in the AFM material lie on an axis that is orthogonal to the FM moment at the time of cooling through T_N (Figure 8b) [34]. Once the exchange coupling is established by field-cooling through the Néel temperature, a preferred direction of magnetization exists at the interface. On application of a weak or moderate field to an exchange coupled system, the soft FM material will tend to follow the field subject to the coupling at the interface (Figure 8c). The twist of spins occurs at the length of a domain wall.

The additional magnetic field energy required to create an interfacial magnetization twist in the FM material shows up as a shifted M(H) loop (toward negative field in the case shown in Figure 8). For magnetic measurements, approximately 10 MLs of the Co/CoO nanoparticles were deposited onto a Si(100) substrate by drying the appropriate amount of the solution. The center-to-center distance between particles is about 16 nm. Assuming the spontaneous magnetization of fcc bulk Co, the magnetic moment of 7.5 nm Co particle $3.2*10^{-19}$ Am^2 ($1.45*10^4 \mu_B$).

The dipole-dipole interaction energy between two neighbouring particles with magnetic moments aligned along the same axis E_{dd} = $2.5*10^{-21}$ J (15.63 meV). Assuming closed packed arrays of particles in the sample, the volume fraction of the magnetic material is about 21%, and the dipole-dipole interaction cannot be neglected.

The temperature dependence of the dc magnetization was measured after cooling the sample without external magnetic field (zero-field cooling (ZFC)) and after cooling in a magnetic field of 5 T (FC) from 370 K to 10 K. Then a magnetic field 20 G is applied and the magnetization is recorded at increasing temperature. Figure 9 shows ZFC and FC dc magnetization as a function of temperature from 10 to 370 K.

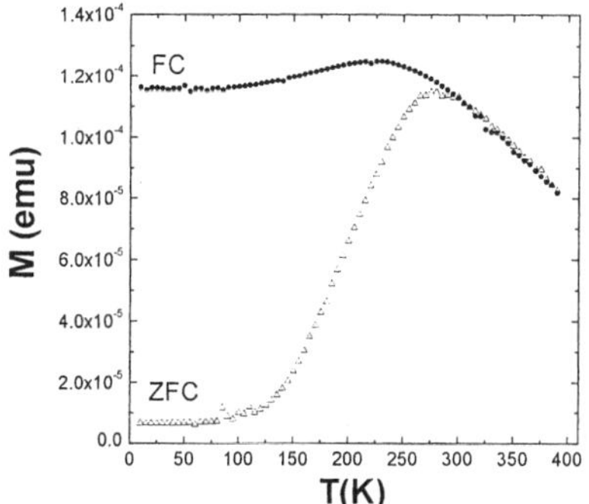

Figure 9. Zero-field-cooled (•) and field-cooled (Δ) magnetization measured at 20 G as a function of temperature for the CoO@Co nanoparticle assembly. The cooling field is 5 T.

The ZFC and FC magnetizations M do not change from 10 to 100 K The ZFC M rapidly increases with temperature above 100 K and reveals a maximum at 280 K. Upon reaching a temperature of 290 K, the FC M coincides with the ZFC M and starts to decrease with increasing temperature behaving like a Curie-Weiss paramagnet. The blocking temperature for the nanoparticle array is estimated to be about 290 K. The FC M shows non-monotonic temperature dependence: it is unchanged below 100 K and slightly increases with temperature revealing a maximum at 230 K. This behaviour of the FC M is attributed to a strong exchange coupling between the Co core and CoO shell resulting in a strong unidirectional exchange anisotropy.

Hysteresis loops of an array of the CoO@Co particles were measured at 10 K both after zero-field cooling (ZFC) and field cooling (FC) in an applied field of 5 T from 370 K to 10 K. The orientation of the magnetic field is parallel to the substrate plane. The direction of field used to measure the hysteresis loop was parallel to that of the cooling field. Figure 10 shows the ZFC and FC loops of the CoO@Co nanoparticles.

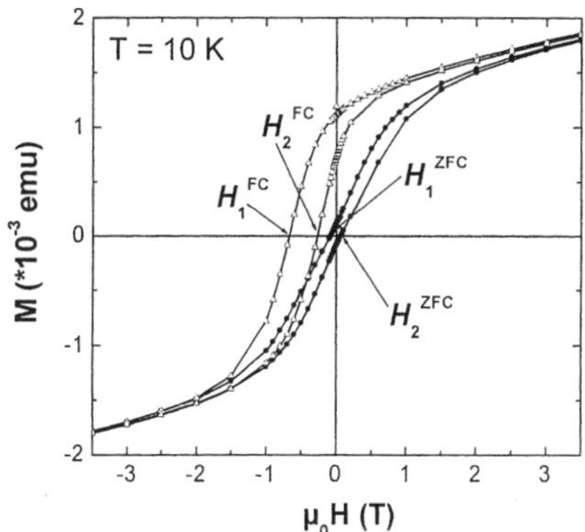

Figure 10. Zero-field-cooled (•) and field-cooled (Δ) hysteresis loops of the CoO-coated Co nanoparticles measured at 10 K. A magnetic field of 5 Tesla was applied during field cooling. Coercive fields for FC and ZFC curves are shown by arrows.

The ZFC loop is symmetrical about the origin. The FC loop is shifted along the applied field direction which shows the presence of UEA in our system. The value of the exchange bias field H_{eb} ($= \left|H_1^{FC} + H_2^{FC}\right|/2$) of 0.46 T indicates a strong UEA in the sample. The FC coercivity is increased ($\left|H_1^{FC} - H_2^{FC}\right|/2 \approx 0.2$ T) in comparison to the ZFC case ($\left|H_1^{ZFC} - H_2^{ZFC}\right|/2 \approx 0.07$ T) where the UEA is randomly distributed. H_1 and H_2 are marked in Figure 10. The increase of the coercivity in the FC sample indicates UEA increases of the effective uniaxial anisotropy of the particle.

To determine the correlation of the hysteresis loop shift with the ordering temperature of antiferromagnetic CoO, the following measurement was performed: the demagnetized sample was cooled from 370 K in the absence of an external magnetic field to a temperature T_f at which a magnetic field of 5 T was applied and the sample was further cooled to 10 K where the hysteresis loops were recorded. The exchange bias field H_{eb} and the coercivity H_c (measured at 10 K) are shown in Figure 11 as a function of the temperature T_f.

H_{eb} and H_c exhibit a similar temperature dependence: the values increase monotonically with increasing the T_f up to 150 K, at which the exchange bias field and the coercivity values achieve saturation and do not change anymore.

At T_f = 150 K, below which the exchange interaction decreases, indicates the thermal instabilities of AFM state in small CoO grains. This temperature is much lower that the Néel temperature (T_N = 293 K) of the bulk CoO. This result is an agreement with temperature dependent measurements of the exchange bias in oxide coated Co nanoparticles with diameters from 6.4 up to 27.5 nm. The thickness of the oxide layer was about 2 nm for different sizes. The shift of the hysteresis loop vanished at 150 K independently of the nanoparticle size [35,36]. That result was attributed to a superparamagnetic behavior of the AFM CoO crystallites (2 nm size) in the shell having a blocking temperature of 150 K. The appearance of this superparamagnetism in the AFM annoparticles has been discussed in terms of uncompensated surface spins. In the case of our core-shell particles, the decrease of the UEA (Figure 11 a) should be due to uncompensated and disordered spins at the core-shell interface [37].

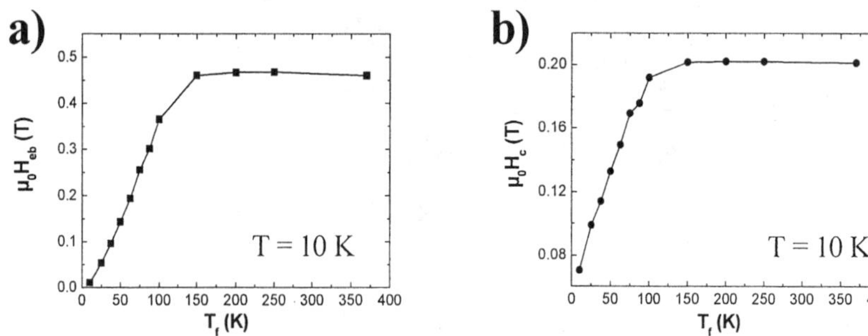

Figure 11. Exchange bias field H_{eb} (a) and coercivity H_C (b) at 10 K as a function of a characteristic temperature T_f below which a magnetic field of 5 T was applied during cooling from 370 to 10 K.

7. Self-Assembled Magnetic Nanoparticle Arrays

Colloidal, surfactant-coated nanoparticles self-assemble from solutions, forming two-dimensional (2D) and/or three-dimensional (3D) ordered structures on a substrate as the dispersing solvent evaporates [38]. These new mesoscopic materials are called colloidal crystals and are of growing interest, since they have unique optical, electronic and magnetic properties. The properties of these arrays strongly depend on their structure and composition. Hence, a lot of effort is directed at predicting and controlling the nanoparticle condensation and their crystallization. Preparation of the particles with a narrow size distributions is critical to achieve long-range order in a nanoparticle assembly: only for size distributions of $\sigma \leq 5$ % long-range order can be achieved [39]. There are many other parameters which influence the nanoparticle condensation. It is important to prolong the sedimentation time, that is reducing the evaporation rate of the solvent: rapid drying (< 60 s) frustrates the formation of long-range order while lowering the evaporation rate by drying in a controlled solvent atmosphere and by using solvents with lower boiling points [40] improves the degree of order. The colloidal nanoparticles can form different superlattices. The preference for a given phase depends on the ratio of the surfactant layer thickness L to the metal core radius R. For $L/R < 0.60$, close-packed fcc or hcp structures were observed, whereas a thick surfactant layer favors more open bcc or bct structures [41]. These phenomena are related to the softness of surfactant layer. If nanoparticles interact like hard spheres (short-range repulsion), the most stable colloidal crystal structure is fcc. For soft spheres (long-range repulsion, $L/R >> 0.6$) the bcc phase is favourable [42].

The wetting properties of the substrate influence of the nanoparticles crystal morphology similar to molecular beem epitaxy in thin film growth. If the nanoparticle dispersion wets the substrate, layer-by-layer growth of the colloidal crystal is observed. If the nanoparticle solution does not wet the substrate, the nanoparticle crystallization process proceeds in a more three-dimensional way [43].

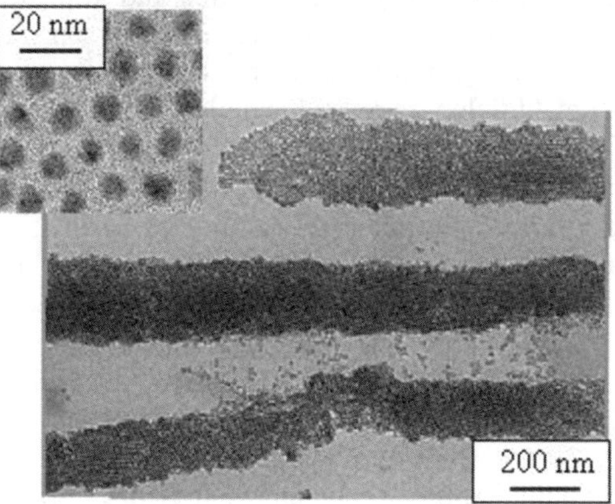

Figure 12. TEM micrographs of Co nanocrystals deposited on a carbon-coated Cu grid in a magnetic field of 0.35 T parallel to the grid plane. Insert: The higher resolution TEM image recorded near a stripe edge (one layer of Co nanoparticles) shows a regular fcc 111 plane ordering of the particles.

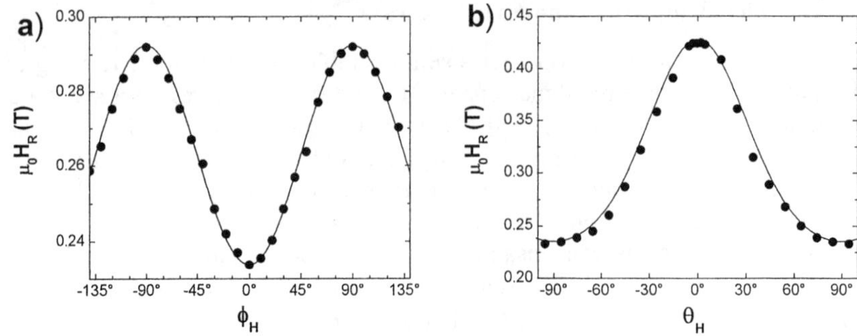

Figure 13. Dependence of H_R on the direction of the external magnetic field: (a) in plane φ_H (measured from the stripe long axis) and (b) out-of-plane θ_H (measured from the normal to the sample). The error bar is on the order of the symbol size. The full lines are fits as described in the text.

In magnetic nanoparticle systems additional magnetostatic forces exist due to dipolar interaction. A colloidal crystal formation can be directed by magnetic forces [28,44,45]. By applying a magnetic field during the nanoparticle deposition one can change the morphology of the growing colloidal crystal.

An example is discussed in the following: Regular arrays of Co nanocrystals were formed by drying the solution in a magnetic field. The TEM micrographs of the dried structure (Figure 12) were recorded after drying a drop of the solution on a carbon-coated Cu grid in a magnetic field of 0.35 T parallel to the grid plane. Stripes consisting of a regular triangular in-plane lattice of the Co nanocrystals with a width of 200 - 250 nm and a length of 1.5 - 3 μm are formed along the applied field direction. Mostly, the stripes consist of 3-4 layers of Co nanocrystals forming a nearly perfect three dimensional fcc lattice with the (111) planes parallel to the substrate. The nearest-neighbour distance is about 16 nm.

Ferromagnetic resonance (FMR) experiments [6] were carried out at 9.5 GHz and 300 K to obtain values for the effective magnetization and in-plane anisotropy field. We have performed two types of angular-dependent measurements: azimuthal angular dependence when the dc field was rotated in the sample plane; and polar angular dependence when the applied field was rotated in a plane normal to the film plane. The results are shown in Figure13. For in-plane measurements (Figure 13a), the azimuthal angle ϕ_H was measured from the direction parallel to the stripes. The polar angle θ_H (Figure 13b) was measured from the normal to the sample plane.

The lowest resonance field 0.233 T was observed when the magnetic field was applied in the film plane and along the stripes. It is smaller than the paramagnetic resonance field ω/γ = 0.3085 T, which shows that there is an additional intrinsic magnetic field due to an effective magnetization of all particles and that the easy axis of the magnetization is in the film plane and parallel to the stripes.

One has to consider three contributions in the analysis of H_R : (a) shape anisotropy due to stripes; (b) configurational magnetic anisotropy originating from the interparticle magnetostatic coupling of the particles ordered in a fcc-like lattice inside of the stripes; (c) the effective magnetic anisotropy of the individual particles which includes shape, volume, and surface anisotropy contributions. Assuming in a first step that all anisotropies due to spin-orbit coupling vanish and only shape anisotropy is present, the resonance condition is written as:

$$\left(\frac{\omega}{\gamma}\right)^2 = \left(H_R + (N_x - N_z)4\pi M_s\right)\times\left(H_R + (N_y - N_z)4\pi M_s\right) \tag{11}$$

($\omega/2\pi$) is the resonance frequency, γ is the gyromagnetic ratio given by $\gamma = ge/2mc$ (g is the spectroscopic splitting factor). N_x, N_y, and N_z are the demagnetising factors along the axes of an ellipsoid. H_R is the resonance field when the field is applied along z-axis. M_s is the saturation magnetization. The demagnetising factors ($N_x = 0.005$, $N_y = 0.095$, $N_z = 0.9$) were calculated [46] in first approximation assuming an ellipsoidal shape of the stripes with semi-axes $l_x = 1000$ nm, $l_y = 125$ nm, and $l_z = 20$ nm. Theses lengths correspond to the average dimensions of the stripes as determined by TEM. We find that $H_R(\theta_H, \phi_H)$ (Figure 13) cannot be described by shape anisotropy (Eq. (11)) only. Other anisotropy contributions exist in our samples.

To fit the experimental data of Figure 13a we use the resonance conditions for a homogeneous thin film with cubic symmetry and an additional in-plane uniaxial anisotropy [47], which includes the small in-plane shape anisotropy (N_x, $N_y << N_z \approx 1$).

$$\begin{aligned}\left(\frac{\omega}{\gamma}\right)^2 &= \left[H_R + 2H_{an}^{4\|}\cos 4\phi - H_{an}^{2\|}\cos 2(\phi - \phi_u)\right] \\ &\quad\times\left[H_R + H_{eff} + H_{an}^{4\|}\left(2 - \sin^2 2\phi\right) - H_{an}^{2\|}\cos^2(\phi - \phi_u)\right]\end{aligned} \tag{12}$$

Resonance is observed at the magnetic field H_R which depends on the equilibrium angle ϕ (in this case identical to the angle ϕ_H of the applied magnetic field) of the magnetization $M(H)$ and the angle ϕ_u of the magnetic field H with respect to the axis of uniaxial in-plane anisotropy. $H_{an}^{2\|}$ is an effective in-plane uniaxial anisotropy field and $H_{an}^{4\|}$ is a fourfold in-plane anisotropy field. The effective anisotropy field is defined as $H_{eff} = -2K_{2\perp}/M_S + 4\pi f M_S$ with the perpendicular anisotropy energy $K_{2\perp}$, the volumetric filling factor of the stripes f and the bulk-like saturation magnetization M_s of individual Co nanocrystals. In Figure 13a we show the best fit (solid line) to the experimental data. The fit according to Eq. (12) yields $H_{an}^{4\|} = 0$, $H_{eff} = 0.127$ T and $H_{an}^{2\|} = 0.037$ T, that is no cubic contribution and only an in-plane uniaxial anisotropy.

In Figure 13b we show the polar angular dependence of H_R when the applied dc field was rotated in a plane normal to the film plane starting parallel to the in-plane easy axis ($\phi_H = 0°$). Note that $H_R(\phi_H = 0°)$ of Figure 13a is the same as $H_R(\theta_H = 90°)$ in Figure 13b. The resonance condition

$$\left(\frac{\omega}{\gamma}\right)^2 = \left[H_R\cos(\theta - \theta_H) - H_{eff}\cos 2\theta\right]\times\left[H_R\cos(\theta - \theta_H) - H_{eff}\cos^2\theta + H_{an}^{2\|}\right] \tag{13}$$

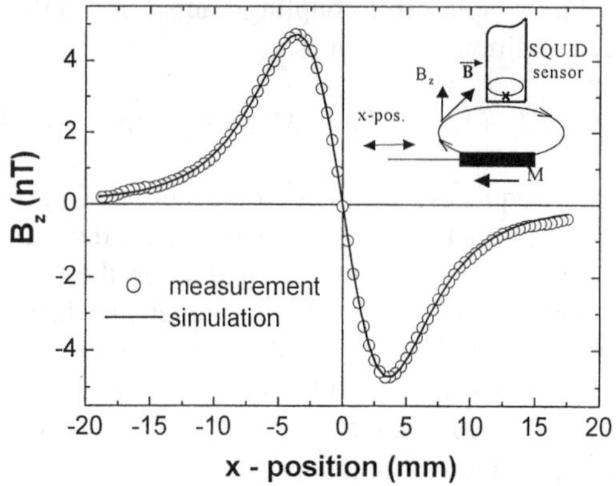

Figure 14. Vertical component B_z of the dipolar magnetic stray field of the magnetized Co nanocrystal array (symbols) and the simulation as described in the text (solid line). The geometry of the measurement is shown in the inset.

yields $H_{eff} = 0.13$ T and $H_{an}^{2//} = 0.037$ T. Again an excellent fit is obtained with the same $H_{an}^{2//}$ without the need to include fourth-order contributions. This uniaxial anisotropy is the result of shape anisotropy of the stripes and a possible alignment of crystalline anisotropy axes of the individual crystals. H_{eff} cannot be explained by assuming a bulk-like magnetization per Co particle ($4\pi\ M_s = 1.79$ T) alone. The Co volume fraction f inside the stripes is about 0.21 (7.5 nm diameter Co particles covered with 2 nm thick CoO shell ordered in fcc structure with the center-to-center nearest-neighbour distance of about 16 nm) and the stripes cover about 40% of the substrate. This yields the smallest possible average magnetization $4\pi\, fM_s = 0.15$ T which is still larger than H_{eff} obtained from the FMR analysis. To explain the difference one has to include a small perpendicular anisotropy field $2K_{2\perp}/M_S$.

To gain further insights into the origin of H_{eff} we performed remanent magnetization measurements with a home-built HTS-SQUID magnetometer at 300 K [48]. In such a set-up (see Ney et al. [49] for details) the component B_z^{tot} of the dipolar magnetic stray field of the in-plane magnetized sample is measured at constant height as a function of lateral position of the SQUID sensor over the sample (inset Figure 14). Before the SQUID measurement the sample was magnetized at 1.3 T parallel to the stripes. Figure 14 shows the measured and the calculated stray-field B_z when the sample was scanned along the x-direction. B_z can be precisely simulated [50] by using the model of an homogeneously magnetized film with the area a^2, the thickness d of our sample and an unknown M parallel to the x-axis.

We find: firstly, the excellent agreement between experimental data and theoretical model shows that the approximation of a quasi – continuous layer with an average magnetization is a very reasonable assumption in the FMR analysis. Secondly, the fit yields the remanent magnetization $M = 0.059$ T. This is only 40 % of the value one would expect for f = 0.084 when the 7.5 nm diameter particle carries a bulk magnetic moment assuming that the CoO outer shell of 2 nm thickness carries no net magnetic moment. With the experimental SQUID value M and H_{eff} = 0.13 T one can calculate M_s = 0.038 T and $K_{2\perp} \approx 1$ μeV/atom. This is a very small value compared to bulk hcp Co, which can be explained by the cubic structure of the particles and the thermal excitations inherent to the dipolar coupled system close to its blocking temperature.

8. Summary

We presented a detailed structural and magnetic investigation of colloidal Co@CoO nanoparticles of 11.5 nm diameter. HRTEM and EELS have shown that the nanoparticles contain a bulk-like fcc Co core covered with a 2-2.5 nm thick CoO shell. The nanoparticles show a large unidirectional exchange anisotropy due to the exchange coupling of the magnetic moments of the FM Co core and the AFM CoO shell. The ordering temperature for AFM CoO grains was estimated to be 150 K. Regular arrays of the nanoparticles were formed by drying the solution in a magnetic field. An angular dependent FMR analysis of the 2D self-assembled Co nanoparticle arrays at room temperature shows that the magnetic response of the ordered layers can be well approximated by assuming a homogeneously in-plane magnetized film with a volumetric filling factor obtained by TEM analysis.

In collaboration with M. U. Wiedwald, R. Ramchal, T. Radetic, U. Dahmen, M. Hilgendorff, and M. Giersig, and supported through Deutsche Forschungsgemeinschaft, EC contract no. HPRN-CT-1999-00150, and EC –ARI program. We especially acknowledge the help of the National Center of Electron Microscopy, Lawrence Berkeley Laboratory, USA in acquiring the TEM images.

9. References

1. Sorensen, C. M. (2001) *Magnetism in Nanoscale Materials in Chemistry*, ed. by Klabunde, K. J., Wiley-Interscience Publication, New York.
2. Moriarty, P. (2001) Rep. Prog. Phys. **64**, 297.
3. Moser, A., Takano, K., Margulies, D. T., Albrecht, M., Sonobe, Y., Ikeda, Y., Sun, S., Fullerton, E. (2002) J. Phys. D: Appl. Phys. **35**, R157.
4. Caruso, F., Spasova, M., Susha, A., Giersig, M. and Caruso, R. A. (2001) Chem. Mater. **13**, 109.
5. Sobal, N. S., Hilgendorff, M., Möhwald, H., Giersig, M., Spasova, M., Radetic, T. and M. Farle (2002) Nano Letters **2**, 621.
6. Farle, M. (1998) Rep. Prog. Phys. **61** 755.
7. Bovensiepen, U., Hyuk J. Choi and Qui, Z. Q. (2000) Phys. Rev. B **61**, 3235.
8. Kawakami R. K., Escorcia-Aparicio Ernesto, J. and Qiu, Z. Q. (1996) Phys. Rev. Lett. **77**, 2570.
9. Chen Jian and Erskine, J.-L. (1992) Phys. Rev. Lett. **68**, 1212.
10. Gradmann, U., Dürkop, T. and Elmers, H. J. (1997) J. Magn. Magn. Mater. **165**, 56
11. Bruno, P. (1988) J. Phys. F **18**, 1291.
12. Heinrich, B. and Bland, J. A. C. (eds.) (1994) *Ultrathin magnetic structures,* Springer, Berlin.
13. Hillebrands, B. (2000) *Brillouin Light Scattering from Layered Magnetic Structures* in Topics in Applied Physics, Vol. 75, Springer, Berlin, 175 – 289.
14. Johnson M-T, Bloemen P-J-H; den Broeder F-J-A and de Vries J-J , (1996) Rep. Prog. Phys. **59**, 1409.
15. O'Handley, R.C. (2000) *Modern Magnetic Materials: Principal and Applications*, Wieley-Interscience Publication, New York.

16. Stoner, E. C. and Wohlfarth, P. (1948) Philos. Trans. R. Soc. London, Ser. A **240**, 599.
17. Néel, L. (1949) Ann. Geophys. **5**, 99; Brown, W. F., Jr. (1963) Phys. Rev. **130**, 1677.
18. Jacobs, I. S. and Bean, C. P. (1963) Rado, G. T. and Shull, H. (eds.) *Magnetism III.*, Academic. New York.
19. Hansen, M. F. and Mørup, S. (1998) J. Magn. Magn. Mater.**184**, 262 and references cited therein.
20. Dormann, J.L., Bessais, L. and Fiorani, D. (1988) J. Phys. C: Solid State Phys. **21**, 2015.
21. Mørup, S. and Tronc, E. (1994) Phys. Rev. Lett. **72**, 3278.
22. Dormann, J. L., D'Orazio, F., Lucari, F., Tronc, E., Prené, P., Jolivet, J. P., Fiorani, D., Cherkaoui, R. and Noguès, M. (1996) Phys. Rev. B **53**, 14291.
23. Dormann, J. L., Fiorani, D. and Tronc, E. (1997) Adv. Chem. Phys. **98**, 283.
24. Hansen, M. F. and Mørup, S. (1998) J. Magn. Magn. Mater.**184**, 262.
25. Romano, S. (1994) Phys. Rev B **49** 12287 and references cited therein.
26. Mørup, S., Christensen, P. H., Clausen, B. S. (1987) J. Magn. Magn. Mater. **68**, 160.
27. Luo, W., Nagel, S. R., Rosenbaum, T. F. and Rosensweig, R. E. (1991) Phys. Rev. Lett. **67**, 2721.
28. Spasova, M., Wiedwald, U., Ramchal, R., Farle, M., Hilgendorff, M. and Giersig, M. (2002) J. Magn. Magn. Mater. **240**, 40; Wiedwald, U., Spasova, M., Farle, M., Hilgendorff, M. and Giersig, M. (2001) J. Vac. Sci. Technol. A **19**, 1773.
29. Yin, J. S. and Wang, Z. L. (1997) Phys. Rev. Lett. **79**, 2570.
30. Murray, C. B., Shouheng Sun, Gascher, W., Doyle, H., Betley, T. A. and Kagan, C. R. (2001) IBM J. Res. & Dev. **45**, 47.
31. Chantrell, R. W., Popplewell, J. and Charles, S. W. (1977) Physica **86-88B**, 1421.
32. Spasova, M., Radetic, T., Sobal, N. S., Hilgendorff, M., Wiedwald, U., Farle, M., Giersig, M. and Dahmen, U. (2002) Mat. Res. Soc. Symp. Proc. **721** in press.
33. Meiklejohn, W. H. and Bean, C. P. (1957) Phys. Rev. **105**, 904.
34. Jungblut, R., Coehorn, R., Johnson, M. T., aan de Stegge, J. and Reinders, A. (1994) J. Appl. Phys. **75**, 6659; Jungblut, R., Coehorn, R., Johnson, M., Sauer, Ch., van der Zaag, P. J., Ball, A. R., Rijks, Th. G. S. M., aan de Stegge, J. and Reinders, A. (1995) J. Magn. Magn. Mater. **148**, 300.
35. Peng, D. L., Sumiyama, K., Hihara, T., Yamamuro, S. and Konno, T. J. (2000) Phys. Rev. B **61**, 3103.
36. Gangopadhyay, S., Hadjiipanayis, G. C., Sorensen, C. M. and Klabunde, K. J. (1993) J. Appl. Phys. **73**, 6964.
37. Peng, D. L., Sumiyamama, K., Hihara, T. and Yamamumo, S. (1999) Appl. Phys. Lett. **75**, 3856.
38. Andres, R. P., Bielefeld, J. D., Henerson, J. I., Janes, D. B., Kolagunta, V. R., Kubiak, C. P., Mahoney, W. J. and Osifchin, R. G. (1996) Science **273**, 1690.
39. Murray, C. B., Kagan, C. R. and Bawendi, M. G. (1995) Science **270**, 1335.
40. Murray, C. B., Shouheng Sun, Doyle, H. and Betley, T. (2001) MRS Bulletin **26**, 985.
41. Korgel, B. A. and Fitzmaurice, D. (2000) Phys. Rev. B **59** 14191.
42. Yamamuro, S., Farrell, D. F. and Majetich, S. A. (2002) Phys. Rev. B **65**, 224431.
43. Korgel, B. A. and Fitzmaurice, D. (1998) Phys. Rev. Lett. **80**, 3531.
44. Golosovsky, M., Saado, Y. and Davidov, D. (1999) Appl. Phys. Lett. **75**, 4168.
45. Bizdoaca, E. L., Spasova, M., Farle, M., Hilgendorff, M. and Caruso, F. (2002) J. Magn. Magn. Mater. **240**, 44.
46. Osborn, J. A. (1945) Phys. Rev. **67**, 351.
47. Heinrich, B., Purcell, T., Dutcher, J. R., Cochran, J.,F. and Arrott, A. S. (1988) Phys. Rev. B **38**, 12879.
48. Ramchal, R. Diploma thesis, Technische Universität Braunschweig 2001, unpublished.
49. Ney, A., Poulopoulos, P., Farle, M. and Baberschke, K. (2000) Phys. Rev. B **62**, 11336.
50. Snigirev, O. V., Andreev, K. E., Tishin, A. M., Gudoshnikov, S. A. and Bohr, J. (1997) Phys. Rev. B **55**, 14429.

CIRCULAR MAGNETIC ELEMENTS: GROUND STATES, REVERSAL AND DIPOLAR INTERACTIONS

U. EBELS[1], M. NATALI[2], L. D. BUDA[1], I. L. PREJBEANU[1]
[1]*URA CEA-CNRS SPINTEC*
CEA-Grenoble, 17 Avenue des Martyrs, 38054 Grenoble, France
[2]*ICIS-CNR, Corso Stati Uniti 4, 35128 Padova, Italy*

1. Introduction

The continuous reduction of the system sizes of magnetic materials into the nanometer regime has kept magnetism an attractive field of research for the past 15 years. New phenomena have been put into evidence, when the system size becomes comparable to typical magnetic or electronic length scales [1-3]. Besides being conceptually attractive to the physicist, these phenomena promise the development of new devices, called spin-electronic devices [4,5]. Spin-electronics refers to the fact that in ferromagnets the conduction electrons can be distinguished depending on whether their spin is aligned parallel or antiparallel to the magnetization. Via spin dependent scattering, the electron spin orientation can then be used to control the electronic signal. For example, in spin-valve [5-7] and tunnel junction [8,9] devices, two ferromagnetic layers are separated by a non-magnetic (metallic or insulating) spacer layer. The resistance across this multilayer stack is higher when the magnetization in the two layers is aligned antiparallel as compared to parallel. Such stacked elements are promising candidates to be used as the basic logic cell for a new generation of Magnetic Random Access Memories (MRAM) [4,5,7,9]. The perpetual increase in storage density as well as the increase in access time (now reaching the GHz regime) define two tasks for the physicist: (i) finding a system of reduced size such that the switching between two well defined micromagnetic configurations (corresponding to a "0" and "1") is stable and repetitive, (ii) finding a mechanism by which this switching occurs on the nanosecond or sub-nanosecond time-scale.

This chapter treats some general aspects of the static micromagnetic configurations in submicron magnetic elements for the particular case of circular flat discs. There are a variety of methods by which small magnetic elements can be fabricated, as evidenced by a number of contributions in this book as well as in the literature [10]. The elements presented here have been fabricated mostly by electron beam lithography and lift-off. They are arranged in 2D square arrays which contain (almost) identical particles, whose size and shape distribution is negligible. While the thickness can be reduced to about 5 nm or less, the lateral sizes are typically restricted to above 100 nm, yielding flat elements. Furthermore, upon variation of the separation between the elements, the influence of dipolar interactions can be investigated systematically.

Before going into details, several remarks will be made in view of the title of this issue.

L.M. Liz-Marzán and M. Giersig (eds.),
Low-Dimensional Systems: Theory, Preparation, and Some Applications, 193–211.

2. Magnetic Length Scales

Firstly, only the static magnetization configuration will be considered here with the justification that the static properties need to be known before understanding or controlling the dynamic properties. These static configurations and the quasi-static reversal in a slowly varying applied field, depend critically on the sample shape, the system size, the separation between elements as well as on the material parameters such as: exchange energy constant A_{ex}, saturation magnetization M_s and magnetic anisotropy energy constant K.

The dependence of the micromagnetic configuration on the system size is related to the magnetic lengthscales, which brings us to the second remark. Although magnetism as such is a quantum mechanical effect, the description of the magnetization distribution and of the magnetization dynamics occurs in terms of a classical angular-momentum-type equation of motion [3,11]. The quantum mechanical exchange interaction, which keeps neighboring spins parallel, is very strong and does not allow for a rapid rotation of the spin orientation. The minimum lengthscales over which a substantial reorientation of the magnetization vector can occur are much larger than atomic distances and are given by the competition between the exchange energy and (i) the magneto-crystalline anisotropy energy and/or (ii) the magneto-static energy. This defines respectively (i) the domain wall width parameter δ_w and (ii) the exchange length λ_{ex}. Here, magneto-crystalline energy denotes the fact, that via spin-orbit interaction the spins feel the crystal symmetry and can lower their energy when aligning into particular lattice directions. Magneto-static energy occurs whenever there is a divergence of the magnetization div$\underline{M} \neq 0$ (volume and surface "charges"). This dipolar energy is mathematically the most cumbersome to deal with, since it is long-range (non-local). On the other hand it is at the origin of the large variety of magnetic configurations [3].

For example, when the magnetization points perpendicular to a sample surface, magnetic surface charges are created which give rise to dipolar (demagnetization) fields. In order to reduce the associated magneto-static energy, the system will break up into domains such as shown in Figure 1a, at the cost of domain wall energy. Exchange impedes a rapid rotation of the magnetization from the up to the down domain, and ideally would like to spread out this rotation over the largest possible distance. This however would imply an increase of the magneto-crystalline energy (denoted by K in Figure 1a). Hence, there is a competition between exchange energy, A_{ex} and magneto-crystalline energy K, which defines the wall width parameter δ_w as $\delta_w = \sqrt{A_{ex}/K}$. Typical values for the domain wall widths range from 1 nm to 100 nm.

An example for the exchange length λ_{ex}, where the exchange energy is in competition with the magneto-static energy is represented by vortices. These are treated in detail throughout this chapter for the special case of circular flat elements such as shown in Figure 1b. In order to avoid magneto-static surface charges, the magnetization will stay in-plane and keep parallel to the side faces, forming a circular magnetization path. However, in the center of the disc the magnetization cannot continue to turn in an arbitrarily small circle. One solution might be that the magnetization vanishes ($\underline{M} = 0$), but this would not be consistent with micromagnetic theory which requires that the modulus $|\underline{M}|$ is constant [3,11,12]. This latter condition can be fulfilled, when the magnetization turns out of the film plane in the center of the dot, as deduced from micromagnetic calculations, shown in Figure 1b, c.

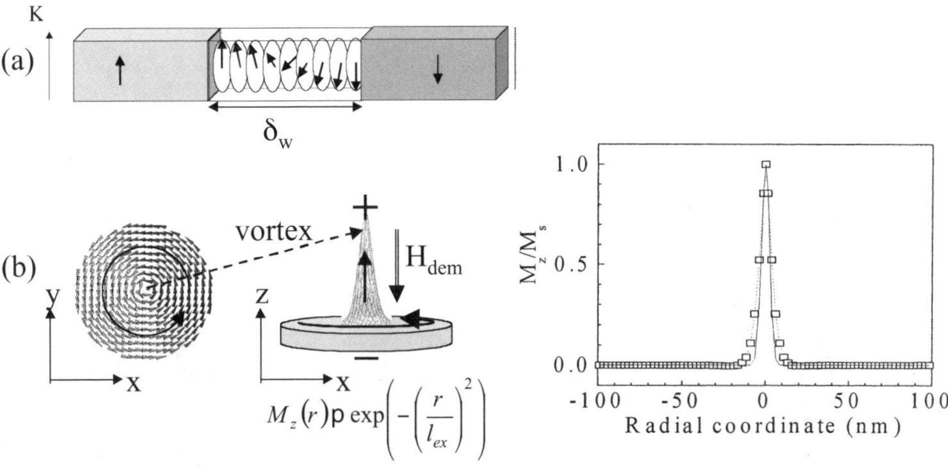

Figure 1. (a) Sketch of a two domain state, separated by a domain wall of width δ_w; (b) The vortex state of a circular element, as deduced from micromagnetic simulations [19]. Magnetization vector plot (left), 3D profile (center) and 1D scan for the out of plane component M_z (right), (full line = analytic, open square = simulation).

The cross sectional scan of the out-of plane magnetization component reveals that the magnetization rotates from in-plane to out of plane back to a reversed in-plane direction, hence creating a large amount of exchange energy. Moreover, in the region where $\underline{M}$ is pointing out of plane, magnetic surface charges are produced (at the top and bottom faces). A reduction of the associated dipolar field energy can be obtained by decreasing the vortex diameter. However, this is in competition with the exchange energy, leading to a vortex diameter which scales with the exchange length $\lambda_{ex} = \sqrt{A_{ex}/2\pi M_s^2}$. Typical values of λ_{ex} are 3 – 10 nm. The corresponding vortex diameter is in the range of 30-50 nm.

For any system whose dimensions are of the order of or below these lengthscales δ_w and λ_{ex}, multidomain structures are suppressed. In sub-micron elements this results in a finite number of magnetization configurations which are either of the flux closure type or of the (quasi-) single domain or high remanence type. In the following we will discuss the occurrence of these states for circular flat dots and define the ground state diagram as a function of dot thickness and dot diameter as well as its reversal properties in applied fields.

3. Observation of the vortex and the single domain state

Vortex State: The first observation of the circular flux closure structure has been made for triangular Co dots using electron holography [13]. This structure has been confirmed recently using a related technique, Lorentz microscopy [14], where electrons are deflected by the Lorentz force due to the in-plane magnetization either away from or towards the disc center, depending on the circulation sense of the magnetization. This gives rise to a decreased (black) or an enhanced (white) contrast at the center of the disc, see Figure 2a. These contrasts confirm that the magnetization is confined to lie in the film plane and that it follows a circular magnetization path. However, they do reveal no information about the spin arrangement in the center.

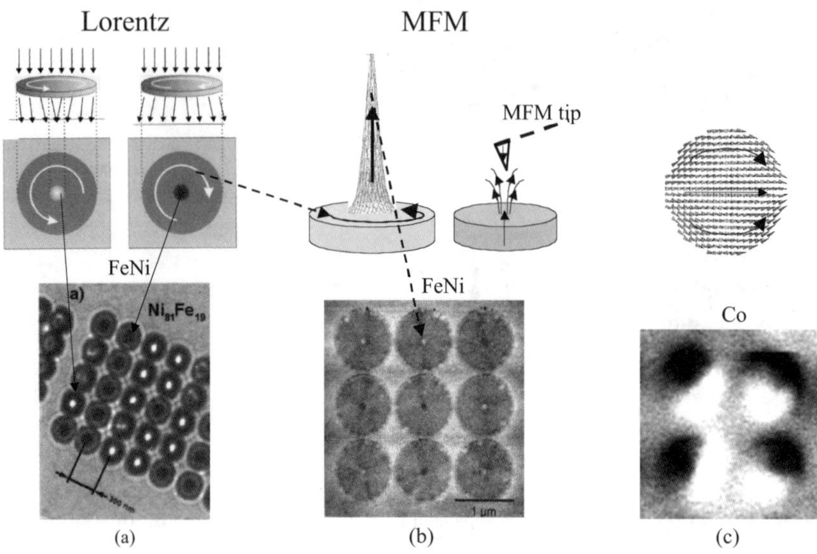

Figure 2. (a) Lorentz microscopy image of the vortex state, shown for D = 200 nm and t = 43 nm FeNi discs, from ref. [14]. (b) MFM image of the vortex state shown for D = 200 nm , t = 15 nm FeNi discs, from ref. [15]. The inset to the top right shows the interaction of an MFM tip with the vortex stray field. (c) MFM image of the single domain state for D = 200 nm, t = 15nm Co discs [17], and the magnetization vector plot as obtained from micromagnetic simulations [19].

In contrast it has been possible to determine unambiguously the direction of the magnetization of the vortex core pointing out of plane, by using magnetic force microscopy (MFM) [14-16]. MFM makes use of the interaction between a perpendicularly magnetized tip with the stray field of the sample. This interaction gives rise to either a black or a white contrast in the center, as shown in Figure 2b, depending on the orientation of the magnetization in the vortex core. In total there are four degenerate flux closure configurations: the vortex can point up or down combined with a clockwise or counter clockwise circulation path of the in-plane magnetization.

Single Domain State: Similar observations on the vortex state have been made on circular Co dots at remanence after saturating the sample in a field perpendicular [17] or parallel [18] to the dot plane. For the same dots, the single domain (SD) state could be stabilized at in-plane remanence, such as shown by the strong dipolar contrast in the MFM images in Figure 2c. The magnetization vector plot in Figure 2c, obtained from micromagnetic calculation [19,20], shows that in fact the denomination of 'single domain' state is not quite correct. The magnetization still tries to align parallel to the dot border, allowing deviations from the average magnetization direction. A proper notation therefore is a *quasi*-SD state, which is what we will refer to in the following.

Since in general the remanent state depends strongly on the magnetic history, the magnetic states observed in Figure 2 may not necessarily represent the ground state configuration for the given dot dimensions. In the following we will present a calculation of the energy density of these two configurations to deduce the ground state diagram [21,22] as a function of the dot thickness t and the dot diameter D.

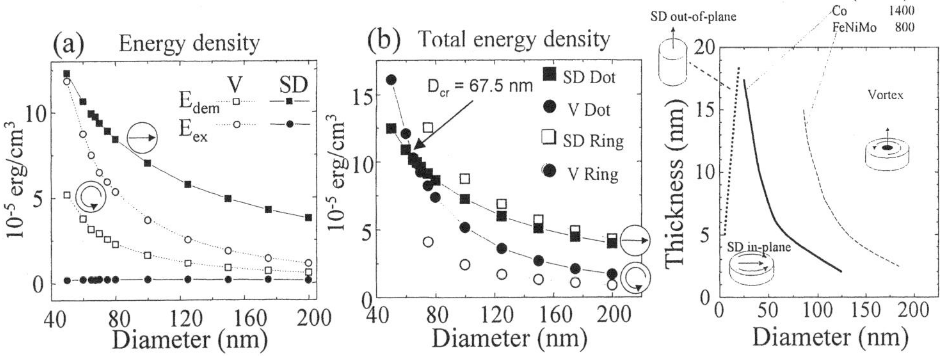

Figure 3. Micromagnetic simulations for (a) the demagnetization field energy density E_{dem} and the exchange energy density E_{ex} as a function of diameter for the Vortex (V) and single domain (SD) state of a Co dot (t = 10 nm); (b) the total energy density of the SD and V state for Co dots (full symbols, t = 5 nm) and Co rings (open symbols, t = 5 nm); (c) The state diagram of the vortex, in-plane and out of plane single domain states.

4. Energy considerations and zero field state diagram

First experiments to deduce the state diagram for circular dots have been performed in ref. [16] using MFM. Further studies using conventional magnetometry techniques, MFM or micromagnetic simulations have been performed for circular flat dots in references [19,20,23-28], for flat squares in ref. [29] and for cylindrical pillars of varying aspect ratio in ref. [30]. Furthermore, a variety of experiments have been performed for elliptical or rectangular elements in refs. [27,28,31,32].

For circular flat elements, in the absence of any magnetic anisotropy energy, the only two energy contributions are: the demagnetization field energy, dominating in the SD state and the exchange energy, dominating in the vortex state. The corresponding calculated energy densities as a function of the dot diameter are shown in Figure 3a. Due to the out-of plane orientation of the vortex core, there is a dipolar field contribution besides the exchange energy in the vortex state. Both energy terms act only locally, over the diameter of the vortex core and the total energy is thus stored only inside this region. Hence, the total energy of the vortex state is independent of D, while the total energy density (energy per dot volume) increases with decreasing D [19,20], see Figure 3b. For large diameters, the total energy density of the vortex state is smaller than that of the SD state and the vortex is the ground state. Below D_{cr} the vortex state transforms into the energetically more favorable SD state. From Figure 3b, D_{cr} is 67.5 nm as calculated for flat soft Co dots of 5 nm thickness [19,20]. This is about twice the vortex diameter (compare Figure 1b).

Figure 3c summarizes the calculations in a ground state diagram as a function of thickness and diameter for circular flat Co dots [20]. Since the demagnetization field energy of the SD state scales with the dot thickness (increase of surface charges on the side faces), the critical dot diameter D_{cr} of the vortex to in-plane SD transition decreases with increasing thickness. For comparison, Figure 3c includes the boundary (dashed line) deduced in ref. [23] for NiFeMo circular dots. Due to the lower value of the saturation magnetization M_s of NiFeMo, the SD state has a lower energy than for Co dots for the same thickness and diameter, shifting the critical boundary of the vortex to SD transition towards larger values of t and D.

4.1 INFLUENCE OF HIGH ASPECT RATIO AND OF SHAPE

So far, the state diagram of Figure 3c considered only flat soft circular dots, where flat refers to a low aspect ratio R < 1 (R = thickness t / diameter D) and soft to negligible magnetic anisotropy. The boundaries between different states can change and further boundaries may occur when considering high aspect ratio particles (t / D > 1), non-circular elements or by including magnetic anisotropy energy (uniaxial, cubic...).

High aspect ratio: Demagnetization field considerations indicate that when the aspect ratio is larger than one (t > D), the in-plane SD state transforms into an out of plane SD state. The calculated boundary of this transition for Co is indicated in the state diagram of Fig 3c by the dotted line. This transition has been deduced in ref. [30] using micromagnetic simulations, giving a critical transition ratio of $(t / D)_{cr} = 0.9$ (for K = 0) in agreement with direct calculations for homogeneously magnetized cylinders [33]. It is furthermore shown in refs. [30,34], that also the out of plane SD state for elongated pillars (t / D > 0.9) transforms into a vortex state above some critical diameter. For very large aspect ratios, this vortex however is confined to the two extremities with the central part of the pillar remaining in the quasi-SD state.

Rings: One way to almost completely suppress the transition to the SD state in flat circular elements, is to remove the central part of the dot, producing a ring structure [35-37] as shown in Figure 4. For the vortex state, it is this central part which contains most of the energy. The costly vortex energy is removed, thus stabilizing the flux closure state to much lower diameter values, see Figure 3b [35,38]. It should be noted though, that this can be correct only as long as the inner diameter is larger than the vortex diameter. Otherwise, a transition to the SD state can still occur [35].

Squares: A phase diagram similar to Figure 3c has been deduced in ref. [29] for flat square elements. In this case, four different configurations have been considered: the flux closure, buckling, leaf and flower state. Of particular mention are the latter two which correspond two quasi-SD domain configurations [39], with the average magnetization aligned either along the border (flower) or along the diagonal (leaf), see Figure 5.

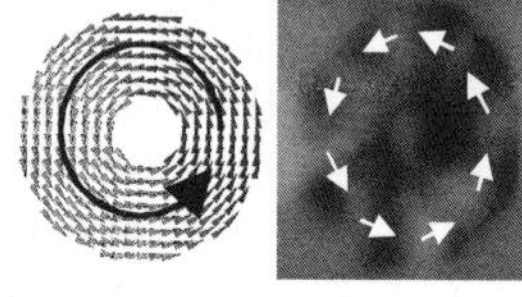

Figure 4. Flux closure configuration in a ring Left: calculation (t = 5 nm D = 200 nm); right: MFM imaging (t = 20 nm, D= 800 nm).

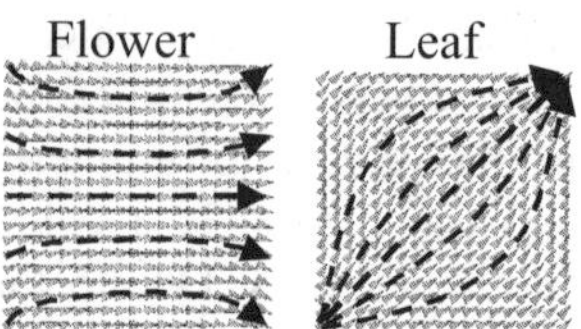

Figure 5. The two degenerate quasi-single domain configurations for square elements.

For a square that is magnetized homogeneously along the border or the diagonal the (dipolar) energy is the same. However, the demagnetization fields inside the square are not homogeneous and thus induce an inhomogeneous magnetization configuration in zero-field, giving rise to additional exchange energy contributions in competition with the dipolar energy. As a result the energies of the flower and leaf configurations are not degenerate. This dependence of the system energy on the magnetization pattern is called configurational anisotropy [39]. It should not be confused with the shape anisotropy (which is zero in the case of squares). This configurational anisotropy is present for all non-circular shaped elements such as squares, triangles, pentagons etc [40].

5. Magnetization Reversal of the Vortex State under in-plane fields

The macroscopic magnetization reversal under in-plane fields has been studied systematically as a function of diameter and thickness in ref. [23] for circular flat NiFeMo dots. A state diagram similar to Figure 3c has been derived, where the SD domain state was deduced from magneto-optic Kerr effect (MOKE) hysteresis loops, with an almost square like shape, similar to the one shown in Figure 8 further below. The corresponding reversal is interpreted as a coherent rotation of the total average moment into the reversed field direction. Similarly, the formation of the vortex state has been deduced from a very characteristic shape of the hysteresis loop, see Figure 6d, which is discussed in the following.

Hysteresis Loop: The details of the reversal process corresponding to the loop shown in Figure 6d have been identified using Lorentz microscopy [41] (Figure 6a), MFM imaging [18,31,42] (Figure 6c) as well as by micromagnetic simulations [23,43-45] (Figure 6b). At saturation, the system is first in a single domain state, with the magnetization pointing into the saturation field direction (point (i) in Figure 6). In the Lorentz image there is no contrast visible, while the MFM shows a strong dipolar contrast. Upon lowering the field, there is a sudden drop in the magnetization in the loop at a critical field called the vortex nucleation field H_n (point (ii) in Figure 6), where a vortex is nucleated at a dot border. As can be seen from the Lorentz microscopy images, (point (ii), Figure 6a), the vortex then moves from the dot border towards the dot center in a direction perpendicular to the applied field direction. In the corresponding MFM images, the dipolar contrast is pushed towards one border (lower border, point (ii), Figure 6) which indicates the location of the vortex core. In zero field the vortex is stabilized in the center (Figure 6, point (iii)) and the dipolar contrast in the MFM vanishes, leaving a localized contrast in the center indicative of the vortex core position. Upon field reversal, the vortex moves to the opposite dot border (Figure 6 points (iv) and (v)) and is expelled at a second critical field, called the vortex annihilation field H_a. The dipolar contrast in the MFM is then reversed.

Initial Susceptibility: The vortex motion perpendicular to the applied field occurs because the region of the flux closure structure which is aligned parallel to the applied field will increase in size, while the region opposite to the field decreases. In absence of defects, the vortex should move freely and thus reversibly inside the dot in response to the applied field [46]. This reversible motion can be described by a low field susceptibility [45] χ_v. It is given by the balance between the applied field, pushing the vortex towards the dot border, and the dipolar field, driving the vortex back to the dot center. The dipolar field strength increases with the displacement of the vortex from the dot center due to the appearance of magnetic charges on the side faces, see MFM contrast of points (ii, iv, v) in Figure 6c.

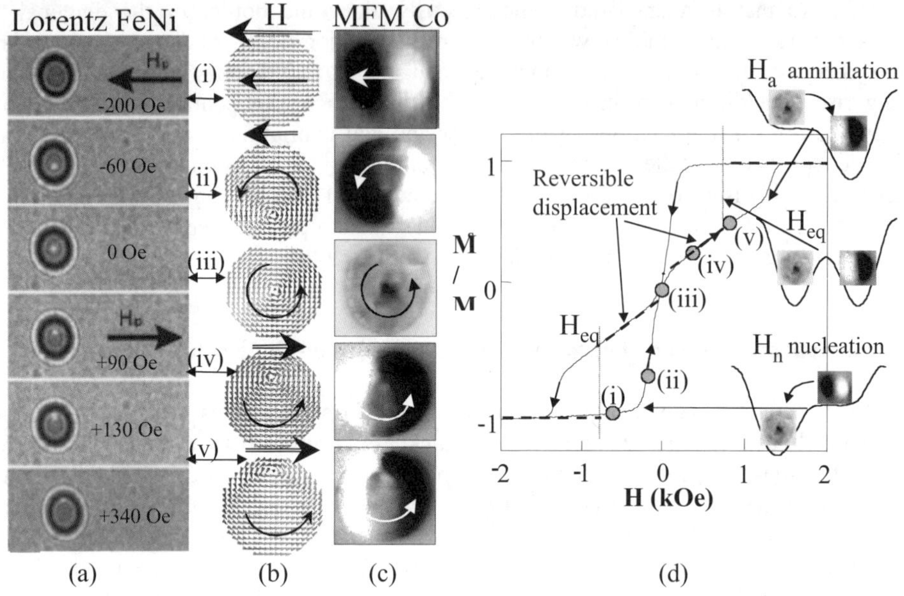

Figure 6. Vortex reversal as a function of an in-plane field, investigated by (a) Lorentz microscopy for a FeNi disc t = 15 nm D = 340 nm, from ref. [41] (b) micromagnetic simulation [18] (c) MFM imaging [18] for a Co dot t = 30 nm D = 500 nm (d) MOKE magnetometry for dots in (c) [18,48]. The thin vertical dashed lines indicate the equilibrium field H_{eq}. The thick dotted lines indicate, that at these field values the system is in its lowest energy state, while along the rest of the loop it is in a metastable minimum [23,44,45].

Nucleation: At saturation the dipolar field of the SD state is strongest, opposing the applied field. It thus acts as the driving field for the nucleation of the vortex. The nucleation itself is a more complex mechanism and can depend on the surface defects of the dot border. It involves the rotation of the vortex spins out of the film plane, thus creating local dipolar fields. There is hence some energy barrier to overcome for the vortex nucleation. Crossing this barrier is facilitated by an initial deformation of the SD state, where different inhomogenous structures might occur just below saturation [45,47]. For comparison, in Fig, 6d, the equilibrium field H_{eq} at which the single domain state energy is equal to the vortex state [23,44,45] energy is indicated schematically by the vertical thin dashed line. Due to the barrier, the field 'undershoots' the equilibrium field when coming from saturation, before the transition to the vortex state occurs.

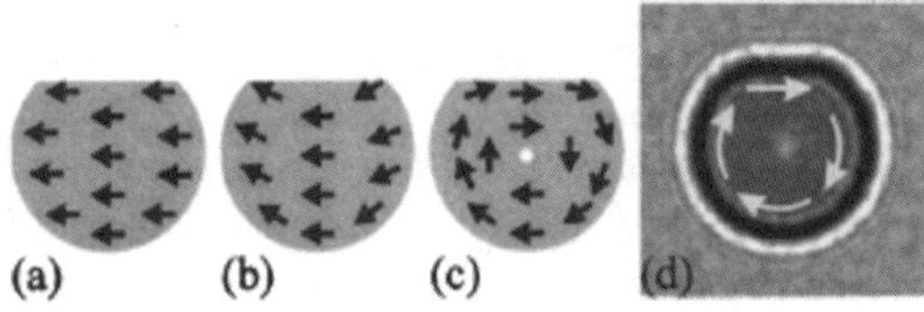

Figure 7. Schematic and Lorentz microscopy image for the vortex nucleation in dots with one flat border, from ref. [49].

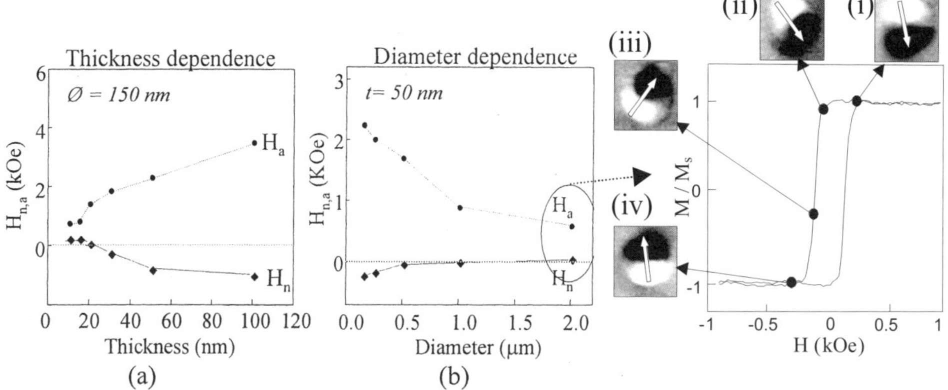

Figure 8. Dependence of the nucleation H_n and annihilation H_a for Co dots field as a function of (a) thickness with D = 150 nm, (b) diameter with t = 50 nm Co dots [48]. (c) MOKE hysteresis loop and corresponding MFM images at selected field values for a t = 10 nm D = 500 nm Co dot array [18,48].

Due to symmetry, two vortices can nucleate at opposite dot borders. However exchange will not allow to develop two stable vortices for the diameter range of interest. Defects and inhomogeneities will decide which of the two vortices will form. Hence the nucleation is rather arbitrary. Introducing however an asymmetry in the dot, see Figure 7, such as done in ref. [49] with one flattened border, it could be shown that the vortex nucleates always at the flat border. Since the magnetization tries to follow the dot border, then by decreasing the applied field coming from saturation, the magnetization first bends along the curved border, thereby determining the sense of rotation of the vortex and its nucleation at the flat border. This provides a very simple way for controlling the rotation sense of the vortex by simply changing the saturation field direction.

Annihilation: Similar to nucleation, the vortex is expelled at the annihilation field once the external field overcomes the dipolar field. It is noted that the annihilation field is always larger than the nucleation field [44]. Furthermore, the equilibrium field H_{eq} lies between H_a and H_n : $H_n < H_{eq} < H_a$ [see refs. 44,45]. Upon, increasing the field from the vortex state, the field 'overshoots' the equilibrium field before the transition to the SD state occurs.

Dependence of the nucleation and annihilation field on dot dimensions: As described in the paragraphs above, the demagnetization field of the saturated SD state or the displaced vortex state is (i) the stabilizing force for small vortex displacements from the center, (ii) the driving mechanism for the vortex nucleation, and (iii) the opposing field for the vortex annihilation. The demagnetization field in turn increases with increasing dot thickness and decreasing dot diameter, scaling roughly with the aspect ratio t/D. Hence, there is a strong dependence of the low-field susceptibility and the two critical fields H_a and H_n on dot dimensions. This dependence is confirmed by a large variety of experiments, on circular [23,43,47,48] and elliptical [31,50,51] dots and has furthermore been derived more quantitatively by calculation in ref. [45]. As an example, Figure 8 shows for Co dots [48] that, the absolute value of H_a and H_n increases with decreasing diameter and increasing thickness. Similarly, the low-field susceptibility increases with decreasing t/D [45].

For decreasing t/D and hence for decreasing demagnetization fields, it may occur, that the vortex nucleation field reverses sign [18,48].

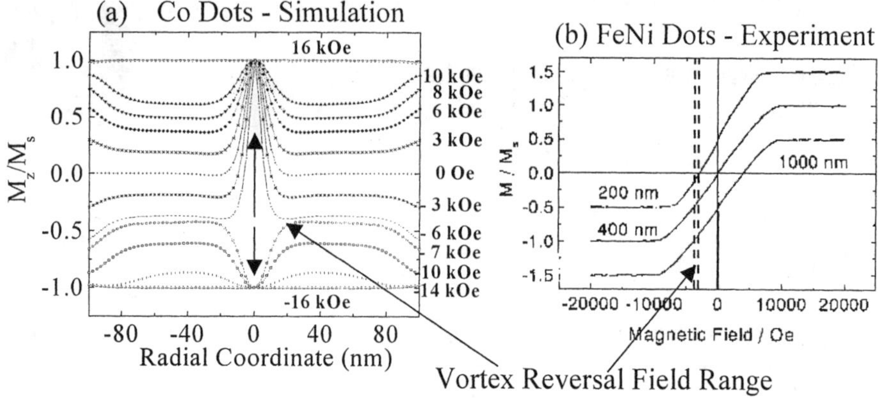

Figure 9. (a) 1D line scans across a dot of the out of plane magnetization component as calculated for K = 0 Co dots (D = 200 nm, t = 15 nm) as a function of a perpendicular field. (b) Out of plane hysteresis loops of FeNi arrays (t = 50 nm, D as indicated) from ref. [52]. The loops for different D have been offset vertically in ref. [52] for clarity. The vertical line indicates the field range in which the vortex reversal has been observed by MFM.

In this case, the SD state persists at remanence (even though the vortex state is the ground state) and the reversal may take another route. The vortex nucleation barrier can then be bypassed by a coherent rotation-like reversal process, yielding a square hysteresis loop such as shown in Figure 8c.

6. Magnetization Reversal of the Vortex State under out of plane fields

The Lorentz microscopy images in Figure 7 for dots with one flat border, indicate that the magnetization circulation sense can be controlled by an external in-plane field. However, the vortex core orientation may still be either up or down. This orientation can be furthermore controlled by simultaneously applying an out of plane field as has been studied in ref. [52] for the vortex reversal of FeNi dots as a function of the applied field angle.

Perpendicular reversal Theoretical investigations of the vortex state reversal in a field applied perpendicular to the dot plane (± 5°) have been carried out in ref. [38], for the case of Co dots. As seen in Figure 9a, from the 1D line scans for the out of plane magnetization component, the vortex is nucleated at some critical field (above 10 kOe) in the dot center and develops gradually to its zero field configuration. Upon reversing the field, the vortex core remains in the dot center, but its diameter is compressed, up to a critical value [38] (3 to 4 times λ_{ex}), where the vortex core reverses orientation. The exact mechanism how this reversal takes place is beyond the resolution of the calculation performed and needs further investigation. To be noted however, is that this critical reversal field is rather larger (6-7 kOe for Co [38] and 3-4 kOe for FeNi [52]), but it is still smaller than the saturation field. This latter point, derived from modeling, is essentially confirmed by the combined MFM and SQUID studies in ref. [52], see Figure 9b.

Field at an angle: Further angular dependent studies of ref. [52] reveal, that the reversal process is of the form shown in Figure 9, if the field angle Θ_H with respect to the dot normal is smaller than some critical angle (15° for t = 50 nm and D = 400 nm FeNi dots). Above this critical angle, the hysteresis loops are of the in-plane type, such as shown in Figure 6.

The difference being that the critical fields for the vortex nucleation and annihilation are larger and follow a $H_{a,n}(\Theta_H) = H_{a,n}(90^o)/\cos(\Theta_H)$ dependence. This means that when the in-plane component of the field applied at Θ_H, is equal to the in-plane annihilation or nucleation field $H_{a,n}(90^o)$, the vortex is expelled from the dot and reenters the dot, with the opposite orientation upon reducing the field. Below the critical angle, the annihilation field $H_a(\Theta_H)$ is larger than the perpendicular reversal field, and thus the process for reversal does take place by a direct vortex reversal, corresponding to loops of Figure 9b and not by vortex expulsion.

7. Dipolar Interactions

In the above discussions of sections (3) to (6), only isolated dots have been considered. In the experiment this is realized by large inter-dot spacing S, with the spacing at least one dot diameter. Upon reducing S, magneto-static interactions will come into play and result in a number of effects such as magnetic ordering of individual elements, alteration of the switching fields and their distribution, collective switching or induced anisotropies. Some general aspects of magnetostatic interactions are the following:

1) Magnetic dipole fields are proportional to (i) the amount of magnetic charges distributed inside the sample or on its surface and (ii) the magnetic moment of the dots. They decay as $1/r^3$ with distance. Therefore dipolar interactions will be enhanced in materials with high saturation magnetization density or for decreasing dot separations.

2) The charge distribution inside an element and the resulting interaction field (strength) is determined by its magnetization configuration. For the circular elements one should therefore distinguish the interactions occurring in arrays of in-plane SD dots, of out of plane SD dots and of dots in the vortex state. For the latter, the only magnetic charges are those associated with the perpendicular magnetization of the vortex core. Since this is confined to a small region, see Figure 1b, this interaction is negligible.

3) Besides the charge distribution inside a single element, the geometric arrangement of the dots inside an array and their coordination number are important. For instance, for square or honeycomb arrays of in-plane magnetized SD dots, the dipole interaction field is predicted to be angularly isotropic in the plane [53] while for in-plane magnetized SD dot chains a uniaxial anisotropy with easy axis along the chain is expected [45,54,55].

4) In many cases the dipolar interaction can be approximated by replacing the charge distribution by a single point dipole located inside the center of the element. Although useful for a qualitative understanding of a number of effects (see 4.2.1 - 4..3.2), care must be taken when quantitative evaluations are needed [56]. In some cases this point dipole interaction picture can even be wholly inadequate to describe even qualitatively the observed phenomena (see 4.3.3). Multipolar terms [57,58] or micromagnetic simulations must then be considered [59].

7.1 MAGNETIC ORDERING

Interactions between magnetic elements produce a tendency of the elements to establish a magnetically ordered state. This is best illustrated for a chain of in-plane SD dots. In ref [54] a qualitative phase diagram has been described for such SD dots in the presence of a uniaxial anisotropy and an applied field, both oriented in-plane and perpendicular to the chain axis.

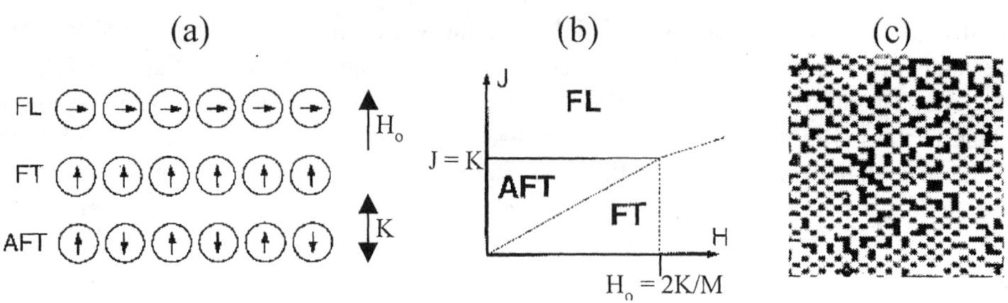

Figure 10. a) Schematic illustration of different ordered states in chains of magnetic SD dots and (b) a corresponding phase diagram, in terms of the coupling strength J and the field H_o applied perpendicular to the chain, from ref. [54], (c) Antiferromagnetic order in an array Pt/Co dot array with perpendicular anisotropy, from ref. [60]

As shown in Figure 10a, three different configurations can develop depending on the strength of the dipolar coupling J, the perpendicular uniaxial anisotropy K and the strength of the applied field H_o: 1) dots are magnetized along the direction of the chain, corresponding to a ferromagnetic longitudinal order (FL), 2) the dots are magnetized perpendicular to the chain, corresponding to a ferromagnetic transverse order (FT) and 3) the dots are magnetized alternatively in opposite transverse directions, corresponding to an antiferromagnetic transverse order (AFT).

In the FL state the dipole interaction field of each dot is aligned with the magnetization of neighboring dots, thereby stabilizing the dipoles in the direction parallel to the chain axis. This configuration is the ground state configuration at zero field for weak or negligible perpendicular anisotropy ($K < J$), see Figure 10b. In the FT state, the dipole interaction field is opposite to the dots magnetic moment and tends to destabilize the orientation of the dipole moments. In consequence the AFT state is the zero-field ground state when the perpendicular anisotropy field is stronger than the dipolar interaction field ($K > J$). The FT state is induced in both cases by applying a sufficiently large perpendicular field.

Similarly, for 2D arrays, the transition of an out of plane SD state to an in-plane SD state as a function of aspect ratio, anisotropy and dipolar interaction has been treated theoretically in ref. [33]. Experimental observations of an antiferromagnetic ordered state of several systems has been reported in refs. [60-62]. Of particular mention are the 2D arrays of flat square ultrathin Co/Pt dots [60] with perpendicular out of plane anisotropy. A "checkerboard" configuration is obtained after demagnetization as shown in Figure 10c. Ordered regions are delimited by disordered boundaries, similar to antiphase boundaries in crystals, that result from frustration effects when two incompatible ordered domains grow and meet.

7.2 SWITCHING FIELD AND COLLECTIVE SWITCHING: SINGLE DOMAIN DOTS

An important consequences of dipolar interactions in magnetic dot arrays, is the alteration of switching fields, involving both their average value and their distribution.

Considering first the reversal of in-plane SD dots arranged in a FL ordered chain, with fields applied opposite to the FL moment (along the chain axis). In this case, the dipolar interaction fields counteract the reverse applied field, trying to keep the FL order. Hence, the (reversed) switching field is shifted towards higher values as compared to isolated dots.

In the simplest case this shift is proportional to the dipole interaction field [59,63-65].

Once the first dot reverses, its magnetic moment is aligned head to head with respect the nearest neighbor magnetic moments, effectively reducing the interaction field on the neighboring dots. Thus the neighboring dots tend to reverse as well, leading to a switching cascade. The field range (or distribution), necessary to reverse the whole chain, is then reduced by the collective switching. This behavior has been experimentally observed in interacting dot arrays by different authors [55,64-67].

For the case of FT single domain dots in fields applied perpendicular to the chain axis, the dipole interaction fields destabilize the FT order, supporting the reversed field. Hence they should accelerate the onset of reversal with a decrease of the switching field as compared to isolated dots. Once in the intermediate AFT or FL state however, the remaining part of the magnetization reversal is blocked, because the interactions of the AFT or FL state stabilize the configuration. Consequently the switching field distribution is broadened. Such behavior has been observed in a number of systems [67,68], including the Co/Pt array shown in Figure 10c.

7.3 CRITICAL FIELDS OF INTERACTING VORTEX DOTS

For circular flat dots with a vortex ground state, the magnetic flux is closed and the magnetic surface charges from the small vortex core region are negligible. However, as has been shown by the MFM contrast in Figure 6c, when an in-plane field is applied, the vortex is displaced from the dot center towards the border, resulting in an uncompensated net magnetization and the appearance of strong magnetic charges.

For decreasing separation S between dots, the arising dipolar interactions of the displaced vortex state will then alter the vortex nucleation and annihilation field values, similarly to what has been described above for the SD FL chain. Upon coming from saturation, the interaction field of the SD state stabilizes the SD configuration and opposes the nucleation of the vortex, while close to saturation the interaction fields support the expulsion (annihilation) of the vortex. In consequence, the nucleation field H_n and the annihilation field H_a are reduced in absolute value, as shown in Figure 11a, for Co dot arrays.

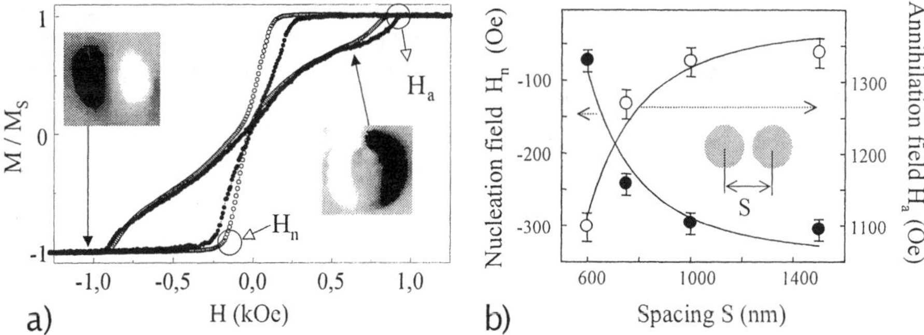

Figure 11. a) MOKE hysteresis loops for in-plane magnetized circular Co dots (D = 550nm, t = 30nm) for two different dot spacings S = 750nm (open circles) and S = 1000 nm (filled circles). The insets show the MFM contrast of a displaced vortex state in an applied field and the SD state at saturation. b) variation of the vortex annihilation and nucleation field with dot spacing S.

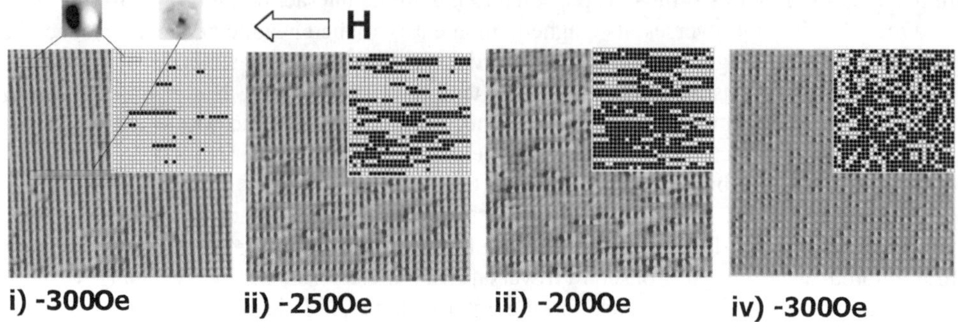

Figure 12. i-iii) MFM images (scan size 30 × 30 μm²) are shown for an array of interacting circular Co dots (D = 550nm, S = 600nm, t = 30nm) for different applied fields obtained after in-plane saturation in a field of −1.5kOe. iv) MFM image of an array of non interacting dots (spacing 1000nm)

Figure11b summarizes the variation of H_a and H_n with dot spacing S for a square lattice of flat circular Co dots of diameter 550nm and thickness 30nm. The data points closely match the theoretical expression [69] $H_{n/a}(S) = H_{n/a}(\infty) \pm H_{int}$ with H_{int}=4.2 M_S*V/S^3 for the dipole interaction of SD dots in the point dipole approximation. Here the plus sign is for vortex nucleation, the minus sign is for vortex annihilation and V is the dot volume.

7.4 COLLECTIVE SWITCHING OF VORTEX DOTS

A further manifestation of the dipolar interaction are switching cascades, as already mentioned in 7.2 for the in-plane SD elements. They have also been observed experimentally for in-plane reversal in arrays of circular flat dots with a vortex ground state in ref. [64] with a theoretical treatment given in refs. [45,64]. When the first dot in a saturated array nucleates a vortex, the interaction field emanating from the vortex dot is reduced. Consequently the switching of the nearest neighbor dots lying along the field direction is favored compared to other dots that are still surrounded by saturated SD dots. These nearest neighbor dots will then next nucleate a vortex and so on, creating a switching cascade along the field direction. Figure 12 shows MFM images in decreasing in-plane field of an array of flat Co dots, after saturation in a negative field of -1.5 kOe. The saturated state, with all dots in the SD state (strong contrast) persists in applied fields down to –325 Oe. By further decreasing the field (-300 Oe), vortices start to nucleate (weak contrast) along chains parallel to the applied field direction, (Figure12 (i-iii)). The insets show graphical elaborations of the MFM images to reveal more clearly the vortex (black) and the SD states (white). To confirm the chain formation, due to the interaction among dots, Figure12(iv) shows the case of an array of widely spaced, non interacting Co dots where vortex states nucleate randomly.

7.5 BEYOND THE POINT-DIPOLE INTERACTION MODEL

While the point dipole approximation can explain qualitatively a number of effects, one has to keep in mind that nanostructured magnetic dots have a finite size. For example, in the SD state, the charges are effectively distributed along the dot border, rather than in the center.

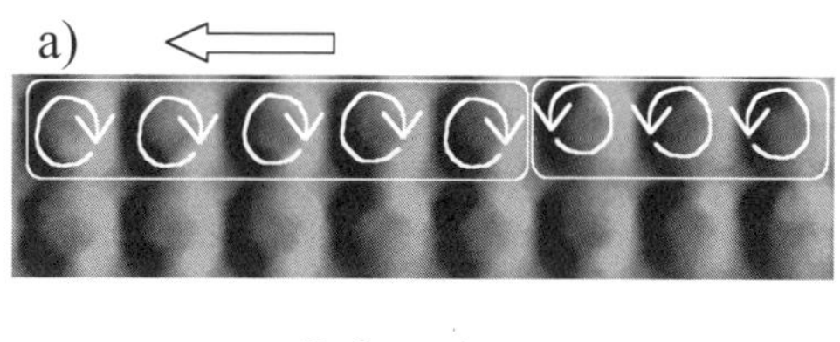

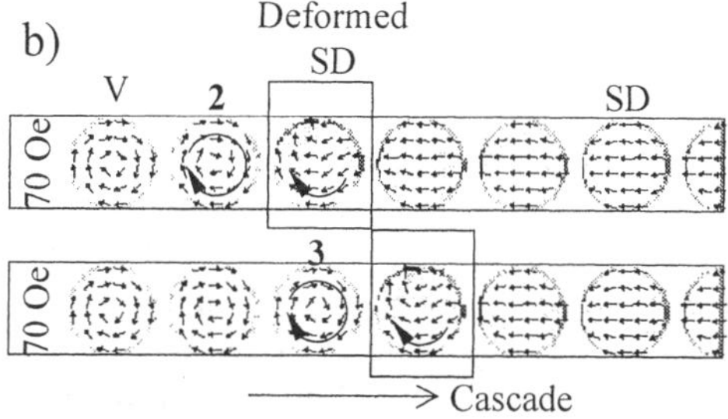

Figure13. a) Zoom onto two vortex chain segments from Figure 11b for the displaced vortex states at H = 250 Oe b) Micromagnetic simulations for a chain of Co dots (D = 170nm, S = 200nm, t = 30nm) showing the development of a correlated vortex state chirality during a nucleation cascade.

This has two consequences: (i) the dipolar field created by a dot is quite different from the one created by a point-dipole, especially at short distances [70,71]. (ii) As already indicated in Figure 2b, in many cases the single-domain dots are not in a perfectly uniformly magnetized state due to inhomogenous self demagnetizing fields. To this adds the non-homogenous interaction field, which further perturbs the magnetization configuration.

In the case of the interacting Co dot arrays of Figure 12, these non-homogeneous interaction fields are supposed to be at the origin of a correlation of the vortex circulation sense (chirality). Such a correlation along chains of vortices has been observed by MFM, [64] after in-plane saturation, as shown in Figure13a. An explanation is proposed using micromagnetic modeling [64], for which a chain of seven circular Co dots is considered (Figure13b). The chain is first saturated in a negative field along the chain axis and then a reversed field is applied. At a reverse field of +30 Oe a first vortex nucleates at the left chain end (the right chain end is fixed as SD). By increasing the field to +70 Oe a nucleation cascade is triggered and vortices nucleate successively in dot 2, dot 3, etc. , see Figure 13b, until all dots are in the vortex state at equilibrium. All vortices have the same chirality, imposed by the first vortex. The origin of this correlation can be understood in terms of magnetostatic interactions as follows: The vortex formation is preceded by a deformation of the SD domain state in the form of a buckling or c-shaped magnetization path [45,47], attempting a flux closure path. The surface charges of the deformed SD state are then concentrated towards the region where the vortex nucleates. The fringing field of these charges induce a perturbation in the neighboring dot such that the resulting deformation of the neighbor dot SD state follows the same circulation sense, which in turn defines the location of vortex nucleation. Simulations on chains with different dot diameters have shown that a correlation of vortex chiralities occurs also for dots with larger (400 nm) and smaller (60 nm) diameters. While the origin of the correlation is related to magnetostatic coupling between the dots it cannot be understood in terms of interaction of simple point dipoles.

8. Summary

In this chapter the static properties of circular flat elements have been summarized. Isolated circular elements are free of shape anisotropy and configurational anisotropy and are therefore a suitable model system to investigate the size dependence on the magnetic ground state configurations as well as the dipolar interactions among dots in closely packed arrays. The size (thickness and diameter) scales the demagnetization field energy, dominating the SD state, which is in competition with the exchange energy, dominating the vortex state. This defines the ground state diagram derived in Figure 3 as well as the dependence of the vortex nucleation and annihilation field for in-plane field reversal.

Dipolar interactions are expected for the SD domain state, while for the vortex state, with its flux closure structure, these interaction are negligible. This is correct only in zero field. Under in-plane field, the vortex displaces and important dipolar interaction fields arise, responsible for the occurrence of switching cascades and the correlation of the vortex circulation sense along chains in square arrays. Further manifestations are the occurrence of higher order anisotropy terms, not discussed here.

While circular dots provide models to study basic concepts of the energy balance and the magnetization reversal, the direct application in spin electronic devices might not be as obvious. Irreversible vortex pinning in stacked layer systems [35], complicated addressing schemes for the dynamics of ring structures [35] and finally the strong dipolar interactions are less suitable for such applications. However, as shown in ref. [55], the occurrence of switching cascades induced by dipolar interactions might be used to transport information over long distances and to create logic cell systems.

9. Acknowledgement

The results on Co dot arrays reported here have been obtained in collaboration between the group of IPCM Strasbourg (now at SPINTEC Grenoble) and the group of L2M Bagneux (now at LPN, Marcoussis). The authors would like to acknowledge both groups, in particular Yong Chen, Amira Lebib, Shunpu Li and Kamel Ounadjela whose contributions have been substantial for the completion of this work. This work has been funded in parts by the European Programs SubMagDev (FMRX-CT97-0147), NanoScaleParticles (HPRN-CT-1999-00150) and Dynaspin (FMRX-CT97-0124).

10. References

1. Heinrich, B. and Bland, J. A. C. (eds.) (1994) *Ultrathin Magnetic Structures,* Springer, Berlin
2. Ziese, M. and Thornton, M. J. (eds.) (2001) *Spin Electronics*, Springer, Berlin
3. Hubert, A and Schäfer, R. (1998) *Magnetic Domains*, Springer, Berlin
4. Wolf, S. A., Awschalom, D. D., Buhrman, R. A., Daughton, J. M., von Molnar, S., Roukes, M. L., Chtchelkanova, A. Y. and Treger, D. M. (2001) Spintronics: a spin-based electronics vision for the future, *Science* **294**, 1488-1495
5. Prinz, G. (1998) Magnetoelectronics *Science* **282**, 1660-1663
6. Dieny, B. Speriosu, V. S., Metin, S., Parkin, S. S. P., Gurney, B. A., Baumgart, P., and Wilhoit, D. R. (1991) Magnetotransport properties of magnetically soft spin-valve structures, *J.Appl. Phys.* **69**, 4774-4779
7. Tehrani, S. Chen, E., Durlam, M., DeHerrera, M., Slaughter, J. M., Shi, J. and Kerszykowski, G (1999) High-density submicron magnetoresistive random access memory, *J. Appl. Phys.* **85**, 5822-5827
8. Moodera J. S., Kinder, L. R., Wong, T. M. and Meservey, R. (1995) Large magnetoresistance at room temperature in ferromagnetic thin film tunnel junctions, *Phys. Rev. Lett.* **74**, 3273-3276

9. Parkin, S. S. P., Roche, K. P., Samant, M. G., Rice, P. M., Beyers, R. B., Scheuerlein, R. E., O'Sullivan, E. J., Brown, S. L., Bucchigano, J., Abraham, D. W., Lu, Y., Rooks, M., Trouillard, P. L., Wanner, R. A. and Gallagher, W. J. (1999) Exchange-biased magnetic tunnel junctions and application to nonvolatile magnetic random access memory, *J. Appl. Phys.* **85**, 5828-5833.
10. for nanofabrication methods see for instance: Chen, Y., Natali, M., Li, S. and Lebib, A. (September 2002) Application of nanoimprint lithography in magnetism, in Sotomayor, C. M. (ed.), *Alternative Lithography*, Kluwer Academic Publishers, Dordrecht.
11. Brown, J. F. Jr. (1963) *Micromagnetics*, J. Wiley and Sons, New York.
12. Miltat, J. (1994) Domains and domain walls in soft magnetic materials, in R. Gerber, C. D. Wright, G. Asti (eds), *Applied Magnetism*, NATO ASI Series, Kluwer Academic Publishers, Dordrecht, pp. 221.
13. Tonomura, A. (1987) Applications of electron holography, *Rev. Mod. Phys.* **59**, 639-669.
14. Raabe, J., Pulwey, R., Sattler, R., Schweinbock, T., Zweck, J., Weiss, D., (2000) Magnetization pattern of ferromagnetic nano-disks, *J. Appl. Phys.* **88**, 4437-4439.
15. Shinjo, T., Okuno, T., Hassdorf, R., Shigento, K., Ono, T. (2000) Magnetic vortex core observation in circular dots of permalloy, *Science* **289**, 930-932.
16. Miramond, C., Fermon, C., Rousseaux, F., Decanini, D., and Carcenac, F., (1997) Permalloy cylindrical submicron size dot arrays, *J. Magn. Magn. Mat.* **165**, 500-503.
17. Demand, M. Hehn, M., Ounadjela, K., Stamps, R. L., Cambril, E., Cornette, A. and Rousseaux, F. (2000) Magnetic domain structures in arrays of submicron Co dots studied with magnetic force microscopy, *J. Appl. Phys.* **87**, 5111-5113.
18. Prejbeanu, I. L., Natali, M., Buda, L. D., Ebels, U., Lebib, A., Chen, Y. and Ounadjela, K. (2002) In-plane reversal mechanisms in circular Co dots, *J. Appl. Phys.* **91**, 7343-7346.
19. Buda, L. D., Prejbeanu, I. L., Demand, M., Ebels, U. and Ounadjela, K. (2001) Vortex states stability in circular Co(0001) dots, *IEEE Trans. Magn.* **37**, 2061-2063.
20. Buda, L. D., Prejbeanu, I. L., Ebels, U. and Ounadjela, K. (2002) Micromagnetic simulations of magnetisation in circular cobalt dots, *Comp. Mat. Science* **24**, 181-185.
21. Most simulation results presented in this chapter have been obtained using the micromagnetics calculations as described in refs. [19, 20]. Some calculations have been performed using the publicly available code OOMMF ref. [22].
22. http://math.nist.gov/oommf/
23. Cowburn, R. P, Koltsov, D. K., Adeyeye, A. O., Welland, M. E.and Tricker, D. M. (1999) Single-domain circular nanomagnets, *Phys. Rev. Lett.* **83**, 1042-1045.
24. Metlov, K. L.and Guslienko, K. Y. (2001) arXiv:cond-mat/0110037.
25. Usov, A. and Peschany, S. E. (1994) Fiz. Met. Metalloved. **12**, 13 and (1993) Magnetization curling in a fine cylindrical particle *J. Magn. Magn. Mat.* **118**, L290-L294.
26. Hanson, M., Johansson, C., Nielsson, B. and Svedberg, E. B., (2001) Magnetic properties of epitaxial Ni (0 0 1) films and sub-micron particles, *J. Magn. Magn. Mat.* **236**, 139-150.
27. Hanson, M., Johansson, C., Nielsson, B., Isberg, P. and Wäppling, R. (1999) Magnetic properties of two-dimensional arrays of epitaxial Fe (001) submicron particles, *J. Appl. Phys.* **85**, 2793-2799.
28. Kazakova, O., Hanson, M., Blomquvist, P. and Wäppling, R. (2001) Arrays of epitaxial Co submicron particles. Critical size for single-domain formation and multidomain structures, *J. Appl. Phys.* **90**, 2440-2446.
29. Cowburn, R. P and Welland, M. E. (1998) Phase transitions in planar magnetic nanostructures, *Appl. Phys. Lett.* **72**, 2041-2043.
30. Ross, C. A., Hwang, M., Shima, M., Cheng, J. Y., Farhoud, M., Savas, T. A., Smith, H. I., Schwarzacher, W., Ross, F. M., Redjdal, M., Humphrey, F. B. (2002) Micromagnetic behavior of electrodeposited cylinder arrays *Phys. Rev. B* **65**, 144417-144424.
31. Fernandez, A. and Cerjan, C. J. (2000) Nucleation and annihilation of magnetic vortices in submicron-scale Co dots, *J. Appl. Phys.* **87**, 1395-1401.
32. Fernandez, A., Gibbson, M. R. , Wall, M. A. and Cerjan, C. J. (1998) Magnetic domain structure and magnetization reversal in submicron-scale Co dots, *J. Magn. Magn, Mat.* **190**, 71-80.
33. Guslienko, K. Y., Choe, S. B. and Shin, S. C. (2000) Reorientational magnetic transition in high-density arrays of single-domain dots, *Appl. Phys. Lett.* **76**, 3609-3611.
34. Usov, N. A. (1999) Magnetization curling in soft type ferromagnetic particles with large aspect ratios, *J. Magn. Magn. Mat.* **203**, 277-279.
35. Zhu, J.-G., Zheng, Y.and Prinz, G. A. (2000) Ultrahigh density vertical magnetoresistive random access memory, *J. Appl.Phys.* **87**, 6668-6673.
36. Li, S., Peyrade, D., Natali, M., Lebib, A., Chen, Y., Ebels, U., Buda, L. D. and Ounadjela, K. (2001) Flux Closure Structures in Cobalt Rings, *Phys. Rev. Lett.* **86**, 1102-1105
37. Rothman, J., Kläui, M., Lopez-Diaz, L., Vaz, C. A. F., Bleloch, A., Bland, J. A. C., Cui, Z. and Speaks, R. (2001) Observation of a bi-domain state and nucleation free switching in mesoscopic ring magnets, *Phys.*

Rev. Lett. **87**, 1098-1101.

38. Buda, L. D., Prejbeanu, I. L., Ebels, U. and Ounadjela, K. (2002) Magnetotransport measurements as a tool to probe the micromagnetic configurations in epitaxial Co wires, *J. Magn. Magn. Mat.* **240**, 27-29.
39. Cowburn, R. P.and Welland, M. E. (1998) Configurational anisotropy in nanomagnets, *Phys. Rev. Lett.* **84**, 5414-5417.
40. Cowburn, R. P. (2000) Property variation with shape in magnetic nanoelements, *J. Phys. D: Appl. Phys.* **33**, R1-R16.
41. Schneider, M., Hoffmann, H. and Zweck, J. (2000) Lorentz microscopy of circular ferromagnetic permalloy nanodisks, *Appl. Phys. Lett.* **77**, 2909-2911.
42. Pokhil, T. Song, D. A. and Nowak. J. (2000) Spin vortex states and hysteretic properties of submicron size Rife elements, *J. Appl. Phys.* **87**, 6319-6321.
43. Novosad, V., Guslienko, K. Yu., Shima, H., Otani, Y., Fukamichi, K., Kikuchi, N., Kitakami, O. and Shimida, Y. (2001) Nucleation and annihilation of magnetic vortices in sub-micron Permalloy dots, *IEEE Trans. Magn.* **37**, 2088-2090.
44. Guslienko, K. Yu. and Metlov, K. L. (2001) Evolution and stability of a magnetic vortex in a small cylindrical ferromagnetic particle under applied field, *Phys. Rev. B* **63**, 100403-100406.
45. Guslienko, K. Yu., Novosad, V., Otani, Y., Shima, H. and Fukamichi, K. (2001) Field evolution of magnetic vortex state in ferromagnetic disks, *Appl. Phys. Lett.* **78**, 3848-3850 and (2001) Magnetization reversal due to vortex nucleation, displacement, and annihilation in submicron ferromagnetic dot arrays, *Phys. Rev. B* **65**, 24414-24424.
46. Shima, H. Novosad, V., Otani, Y. Fukamichi, K., Kikuchi, N., Kitakami, O. and Shimada, Y. (2002) Pinning of magnetic vortices in microfabricated permalloy dot arrays, *J. Appl. Phys.* **92**, 1473-1476.
47. Lebib, A., Li, S. P., Natali, M. and Chen, Y. (2001) Size and thickness dependencies of magnetization reversal in Co dot arrays, *J. Appl. Phys.* **89**, 3892-3896.
48. Schneider, M., Hoffmann, H., Otto, S., Haug, Th. and Zweck, J. (2002) Stability of magnetic vortices in flat submicron permalloy cylinders, *J. Appl. Phys.* **92**, 1466-1472.
49. Schneider, M., Hoffmann, H. and Zweck, J. (2001) Magnetic switching of single vortex permalloy elements, *Appl. Phys. Lett.* **79**, 3113-3115.
50. Girgis, E., Schelten, J., Shi, J., Janesky, J., Tehrani, S. and Goronkin, H. (2000) Switching characteristics and magnetization vortices of thin-film cobalt in nanometer-scale patterned arrays, *Appl. Phys. Lett.* **76**, 3780-3782.
51. Johnson, J. A., Grimsditch, M., Metlushko, V., Vavassori, P., Illic, B., Neuzil, P. and Kumar, R. (2000) Magneto-optic Kerr effect investigation of cobalt and permalloy nanoscale dot arrays: Shape effects on magnetization reversal, *Appl. Phys. Lett.* **77**, 4410-4412.
52. Okuno, T., Shigeto, K., Ono, T., Mibu, K. and Shinjo. T (2002) MFM study of magnetic vortex cores in circular permalloy dots: behavior in external field, *J. Magn. Magn. Mat.* **240**, 1-6.
53. Prakash, S. and Henley, C. L. (1990) *Phys. Rev. B* **42**, 6574-6598.
54. Cowburn, R. P. (2002) Probing antiferromagnetic coupling between nanomagnets, *Phys. Rev. B*, **65**, 92409-92412.
55. Cowburn, R. P. and Welland, M. E. (2000) Room temperature magnetic quantum cellular automata, *Science* **287**, 1466-1468.
56. Joseph, R. I. and E. Schlömann, E. (1965) *J. Appl. Phys.* **36**, 1579.
57. Olive, E. and Molho, P. (1998) Thermodynamic study of a lattice of compass needles in dipolar interaction, *Phys. Rev. B* **58**, 9238-9247.
58. Guslienko, K. Y. (1999) Magnetostatic interdot coupling in two-dimensional magnetic dot arrays, *Appl. Phys. Lett.* **75**, 394-396.
59. Natali, M., Lebib, A., Chen, Y., Prejbeanu, I. L. and Ounadjela, K. (2002) Configurational anisotropy in square lattices of interacting cobalt dots, *J. Appl. Phys.* **91**, 7041-7043
60. Aign, T., Meyer, P., Lemerle, S., Jamet, J. P., Ferre, J., Mathet, V., Chappert, C., Gierak, J., Vieu, C., Rousseaux, F., Launois, H. and Bernas, H. (1998) Magnetization Reversal in Arrays of Perpendicularly Magnetized Ultrathin Dots Coupled by Dipolar Interaction, *Phys. Rev. Lett.* **81**, 5656-5659.
61. Hwang, M., Abraham, M. C., Savas, T. A., Smith, H. I. Ram, R. J. and Ross, C. A. (2000) Magnetic force microscopy study of interactions in 100 nm period nanomagnet arrays, *J. Appl. Phys.* **87**, 5108-5110.
62. Kirk, K. J., Chapman, J. N. and Wilkinson, C. D. (1997) Switching fields and magnetostatic interactions of thin film magnetic nanoelements, *Appl. Phys. Lett.* **71**, 539-541.
63. Grundler, D., Meier, G., Broocks, K. B., Heyn, C. and Heitmann, D. (1999) Magnetization of small arrays of interacting single-domain particles, *J. Appl. Phys.* **85**, 6175-6177.
64. Natali, M., Prejbeanu, I. L., Lebib, A., Buda, L. D., Ounadjela, K. and Chen, Y. (2002) Correlated Magnetic Vortex Chains in Mesoscopic Cobalt Dot Arrays, *Phys. Rev. Lett.* **88**, 157203-157206
65. Gibson, G. A. and Schultz, S. (1993) Magnetic force microscope study of the micromagnetics of

submicrometer magnetic particles, *J. Appl. Phys.* **73**, 4516-4521.
66. Evoy, S., Carr, D. W., Sekaric, L., Suzuki, Y., Parpia, J. M. and Craighead, H. G. (2000) Thickness dependent binary behavior of elongated single-domain cobalt nanostructures, *J. Appl. Phys.* **87**, 404-409.
67. Kirk, K. J., Chapman, J. N., McVitie, S., Aitchison, P. R. and Wilkinson, C. D. W. (2000) Interactions and switching field distributions of nanoscale magnetic elements, *J. Appl. Phys.* **87**, 5105-5107.
68. Ridley, P. H. W., Roberts, G. W., Chantrell, R. W., Kirk, K. J. and Chapman, J. N. (2000) Computational and experimental micromagnetics of arrays of 2-D platelets, *IEEE Trans. Magn.* **36**, 3161.
69. M. Grimsditch, M., Jaccard, Y. and Schuller, I. K. (1998) Magnetic anisotropies in dot arrays: Shape anisotropy versus coupling, *Phys. Rev. B* **58**, 11539-11543.
70. Metlov. K. L. (2000) Micromagnetics and interaction effects in the lattice of magnetic dots, *J. Magn. Magn. Mat.* **215-216**, 37-39.
71. Dunin-Borowski, R. E., McCartney, M. R., Kardynal, B., Smith, D. J. and Scheinfein, M. R. (1999) Switching asymmetries in closely coupled magnetic nanostructure arrays, *Appl. Phys. Lett.* **75**, 2641-26413.

SUBMICRON SIZE PARTICLES OF MAGNETIC FILMS AND MULTILAYERS

M. HANSON AND O. KAZAKOVA
Department of Solid State Physics, Chalmers University of Technology and Göteborg University
SE-412 96 Göteborg, Sweden

1. Introduction

The area between the micro- and macroscopic ranges of magnetism offers an exciting field for research and development. The design of magnets for small-scale applications requires several physical parameters to be simultaneously controlled and matched to each other. First one should control the exchange energy, the crystalline anisotropy and the atomic magnetic moment; material parameters governing e. g. the ordering temperature and magnetization of the material. This may be accomplished by applying modern preparation techniques to make thin films and layered materials, yielding a variety of intrinsic magnetic properties [1]. Second, the demagnetizing effects that are inevitably introduced when the lateral extensions of the material are limited must be incorporated in the design to yield the proper zero-field state as well as dynamic response [2]. There are many questions to answer about the zero-field state of a magnetic particle, for instance how the critical sizes for formation of a single domain (SD) can be reached. SD particles with two possible orientations of their moment in zero field - a binary bit - are suggested as building blocks of a novel magnetic memory [3]. At the same time as a stable zero-field state is obtained, magnetization reversal should occur at an appropriate field with a narrow distribution of switching fields among the particles.

With the aim to study how the magnetic properties change as the size of the magnetic system decreases, we developed and optimized the fabrication methods for different materials, allowing us to produce arrays of submicron size particles with well-defined topography. This yields excellent model systems comprising patterned structures of identical particles, in which the lateral extensions of the particles, the geometry of the arrays,

L.M. Liz-Marzán and M. Giersig (eds.),
Low-Dimensional Systems: Theory, Preparation, and Some Applications, 213–226.

the film thicknesses and the starting materials can be varied independently of each other. We performed systematic studies, of such systems of particles with lateral sizes down to 100 nm. The starting materials were epitaxial films of the classical ferromagnets Fe [4, 5], Co [6] and Ni [7], epitaxial Fe/Co multilayers [8, 9, 10] or polycrystalline Permalloy [11]. In this paper we describe the method for preparation and give account of some of the results obtained from those investigations. In particular we give examples of how the zero-field magnetic domain structure and magnetization reversal processes in the particles depend on the particle shape and size, the magnetocrystalline anisotropy and the inter-particle interactions.

2. Preparation and magnetic characterization of particle arrays made from epitaxial films

To prepare small particles with steep sidewalls and flat top surfaces from epitaxial films, we developed a method in which we apply a combination of electron beam lithography and ion beam milling, using a carbon masking-layer and a NiCr layer serving as a pattern transfer layer. Since the carbon film has a very low sputter rate it can be used to pattern hard materials with high resolution by ion beam milling, where otherwise mask erosion would be a serious problem. In the following we describe the main steps of the preparation process, also referring to Fig. 1a.

2.1. EPITAXIAL MAGNETIC FILMS

The epitaxial films used in the investigations were sputtered onto substrates of single crystals of MgO(001). The films were prepared to yield the epitaxial relationship Fe(001)||MgO(001) and Fe(110)||MgO(100) for bcc Fe [12], and the same for bcc Fe/Co multilayers [13, 14]. For fcc Ni the relationship Ni(001)||MgO(001) and Ni(100)||MgO(100) was obtained [15]. Further details of the film deposition and characterization can be found in the respective references. After deposition the magnetic film is covered with a protective layer of an inert, non-magnetic metal to prevent oxidation. Also polycrystalline Permalloy films on Si substrates were grown by sputtering.

2.2. THE LITHOGRAPHY PROCESS

In case the protective layer on top of the magnetic film is not suitable to serve as a layer for carbon adhesion, a 20 nm thick layer of Au is first deposited by electron beam evaporation onto the film. Then a 50 nm thick carbon layer is deposited by electron beam evaporation. A first layer, about 360 nm, of positive resist with low sensitivity (NANO Copolymer, 10 % in ethyl lactate) is spin coated onto the film and baked 5 min at 170 °C on

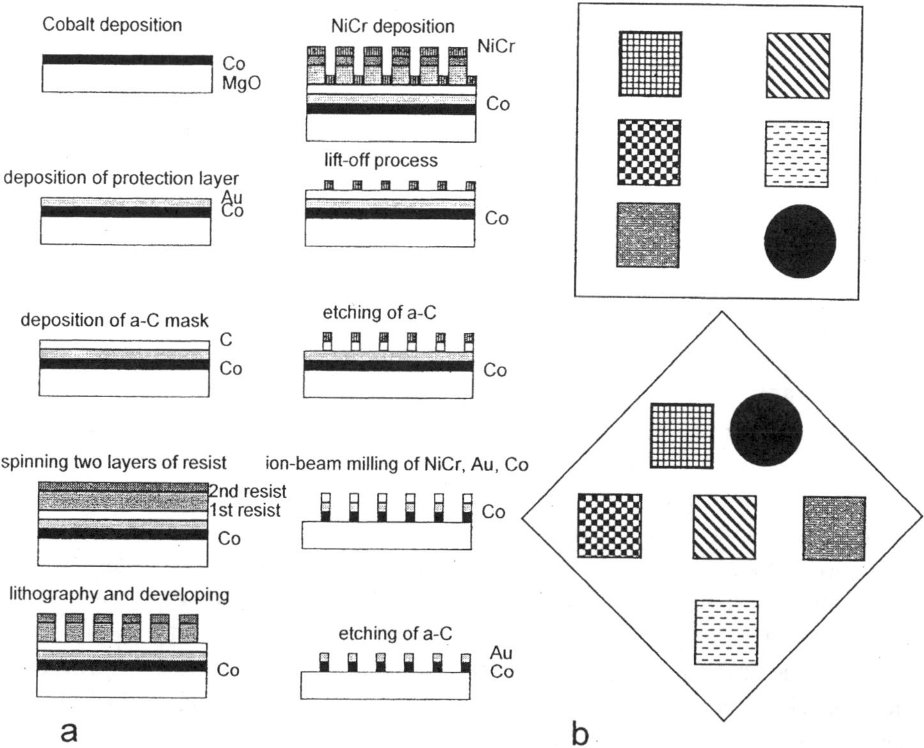

Figure 1. In a) the main steps of the fabrication technique are shown, starting in the left column and continuing in the right one, both read from up to down. b) The layout for the patterning of epitaxial films deposited on 10 mm × 10 mm single crystal MgO substrates, yielding different crystalline orientations of the particles. There are five samples with particle arrays and one reference sample of circular shape.

a hotplate, to remove the residual solvent and to settle the resist polymer. A second layer, about 70 nm, of a positive, high sensitivity resist (NANO PMMA, 2 % in Anisole) is applied and baked in the same way. The layout of the particle arrays is coded by CAD and exposed in a JEOL JBX5D electron beam lithography system, using a beam of diameter 0.04 μm at a current of 300 pA. After the e-beam exposure the resist is developed, removing the exposed resist areas from the surface. Thereby the bottom layer of the resist is removed with an undercut, whereas the top layer retains the shape and size of the written pattern. Subsequently a layer of about 30 nm NiCr is evaporated on the patterned resist. When there is a proper undercut in the resist, the NiCr is deposited into the resist holes and onto the carbon layer, without making contact with the walls of the resist, and with the top of the resist defining the particle sizes. In the next step, a lift-off process, the remaining resist is dissolved in acetone. Then there are three subsequent steps of etching. First the carbon mask is etched in oxygen plasma (Plasma Term Batchtop). Then the wafer is mounted in an ion beam etching system from Oxford Instruments. There the uncovered gold layer and the magnetic film are sputtered away by a beam of argon ions with energy 400 eV. During this process the carbon layer is left nearly intact, while the NiCr buffer layer on top of the carbon mask is removed. After the argon etching the protective carbon layer is removed in oxygen plasma. To ensure proper progress in the fabrication process the samples are checked in an optical microscope or by atomic force microscopy (AFM) between the different steps. A typical set of samples is made of a magnetic film sputtered on a MgO substrate of thickness 0.5 mm and area 10 mm by 10 mm. The film is patterned to yield one reference sample and five samples containing arrays of particles with different sizes, shapes and repetition periods, each covering an area within 3 mm by 3 mm. The particles and the lattice axes of the arrays can be oriented in the desired direction with respect to the crystal directions of the epitaxial film. Fig. 1b shows the layout for rectangular lattices oriented along, or making an angle of 45° with the MgO [100] and [010] directions. AFM and magnetic force microscopy (MFM) studies can be performed on the whole substrate, but for magnetization measurements the wafers are finally diced into rectangular samples using a diamond saw. The processes yield particles of well-defined geometry as shown in the images obtained by AFM (Fig. 2).

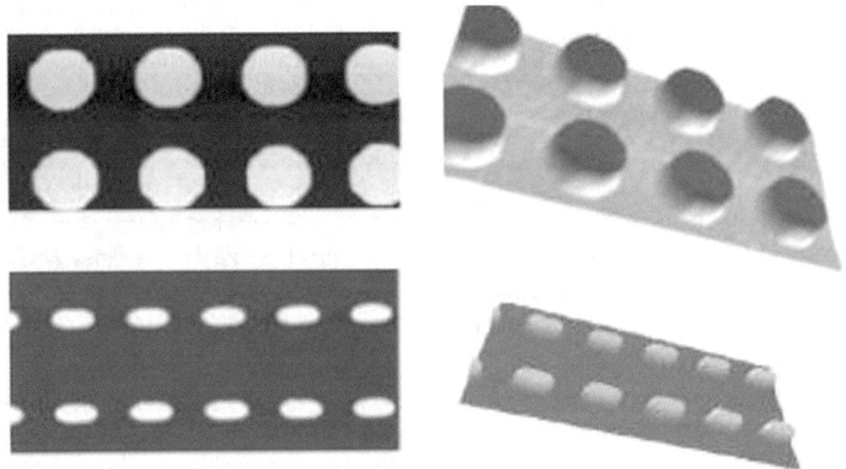

Figure 2. The figure shows the topography of 10 nm thick Fe particles of circular shape (diameter 550 nm) and 20 nm thick Co particles of elliptical shape (axes 150 nm and 450 nm). The particles are imaged by AFM, showing a top view and a 3D representation.

The above methods were successfully applied for preparing arrays of particles with circular, rectangular, elliptical and ring shape from films with thickness t in the range 10 nm $< t <$ 100 nm. The minimum distance for which a proper etching is still obtained between particles positioned in arrays is of the order 50-70 nm. Since the complete process is time consuming it is not apt for large scale production. For such applications one has to find other methods for particle preparation, e. g. nano-imprint for mass production. Such techniques can, however, not be used for patterning of epitaxial films [16].

2.3. MAGNETIC CHARACTERIZATION

Magnetic hysteresis curves were determined using an alternating gradient magnetometer (AGM) from Princeton Measurements Corporation. The samples were measured with the magnetic field, in the range -2 T $< B <$ 2 T, applied in different crystalline directions in the plane of the film or perpendicular to the film plane.

The topography and the local distribution of magnetic moments of the individual particles and arrays were studied by an AFM/MFM scanning probe technique using a Dimension^{TM} 3000 microscope from Digital Instruments (DI). Images were obtained with the instrument operating in the tapping and lift modes using the standard MFM tips delivered by DI. The MFM measurements were performed on the samples in the as-grown state or after different magnetic field histories, for instance ac demagnetization.

3. Magnetic domain structures and magnetization reversal

In order to follow the changes of magnetic properties that are due to the decreasing lateral dimensions, the starting film materials were carefully analyzed. Structural analysis was performed on the as prepared films. The thickness of the films was determined from the deposition rate, in some cases complemented by estimates from the interference fringes (Kiessig fringes) in the reflectivity scans and by Rutherford back scattering. Reference samples with the films covering rectangular (about 2 mm by 3 mm) or circular (diameter about 2 mm) areas, having undergone the same preparation process as the particles, were examined by magnetic measurements. The magnetometer (calibrated with a Ni sample) yields the total moment m of the sample, which is then corrected for the diamagnetic background of the sample holder and substrate. The magnetization M is obtained by dividing m with the volume of magnetic material in each sample. In this way it was verified that no oxidation occured during the sample preparation. The angular dependence of the in-plane magnetization of the reference samples shows the magnetocrystalline anisotropy of the films, and from the area between the hysteresis curves measured in the [110] and [100] directions [17] the first order anisotropy constant K_1 was determined. The coercivity B_c and the ratio between the remanent and saturation magnetizations M_r/M_s are indicators of a good film quality, for instance yielding low B_c and high M_r/M_s when measured in one of the easy directions of magnetization.

In the following we discuss the obtained results treating four aspects: the decrease of lateral extension, effects of varying magnetocrystalline anisotropy inter-particle interactions and ring shaped magnets as elements for weak stray fields.

3.1. EFFECTS OF DECREASING PARTICLE SIZE AND FILM THICKNESS

The magnetic measurements on Fe reference samples showed that the Fe films have the same cubic anisotropy as that of bulk Fe for thicknesses 50, 30 and 15 nm [4, 5]. The measurements on the patterned samples of these films thus were interpreted showing clear effects of the interplay between magnetocrystalline and shape anisotropies, that occur when going from a continuous film to submicron size particles. Fig. 3a shows the low field hysteresis curve of the reference sample of a 50 nm thick Fe film measured in the [110] crystalline direction, which is the intermediate magnetization direction. The ratio M_r/M_s is about 0.7 for the three films, in good agreement with the ratio $1/\sqrt{2}$ expected when the saturation magnetization lying along the easy [100] direction is projected onto the [110] direction. The MFM image of the film in Fig. 3b shows no distinct magnetic contrast, implying that any stray fields are weak.

The arrays of particles with circular shape were designed in order not to introduce any in-plane shape anisotropy. Thus the magnetocrystalline anisotropy together with effects of the finite size are expected to dominate the behaviour. Fig. 3c shows the hysteresis curve for particles with diameter $d = 550$ nm and $t = 50$ nm, made of the same film as above. The coercivity of the particles is the same as that of the film, with the remanence being practically zero. The initial slope can be accounted for by the demagnetizing effects in the individual particles. The overall shape of the magnetization curve is typical for non-interacting magnetic particles that are sufficiently large to accomodate domain walls. In zero field each particle comprises a multi-domain state with zero remanence. The reversible parts (the initial slopes) of the magnetization curves are due to magnetization rotation and domain wall movements and the irreversible parts (the small loops) to annihilation and nucleation of domain walls. Whereas these processes should occur in a sharp step in a single particle, we observe them to take place over a narrow field range. This can be explained by that there is a distribution of annihilation and nucleation fields among the 1.4×10^7 small particles comprising a sample. The MFM image in Fig. 3d shows

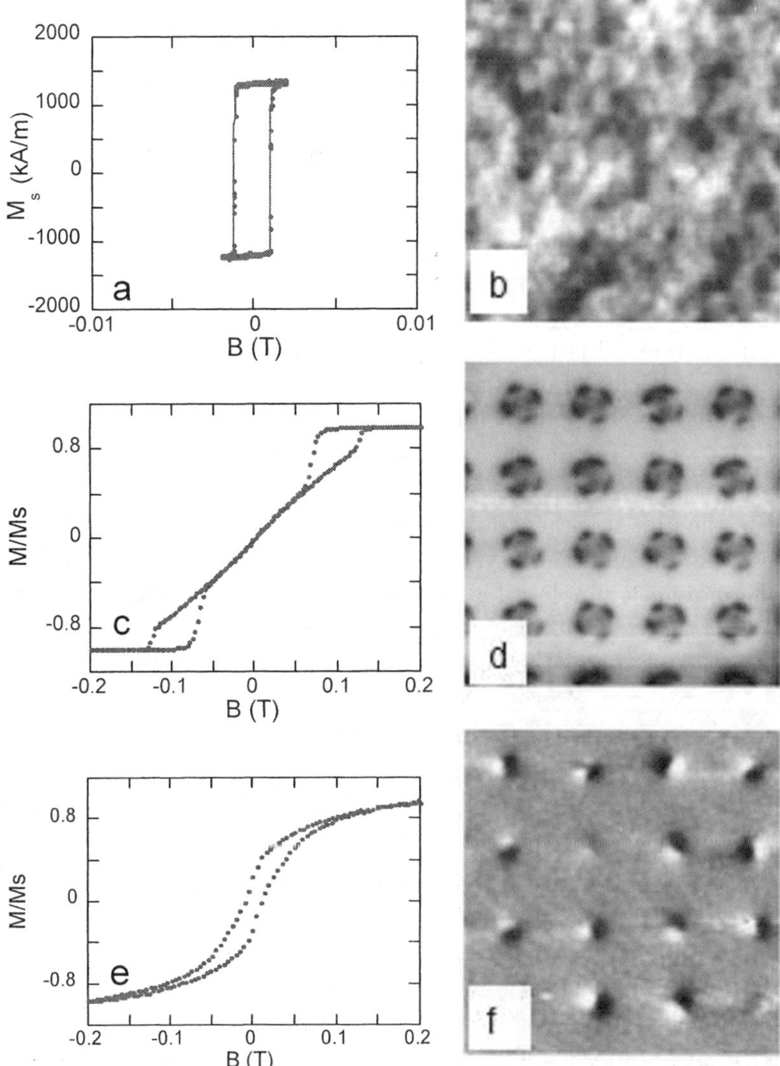

Figure 3. The normalized magnetization versus field of: a) a 50 nm thick Fe(001) film, c) an array of Fe particles of circular shape made of the same film with d = 550 nm, and e) an array of Fe particles of circular shape with d = 200 nm and t = 10 nm. The magnetic field was applied in the plane of the film along the a) [110], c) [110] and e) [100] direction. b) d) and f) MFM images of the samples in a), c) and e) respectively. The scan sizes are 5 μm × 5 μm.

the magnetic domain structure in zero field with the four-fold symmetry characteristic for a closed magnetic structure with the main magnetization oriented along the four equivalent $< 100 >$ directions in the plane of the film. When the lateral dimensions of the ferromagnet are decreased, the magnetostatic energy E_{dem} must be added to the quantum mechanical exchange energy E_{exch} and the magnetocrystalline anisotropy E_{anis}. In order to minimize the total energy E_{magn} at the surface, the spins tend to align along the edges of the particles to form a closed domain structure. In iron E_{anis} is strong enough to force the spins to align mainly along the easy $< 100 >$ directions in the interior of the particle, thereby minimizing the stray fields.

As the thickness dependence of Fe particles was studied [5] it was found that single domain particles were formed in rectangular (300 nm by 900 nm) and elliptical (150 nm by 450 nm) particles only when the film thickness was decreased to 10 nm. Figure 3e and f show the hysteresis curve and an MFM image of Fe particles of circular shape, d = 200 nm and t = 10 nm. Here the particle volume is small enough to lead to a SD state, since the energy required to create domain walls would be too high. The individual magnetic moments of the particles are randomly oriented. This might be explained by that as the film thickness decreases, the the importance of surface effects increases and the magnetocrystalline anisotropy is no longer sufficient to orient the magnetic dipoles. The characteristics of the hysteresis curve show that the magnetization reversal does not occur by coherent rotation, in which case the shape of the curve would have been square.

3.2. EFFECTS OF VARYING MAGNETOCRYSTALLINE ANISOTROPY

Epitaxial multilayers with different combinations of Fe and Co monolayers offer an excellent system for tuning the strength and orientation of the magnetocrystalline energy in arrays of magnetic particles. In a series of bcc Fe/Co(001) multilayer films it was found that the films have a cubic magnetocrystalline anisotropy, with a value of K_1 depending linearly on the Co concentration [10, 14]. An extrapolation of the linear relation yielded K_1 = 0, implying a transition of the easy axis from the initial [100] direction in Fe to [110], at about 23 % Co [10]. We investigated arrays of particles with elliptical shape having axes 150 nm and 450 nm and thickness 20 nm, made of different combinations of layers of Fe and Co [8, 9, 10]. Since the shape anisotropy and the demagnetization effects are practically the same for all of the particles, the dominating variation of the magnetic properties can be ascribed the influence of the magnetocrystalline energy, which changes considerably as the starting material is varied from pure Fe to Co. Clear effects of the varying magnetocrystalline anisotropy can be observed in the remanent state after ac demagnetization along the long axis of the ellipses. This is shown in the MFM images of the particles made of Fe (Fig. 4a), Fe8/Co3 (Fig. 4b), Fe2/Co6 (Fig. 5a) and Co (Fig. 4c). Here the numbers indicate the number of Fe and Co monolayers within each repetition in the

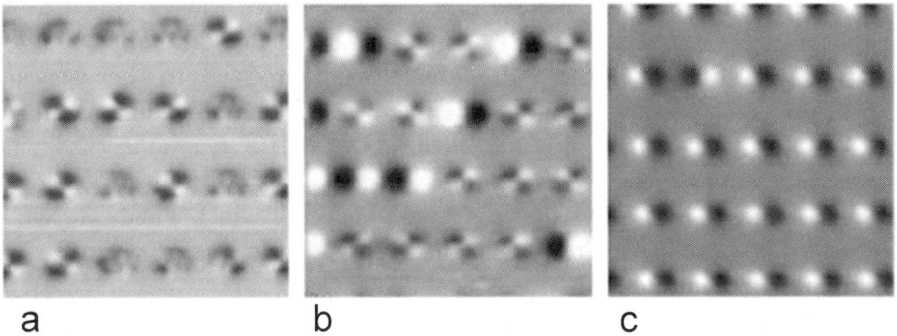

Figure 4. MFM images of particles of elliptical shape (axes 150 nm and 450 nm, $t = 20$ nm) patterned with their axes along the easy magnetocrystalline directions of the film. The particles are made of (a) pure Fe, (b) Fe8/Co3, and (c) pure Co. The scan areas are 5 μm × 5 μm.

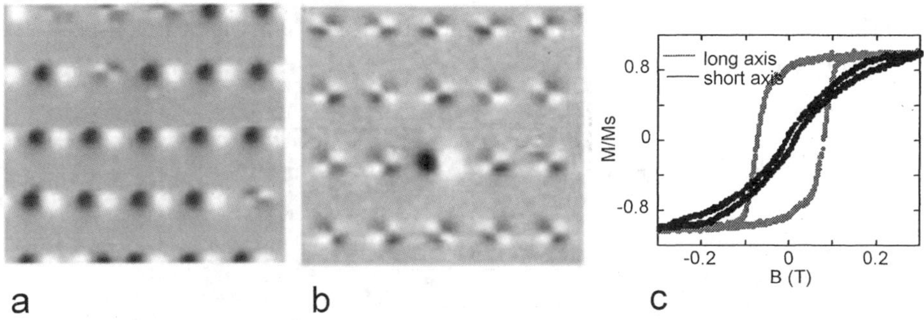

Figure 5. MFM images and hysteresis curves of particles of elliptical shape (axes 150 nm and 450 nm, $t = 20$ nm) patterned with their axes along the easy magnetocrystalline directions of the Fe2/Co6 multilayer film. a) and b) MFM images of the remanent state after demagnetization along the long and short axis, respectively. The scan areas are 5 μm × 5 μm. c) The normalized magnetization versus field applied along the long and short axis according to the notations in the figure.

multilayer. The axes of the ellipses are oriented along the two easy directions (cubic) of magnetization in the plane of the film. In Fe8/Co3 and Fe2/Co6 two different stable zero-field states are observed, one comprising a single domain and another forming a closed structure of mainly two oppositely oriented domains. The observed relative number of SD particles is about 23 % in Fe8/Co3 and 78 % in Fe2/Co6. Only the multidomain states were found in Fe and only SD particles in Co. This shows how the influence of the strong magnetocrystalline anisotropy of Co prevails in the multilayers, thus affecting the magnetic domain structure of the submicron size particles.

As the elliptical particles are positioned with both the short and the long axes along the easy directions of magnetization in the starting film, shape and magnetocrystalline anisotropies co-operate when the field is applied along the long axis, but not along the short. The different behaviour observed in these two cases is demonstrated in Fig. 5. In the remanent state after demagnetization along the long axis the number of SD particles is 78 % (Fig. 5a) to be compared with less than 10 %, as observed when the particles are demagnetized along the short axis (Fig. 5b). This result is reflected also in the hysteresis curves (Fig. 5c) which yield $M_r/M_s = 0.84$ and 0.09, and $B_c = 75$ mT and 9 mT for magnetization with the field applied along the long and short axes, respectively.

The importance of the magnetocrystalline anisotropy is further demonstrated in the switching behaviour of the elliptical particles of the 20 nm thick Fe2/Co6 and Co films. In this procedure the sample was first saturated, leaving all the particles in a SD state, with their magnetic moments oriented in the direction of the field. Then a stepwise increasing field of opposite polarization was applied, and the number of particles that were switched was counted in the remanent state. The switching field distribution is shown in Fig. 6. For comparison data for 10 nm thick Fe ellipses are also included.

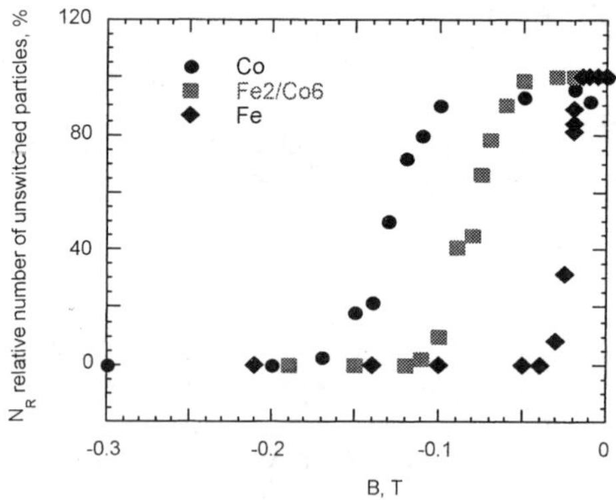

Figure 6. Switching field distribution of elliptical particles (axes 150 nm and 450 nm) made of Co (t = 20 nm), Fe2/Co6 (t = 20 nm) and Fe (t = 10 nm), according to the notations in the figure.

3.3. INFLUENCE OF INTER-PARTICLE INTERACTIONS

To study the influence of interactions between the elements in an array, particles of elliptical shape were positioned in different configurations allowing the angle between the long axes of the ellipses and the distance between the particles to be varied. Effects of interactions were observed in the switching behaviour of the particles as well as in their magnetic domain structure in zero field [9]. An example is shown in Fig. 7 for elliptical particles made

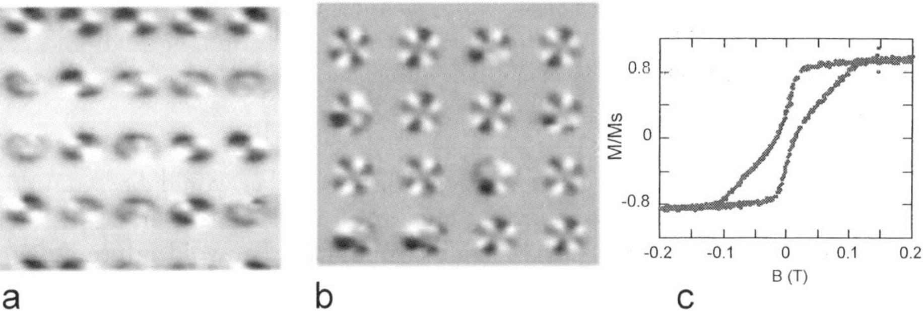

Figure 7. Elliptical particles (axes 150 nm and 450 nm) made of a Fe2/Co6 multilayer film (t = 35 nm). MFM images of the particles positioned in a) a rectangular lattice, array A and b) a square configuration, array B. Both images show the states after demagnetization in the horizontal direction. The scan areas are 5 μm × 5 μm in a) and 8 μm × 8 μm in b). c) The normalized magnetization versus field applied along the long axis of the particles in array A.

of a 35 nm thick Fe2/Co6 multilayer film.

In array A (Fig. 7a) elliptical particles (short axis 150 nm and long axis 450 nm) were positioned in a rectangular lattice with the centre to centre distance between the particles equal to 1.15 μm and 0.90 μm, respectively. In array B (Fig. 7b) subgroups containing four elliptical particles (of the same size as above) were positioned in a square lattice. The centre to centre distance between the subgroups is 1.15 μm. In the subgroups each elliptical particle was positioned with its long axis along one of the sides of a square. The four ellipses in a subgroup thus form a configuration with a minimum distance of 75 nm between the particles.

The particles in array A are multidomains in zero field (Fig. 7a), and the interactions between the particles were found to be weak. As the distance between the particles is decreased, a completely different picture is observed. In array B the majority of the particles relax to a single domain state and their dipolar moments form closed domain structures with a clockwise or anti-clockwise rotation direction (Fig. 7b). All the configurations are not identical, a few less well organized structures may also be seen. The particles in array A have a hysteresis curve with a shape being characteristic for magnetization processes in which domain walls are nucleated, moved and annihilated (Fig. 7c). The coercivity is 14 mT, which is considered to be characteristic for the non-interacting ellipses. The interaction field in configuration B, as estimated for pairs of dipoles, is 32 mT.

From a omparison between these two values it can be understood why the particles in array B relax to a configuration of ordered magnetic dipoles.

3.4. RING SHAPED PARTICLES

When the distance between SD particles decreases, as would be required in a patterned high-density recording medium, the dipolar interactions between the particles become increasingly prominent. To avoid interactions, which make the switching processes more complicated, circular or even ring shaped elements in the vortex state have been suggested [18]. The vortex state appears in circular elements made of materials with low magnetocrystalline anisotropy, within certain ranges of lateral sizes and thicknesses [19]. The state can be described as a flux closed structure, with the spins changing direction gradually in the plane of the circular element. At the centre a core with perpendicular magnetization is formed. The advantage of a ring element is that the core, possessing high magnetic energy, is removed, and the configuration becomes more stable and easy to operate. In an investigation circular and ring shaped particles with outer diameters 0.55 μm and 2.2 μm were prepared of polycrystalline Permalloy films (t = 24 nm and 66 nm). The domain configurations in ring elements were observed to depend on their width [11]. It is of interest to investigate ring elements also in materials possessing higher magnetocrystalline anisotropy. An example is given in Fig. 8, showing the domain structure observed in ring elements,

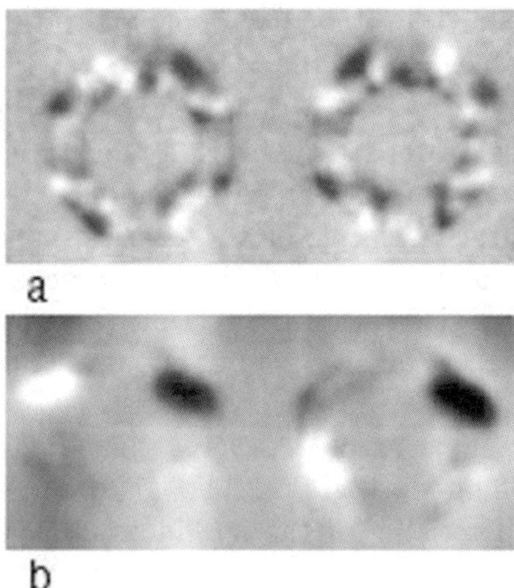

Figure 8. MFM images of ring shaped elements made of a Fe2/Co6 multilayer film, with inner and outer diameters 1.7 μm and 2.1 μm and t = 20 nm. a) shows the state after demagnetization and b) the state after saturation in the horizontal direction. c) a sketch showing the magnetization directions in the "onion" state.

with inner and outer diameter 1.7 μm and 2.1 μm and $t = 20$ nm, prepared of Fe2/Co6. The state after demagnetization is shown in Fig. 8a and the state after saturation in the horizontal direction in Fig. 8b. The particle to the right in Fig. 8b is in an "onion" state [20], but with additional substructures. The "onion" state is characterized by that the ring is split into two domains with their magnetizations oriented in opposite rotation directions (head to head and tail to tail), as shown in the sketch in Fig. 8c.

4. Summary

In this paper we described a method for preparing arrays of submicron size particles of well-defined geometrical shape. The method is applicable for patterning of epitaxial films, thus making it possible to include studies of the interplay between the shape and magnetocrystalline anisotropies. Results obtained from investigations of particles of circular or elliptical shape prepared from epitaxial Fe, Fe/Co or Co films were presented. In these we demonstrated how the magnetic domain structure and magnetization processes were influenced by effects of: decreasing lateral extension, varying shape anisotropy, varying magnetocrystalline anisotropy and dipolar interactions between the particles. New kinds of domain structures were demonstrated in ring shaped particles of Fe2/Co6.

References

1. Bland, J. A. C. and Heinrich, B. (1994) *Ultrathin Magnetic Structures I and II.* Springer, Berlin.
2. See e. g. Aharoni, A. (1996) *Introduction to the Theory of Ferromagnetism.* Clarendon Press, Oxford.
3. See e. g. White, R. L., New, R. M. H. and Pease, R. F. W. (1997) Patterned media: A viable Route to 50 Gbit/in^2 and Up for Magnetic Recording?, *IEEE Trans. Magn.* **33**, 990-995.
   Chou, S. Y. (1997) Patterned Magnetic Nanostructures and Quantized Magnetic Disks, *Proc. IEEE* **85**, 652-671.
   Plumer, M., van Ek, J. and Weller, D. (eds.) (2001) *The Physics of Ultrahigh-Density Magnetic Recording.* Springer, Berlin.
4. Hanson, M., Johansson, C., Nilsson, B., Isberg, P. and Wäppling R. (1999) Magnetic properties of two-dimensional arrays of epitaxial Fe (001) sub-micron particles, *J. Appl. Phys.* **85**, 2793 -2799.
5. Hanson, M., Kazakova, O., Blomqvist, P., Wäppling, R. and Nilsson, B. (2002) Magnetic domain structures in submicron size particles of epitaxial Fe (001): shape anisotropy and thickness dependence, (Unpublished).
6. Kazakova, O., Hanson, M., Blomqvist, P. and Wäppling, R. (2001) Magnetic properties of two-dimensional arrays of epitaxial Co submicron particles: critical size for single-domain formation and multidomain structures, *J. Appl. Phys.* **90**, 2440-2446.
7. Hanson, M., Johansson, C., Nilsson, B. and Svedberg, E. B. (2001) Magnetic propertiees of epitaxial Ni (001) films and sub-micron particles, *J. Magn. Magn. Mater.* **236** 139-150.
8. Kazakova, O., Hanson, M., Blomqvist, P. and Wäppling, R. (2002) Magnetic prop-

erties of submicron size particles made from Fe/Co multilayers, *J. Magn. Magn. Mater.* **240**, 21-23.

9. Hanson, M., Kazakova, O., Blomqvist, P. and Wäppling, R. (2002) Submicron particles of Fe/Co multilayers: Influence of interactions, *J. Appl. Phys.* **91**, 7042-7044.
10. Kazakova, O., Hanson, M., Blomqvist, P. and Wäppling, R. (2002) Interplay between shape and magnetocrystalline anisotropies in patterned bcc Fe/Co multilayers, (Unpublished).
11. Kazakova, O., Hanson, M., Blixt, A. M. and Hjörvarsson (2002) Domain structure of circular and ring magnets, *J. Magn. Magn. Material*, (In press).
12. Isberg, P., Svedberg, E. B., Hjörvarsson, B., Wäppling, R. and Hultman L. (1997) Growth of epitaxial Fe/V (001) superlattice films, *Vacuum* **48** 483-489.
13. Blomqvist, P. and Wäppling, R. (2002) Growth of ultrathin cobalt films on Fe(001) studied by reflection high-energy electron diffraction and x-ray diffraction, *J. Vac. Sci. Technol. A* **50**, 234-238.
14. Blomqvist, P., Wäppling, R., Broddefalk, A., Nordblad, P., te Velthuis, G. E. and Felcher, G. P. (2002) Structural and magnetic properties of BCC Fe/Co (0 0 1) superlattices, *J. Magn. Magn. Mater.* **248** 75-84.
15. Svedberg, E. B., Sandström, P., Sundgren, J. E., Greene, J. E. and Madsen, L. D. (1999) Epitaxial growth of Ni on MgO(002) 1×1: surface interaction vs. multidomain strain relief, *Surface Science* **429** 206-216.
16. See e. g. Lebib, A., Chen, Y., Carcenac, F., Cambril, E., Manin, L., Couraud, L. and Launois, H. (2000) Tri-layer systems for nanoimprint lithography with an improved process latitude, *Microel. Eng.* **53** 175-178, and references therein.
17. Morrish, A. H. (1965) *The Physical Principles of Magnetism.* Wiley, New York.
18. Zhu, J., Zheng, Y., Prinz, G. (2000) Ultrahigh density vertical magnetoresistive random access memory *J. Appl. Phys.* **87** 6668-6673.
19. See e. g. Shinjo, T., Okuno, T., Hassdorf, R., Shigeto, K. and Ono, T. (2000) Magnetic Vortex Core observation in Circular Dots of Permalloy, *Science* **289** 930-932.
20. See e. g. Rothman, J., Kläui, M., Lopez-Diaz, L., Vaz, C. A. F., Bleloch, A., Bland, J. A. C., Cui Z. and Speaks, R. (2001) Observation of a Bi-Domain State and Nucleation Free Switching in Mesoscopic Ring Magnets, *Phys. Rev. Lett.* **86**, 1098-1101.

    Kläui, M., Lopez-Diaz, L., Rothman, J., Vaz, C. A. F., Bland, J. A. C. and Cui, Z. (2002) Switching properties of free-standing epitaxial ring magnets, *J. Magn. Magn. Mater.* **240**, 7-10.

RELAXATION OF HOT ELECTRONS IN SOLIDS OF REDUCED DIMENSIONS

R. PORATH, T. OHMS, M. SCHARTE, J. BEESLEY,
M. WESSENDORF, O. ANDREYEV, C. WIEMANN,
AND M. AESCHLIMANN
Department of Physics, University of Kaiserslautern
Erwin Schroedinger Str. 46, D-67663 Kaiserslautern

1. Abstract

The dynamics of optically excited electrons in solids of reduced dimensions is studied by means of time-resolved two-photon photoemission. In ultrathin films, the lifetime measurements show a clear change by the transition from bulk to layer dimensions smaller than the mean free path of the excited electrons. In silver nano-particles, the electron dynamics does not only depend on the reduced dimension of the systems but rather how the light is coupled into the system (on or off the Mie plasmon resonance). The later effect can be understood in terms of the increased radiation damping in nano-particles.

2. Introduction

A large variety of modern research fields such as fs-photochemistry and (magneto) electronics depend critically on the dynamics of photoexcited hot electrons. In recent years, the dynamics of optically excited electrons in metals, semiconductors and adsorbates has been investigated intensively, in particular by means of the time (and spin) resolved two-photon photoemission technique (TR-2PPE) [1-5]. These investigations have shown that electron dynamics depends critically on the excitation energy, the electron density of states, i.e. the band structure, as well as on the screening of the excited electrons (see Figure 1). As a consequence, the relaxation process of optically excited electrons in transition metals happens on times scales an order of magnitude faster than in noble metals [6]. Additionally, the influence of the spin polarization has been analysed for ferromagnets and semiconductors [4, 7].

The question arises how one can manipulate the lifetime of these hot electrons to design the electrodynamic properties for a specific application. In this paper it will be demonstrated that the specific structuring of the material on a nanometer scale can govern the effective electron dynamics. In order to tune the properties of optically excited electrons, we altered the dimensionality and morphology by using two different approaches:

L.M. Liz-Marzán and M. Giersig (eds.),
Low-Dimensional Systems: Theory, Preparation, and Some Applications, 227–239.

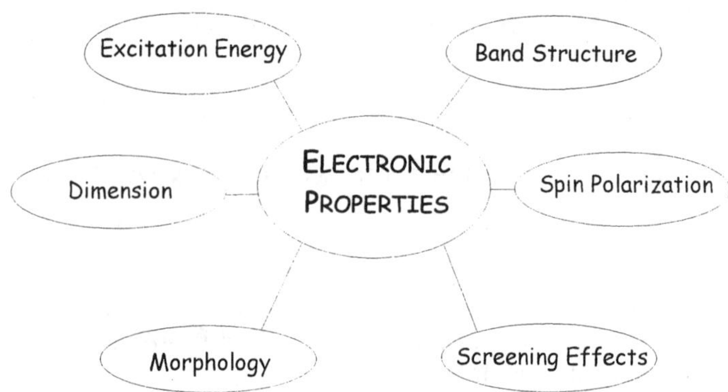

Figure 1. The electronic properties are influenced by different intrinsic (band structure, spin polarization, screening effects) and extrinsic (excitation energy, dimension, morphology) parameters.

i) a reduction from a 3D-bulk to a 2D layered structure and ii) the preparation of defined nanoparticle geometries using the technique of electron beam lithography. By engineering the diameters, heights and the surrounding material the resonance frequency of involved Mie-plasmons can be changed, which considerably influence the electron dynamics.

3. Theory

3.1. THE TRANSITION TO LOWER DIMENSION

The dissipation processes of excited (hot) carriers in metals have been subject of intense theoretical work for several decades. Early theoretical approaches were based on the Landau theory of Fermi liquids (FLT), which treats the excitation as a quasiparticle. The lifetime τ_{ee} of a single particle excitation of an electron gas, as determined by electron-electron interactions, can be calculated from the imaginary part E_I of its self energy E(p), where p is the momentum of the excited electron. Quinn and Ferrell treated this problem in a theoretical study considering a Fermi liquid-like free electron gas [8, 9]. In the case of low excitation energies E(p), and when the effect of plasmon creation can be excluded (E(p) $\ll \hbar\omega_p$), they derived the following simple expression of the lifetime τ_{ee} in a 3D system which is known as Fermi-liquid behavior (FLT):

$$\tau_{ee}^{-1} = \tau_0^{-1} \cdot \left(E - E_F\right)^2 \tag{1}$$

The prefactor τ_0 is primarily determined by the free electron density n and is approximately given by [6]

$$\tau_0^{-1} \approx \frac{\sqrt{3}\cdot\pi^{5/2}}{64}\cdot\frac{e}{\sqrt{m}}\cdot\frac{\sqrt{n}}{1}\cdot\frac{1}{E_F^2}\,. \tag{2}$$

Giuliani and Quinn extended the Fermi-Liquid theory to the case of 2D-systems [10]. Their conclusions are based on theoretical work by M.J. Uren et al. [11], and E. Abrahams et al. [12]. They found a lifetime dependence for excited electrons in a 2D system of the following form:

$$\tau_{ee}^{-1} = \tau_0^{-1}\cdot(E-E_F)^2 \ln\left|E-E_F\right| \tag{3}$$

In comparison to the 3D case, the inelastic lifetime τ_{ee} shows an additional logarithmic dependence of the excitation energy. This is the result of the competing effect of the planar geometry and the conservation of energy and momentum during the electronic collision processes. Furthermore, they emphasized the interesting point, that τ_{ee} does not depend on the electric charge. This is a consequence of the screening by the Coulomb potential. In contrast, in a 3D system τ_{ee} depends on the electric charge e (see Eq. 2).

The quasiparticle behavior in a 1D-system (e.g. nano-wires), especially its inelastic lifetime, was calculated by M. Luttinger [13]. He found that the scattering rate increases linearly with increasing excitation energy:

$$\tau_{ee}^{-1} = \tau_0^{-1}\cdot\left|E-E_F\right| \tag{4}$$

Therefore, the inelastic lifetime of the quasiparticles in a one-dimensional surrounding follows a hyperbolic $1/|E\text{-}E_F|$ curve.

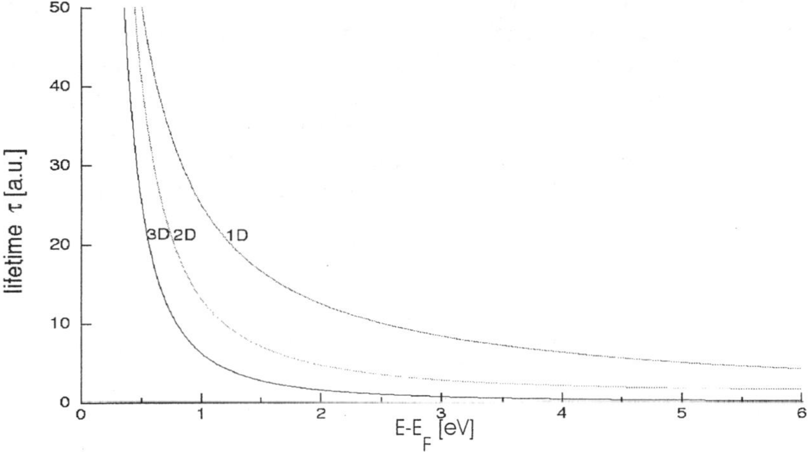

Figure 2. Schematic comparison of the predicted energy dependence of the inelastic lifetimes of excited electrons in a 3D, 2D, and in a 1D system.

Figure 2 shows a plot of the energy dependence of τ_{ee} in a 3D, 2D, and a 1D metal without considering the unknown material dependent prefactor τ_0^{-1}. As expected the reduced phase space for electron-electron scattering events at lower dimension points towards an increased lifetime. But please notice that the prefactor τ_0^{-1} is not included in these theoretical plots, which can considerably change the lifetime ratio between the 3D, 2D and 1D results at certain excitation energies $E - E_F$.

In this paper we will focus on the change of the electron dynamics by a transition from 3D to 2D. We continually reduced the thickness of an ultrathin silver film on a HOPG (highly oriented pyrolitic graphite) substrate and studied the change of the inelastic lifetime τ_{ee} of optically excited electrons.

3.2. PLASMON EXCITATION IN NANOPARTICLES

In mesoscopic systems the dynamical properties of optically excited electrons are not only altered due to the reduced phase space for scattering events but rather how the light is adsorbed in the system. In metal nanoparticles, collective electronic oscillations, so-called Mie-plasmons, can be excited by light and are, therefore, detectable as a pronounced optical resonance in the visible or UV parts of the spectrum. In recent years, several linewidth measurements and time resolved SHG autocorrelation measurements on metallic (Au, Ag) nanoparticles have been published, reporting a dephasing time T_2 (also often called decay or damping time) of the localized particle plasmon excitation in the order of 6-10fs [14, 15, 16].

Much less is known about the physical mechanism underlying the dephasing time T_2. The following mechanisms are possible, depending on the size, size distribution, shape, and dielectric constant of the surrounding medium: First, the plasmon can decay by pure dephasing, e.g. a decay of the fixed phase correlation between the individual electronic excitations of the whole oscillator ensemble, described by a pure dephasing time T_2^*. Common discussed mechanisms are scattering on surfaces or simple decay of the collective mode due to inhomogeneous phase velocities caused by the spread of the excitation energy or the local inhomogeneity of the nanoparticles. Second, the plasmon can also decay due to a transfer of energy into quasi-particles (electron-hole pairs) or reemission of photons (radiation damping and luminescence), described by $T_1 \equiv \tau_{ee}$.

We focus on the study of elliptically-shaped silver nanoparticles with axes of 40 nm by 80 nm and a height of 45 nm. Silver nanoparticles are of particular interest as they can exhibit particularly strong size-dependent optical extinction in the visible spectral range (1.8 eV – 3 eV) due to resonantly driven electron plasma oscillations. Elliptically-shaped metal nanoparticles show two different plasmon resonances, which lie at different wavelengths for light polarized parallel to the short and long axes, respectively (see Figure 3). Tuning the laser wavelength to one of these two resonances allows distinguishing between resonance excitation and off-resonance excitation by simply changing the polarization of the laser pulse. Please notice that in both cases the excitation is the result of an intraband process, since the excitation energies of the Mie-plasmons in those silver nanoparticles lie below the interband transition threshold (~ 4 eV). Consequently, the absorption cross-section in the visible region is dominated by Drude damping [17]. The contribution due to interband transitions is negligible.

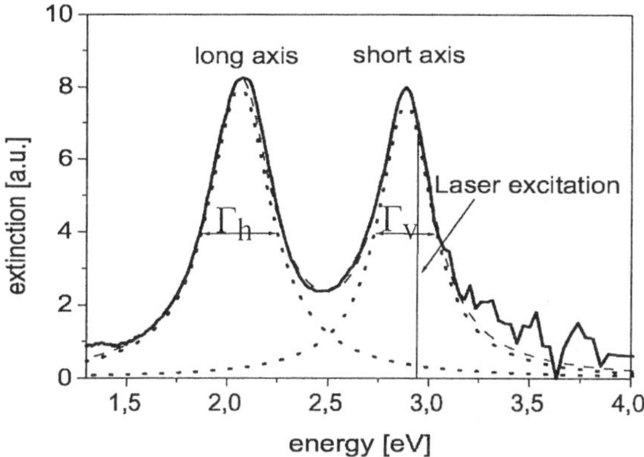

Figure 3. Measured extinction spectrum of the nanoparticle array. The low- and high-frequency peaks correspond to the Mie-plasmon, excited along the long and short axes of the elliptical Ag particles, respectively.

The excited electronic system in large nano-particles (>10nm) can exhibit additional dominant microscopic relaxation channels that are not important in bulk materials, such as radiation damping. Compared with Drude damping the radiation damping is explicitly frequency dependent. In this paper we concentrate on the question whether the same microscopic damping processes (due to photons, phonons, impurities, surface scattering, and electron-electron interactions) are involved when the photoexcitation is close to the Mie-plasma resonance or far from it.

4. The experiments

4.1. TIME-RESOLVED TWO-PHOTON PHOTOEMISSION (TR-2PPE)

The experimental method used to investigate the lifetime of optically excited electrons in a metal is the time-resolved two-photon photoemission (TR-2PPE). This method is known to be a highly accurate method to determine the dynamics of hot electrons in unoccupied and occupied states. The pump-probe technique enables a direct measurement of the dynamical properties in the time domain with a resolution in the range of few femtoseconds [6]. The principle is schematically shown in Figure 4. A pump pulse excites electrons out of the valence band into usually unoccupied states with energies between the Fermi and vacuum level. A second probe pulse subsequently photoemits the excited electrons. The photoemission yield depends on the transient population of these intermediate states as a function of the temporal delay of the probe pulse with respect to the pump pulse. The measured pump-probe-signal contains information on the energy relaxation time T_1 of the population in the intermediate state, as well as its dephasing time T_2.

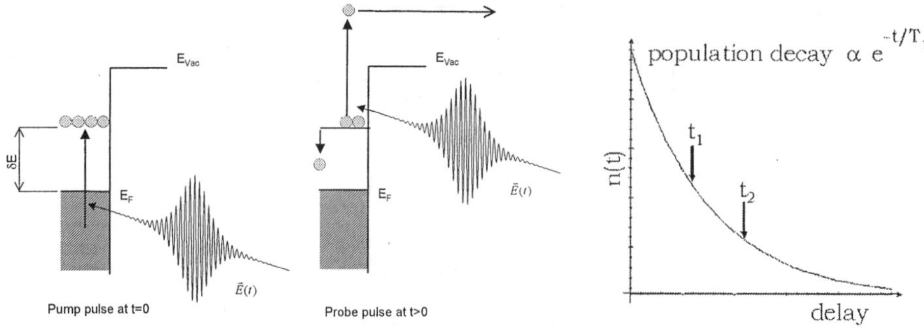

Figure 4. Principal mechanism of the time-resolved two-photon-photoemission (TR-2PPE) and the temporal behavior of the population N(t) in the intermediate state.

The experiments were performed with a femtosecond Kerr lens mode-locked Ti:sapphire laser (Tsunami, SpectraPhysics), pumped by a cw-operated diode pumped solid state laser (Nd:YVO_4, Millennia X, SpectraPhysics). The system delivers transform-limited and $sech^2$ temporal shaped pulses with up to 12 nJ/pulse, a duration of 20 fs – 40 fs at a repetition rate of 82 MHz and a wavelength of 780-830 nm. The linearly polarized output of the Ti:sapphire laser has been frequency-doubled in a 0.2 mm thick beta-barium-borate (BBO) crystal in order to produce UV pulses at about $h\nu = 3.0$ eV corresponding 415 nm. In a Mach-Zehnder-interferometer the pulses are divided by a beamsplitter into equal intensity (pump and probe) pulses. Hereby one path is delayed with respect to the other by a computer-controlled delay stage. Both beams were combined collinearly by a second beamsplitter and focused onto the sample surface. The diameter of the laser spot on the sample is around 100 μm. The plane of polarization can be rotated by means of a half wave plate to any arbitrary angle.
The samples were held in an ultra-high vacuum chamber at a base pressure in the 10^{-10} mbar range. A bias of $U = -4$ V was applied to the sample to eliminate the effects of any stray electric fields. The photoemitted electrons were detected in a cylindrical sector analyzer (CSA, Focus GmbH). The entrance axis of the energy analyzer is 45° with respect to the laser beam. In general, we used a pass energy of 4 eV, leading to roughly $\Delta E = 50$ meV resolution.

4.2. SAMPLE PREPARATION

The ultrahin Ag films were grown onto a HOPG substrate, using an e-beam evaporator at a rate of about 0.1 ML/s, controlled by a pre-calibrated microbalance. Note that one monolayer (1 ML) silver corresponds to a height of 2.045 Å. The base pressure while evaporation was in the 10^{-9} mbar range.

The nanoparticle samples were produced in the group of Professor Aussenegg, Karl-Franzens University in Graz. The two-dimensional array of nearly identical, parallel oriented silver particles are deposited lithographically on a transparent ITO substrate, which itself lies on a glass plate [18]. Figure 5 depicts a SEM picture, and Figure 6 illustrates the geometry, size and the distances of the particles in the array.

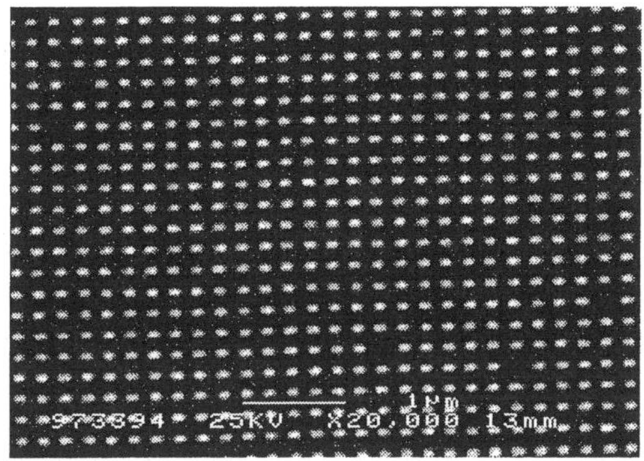

Figure 5. SEM picture of lithographically fabricated nano-particles on ITO

As particle shape and interparticle distances can be varied independently, this method allows us to tailor the optical properties of single particles. Thus the resonance frequency and the strength of particle interactions can be changed separately. Thereby the optical extinction maximum can be tuned to the desired wavelength, e.g. to our illuminating laser wavelength of 415 nm.

Figure 3 shows the two different resonances in the extinction spectrum at $\omega_a = 2.1$ eV and $\omega_b = 2.95$ eV for both axes, respectively. Liebsch's predictions that the Mie-plasmon damping is nearly independent of the resonance frequency [19] are in good agreement with the measured linewidth of the peaks in the extinction spectrum. For the long axis, $\Gamma_a = 0.4$ eV and for the short axis, $\Gamma_b = 0.39$ eV are observed. In the following, the long axis of the elliptic particle is called 'v' and the short axis is called 'h'.

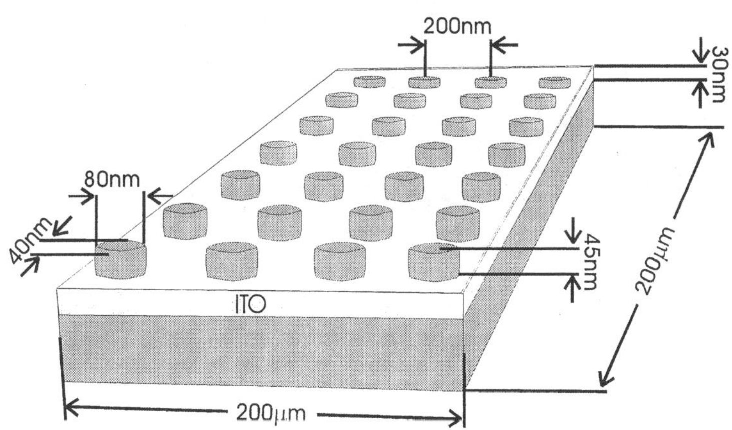

Figure 6. Array of silver-nanoparticles deposited on ITO

5. Results and discussion

5.1. ELECTRON DYNAMICS IN ULTRATHIN SILVER FILM ON HOPG

The relaxation of photoexcited electrons in ultrathin Ag-films at an energy of $E-E_F = 1.24$ eV is shown in Figure 7. Above 25 ML a small reduction in the lifetime τ with increasing film thickness is observed, caused by an increasing transport effect as discussed in [20]. The transport of the photoexcited electrons away from the surface region adds to the system-intrinsic decay due to inelastic scattering and reduces the measured lifetime with respect to the actual one. This transport effect is diminished and finally even eliminated by using films as thin as the penetration depth of the laser light within the investigated material.

For thinner films (<25 ML) a completely anomalous behavior is observed. The lifetime increases from about 25 fs at 25 ML to almost 40 fs at 15 ML silver. The data seem to match the intuitive assumption that the dynamics of the electron relaxation process should differ as soon as the thickness of the metal film becomes smaller than the mean free path of the excited electrons. The electronic structure of metallic thin films grown on a substrate is very different from that of the corresponding bulk crystal. The continuous bands split up into discrete energy levels due to quantum size effects (QSE). This electron confinement should result in a decreased phase space for electron-electron scattering events and, hence, in an increase of the inelastic lifetime τ_{ee} of the optically excited electrons.

An additional increase in the lifetime τ_{ee} is expected at even lower film thickness (< 15ML) due to the transition from a 3D-bulk to a 2D-system as shown in Figure 2. However, the measured data in Figure 7 show a decrease in τ_{ee} at lower coverage.

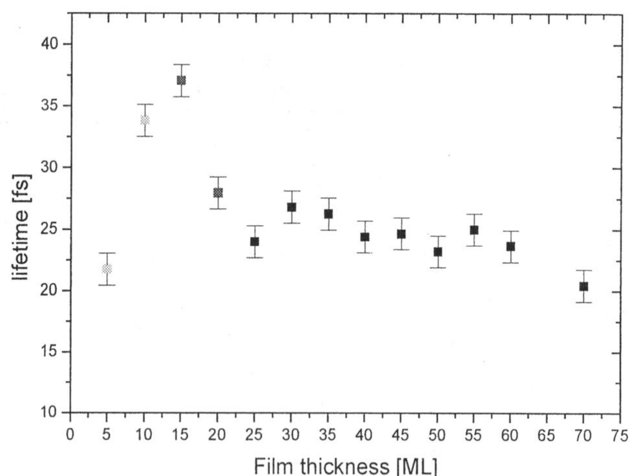

Figure 7. Film thickness dependence of silver on HOPG, at $E-E_F = 1.24$ eV. It is clearly visible that there is a levelling off at thicker layers, and a dramatic change in the region, where the system changes from 3D to 2D.

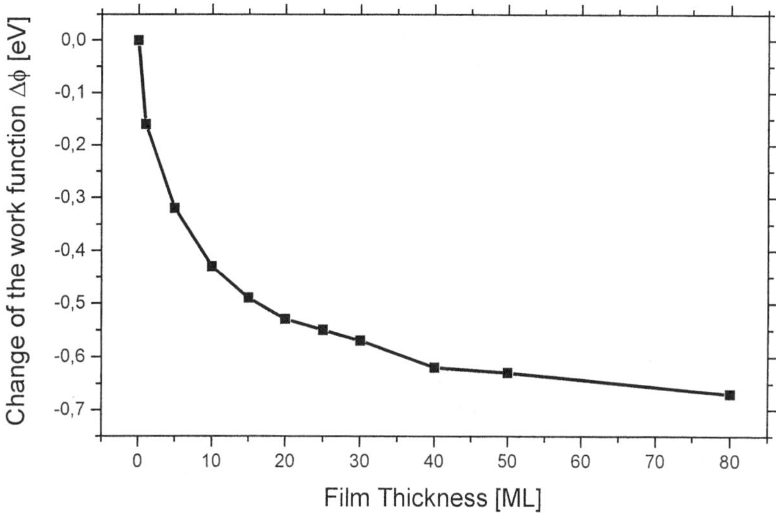

Figure 8. The change of the work function from pure HOPG via small layer thicknesses to an 80 ML silver layer.

Atomic force microscopy (AFM) measurement revealed that the growth mode of silver on is not regularly or even epitactic. The images show a clear tendency to metal clustering below 20 ML. Nevertheless, also small metal clusters should give rise to QSE and cannot explain the observed decrease in the lifetime τ. The reduction in τ can be explained by the fact that the lifetime measurements at lower silver coverage (0 - 15 ML) are influenced by an increasing number of photoexcited electrons from the uncovered HOPG areas. Additional measurements on clean HOPG in the investigated excitation energy range ($E-E_F > 1.0$ eV) showed a very rapid (< 5 fs) relaxation of the photoexcited electrons. This explanation for the reduction of the lifetime τ below 15 ML is confirmed by the measured change of the work function with increasing Ag thickness as shown in Figure 8.

The work function difference between HOPG and a thick Ag film is about 0.7eV. But as presented in Figure 8 the drop of the work function does not happen within a few ML of Ag as expected by an uniformly grown silver film. Therefore, it is assumed that the decrease of the lifetime τ at thicknesses below 15 ML is rather caused due to the increasing number of photoexcited electrons of the clean HOPG surface than due to the transition from 3D-bulk to a 2D layered structure.

To summarize our results on ultrathin Ag "films": The data show a clear change of the dynamics of optically excited electrons as soon as the nominal thickness of the metal film becomes smaller than the mean free path of the excited electrons. In order to obtain a quantitative picture of the change of the dynamics by the transition from 3D-bulk to a 2D layered structure, a more uniform grown thin film on a dielectric substrate (e.g. MgO) is required. Further investigations in this project are in progress.

5.2. THE LIFETIME DIFFERENCE OF MIE-PLASMONS AT AND OFF-RESONANCE

Let us now examine the electron relaxation in the silver nanoparticles, deposited as an array on an ITO substrate. Figure 9 shows the photoemission yield as a function of the polarization angle of the laser pulse. Clearly visible is the strong enhancement in the 2PPE yield in the case of a resonant plasmon excitation (denoted by 'h') compared to off-resonant excitation (denoted by 'v'). The shapes of the 2PPE photoemission spectra, however, do not show distinct differences by an excitation in the 'v' or 'h' direction (not shown). The resonant enhancement of the photoemission yield can be explained by the enlarged absorption of light by an extinction in the 'h' direction and by the lightning rod effect [21] meaning that the field enhancement can occur simply due to a concentration of the electric field at positions of maximum curvature (tip effect).

Figure 10 shows the FWHM determined from the autocorrelation traces of the TR-2PPE measurement as a function on the polarization angle of the incoming light. Each data point is represented by the FWHM of a sech^2-shaped curve fitted into the autocorrelation measurement. A typical autocorrelation measurement is shown in the inset.

For comparison reference autocorrelation traces of polycrystalline tantalum were measured at an intermediate state energy $E\text{-}E_F = 2.8$ eV. In this energy range, the lifetime of excited electrons in a transition metal is less than 1 fs [6] and hence, the traces obtained for tantalum represent the pure laser autocorrelation. The FWHM data for tantalum do not depend on the polarization angle of the incoming light. Therefore, any effect caused by an increase of dispersion due to rotation of the half wave plate can be excluded.

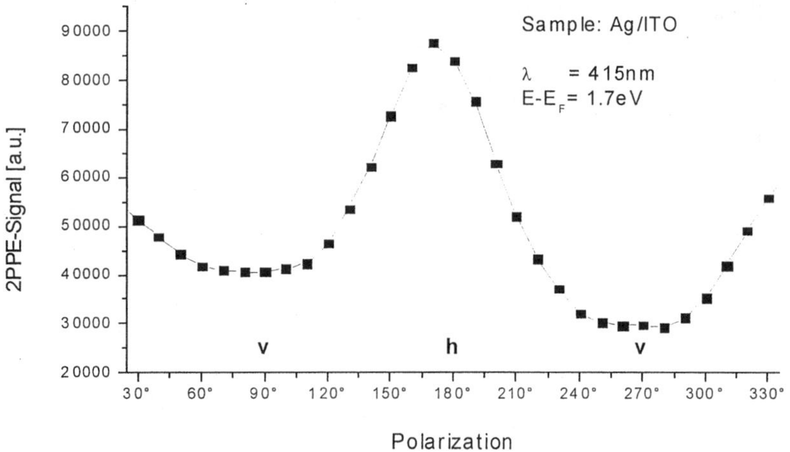

Figure 9. 2PPE yield versus polarization of the incident blue light. In case of the plasmon excitation ('h') a strong rise in the 2PPE-signal occurs.

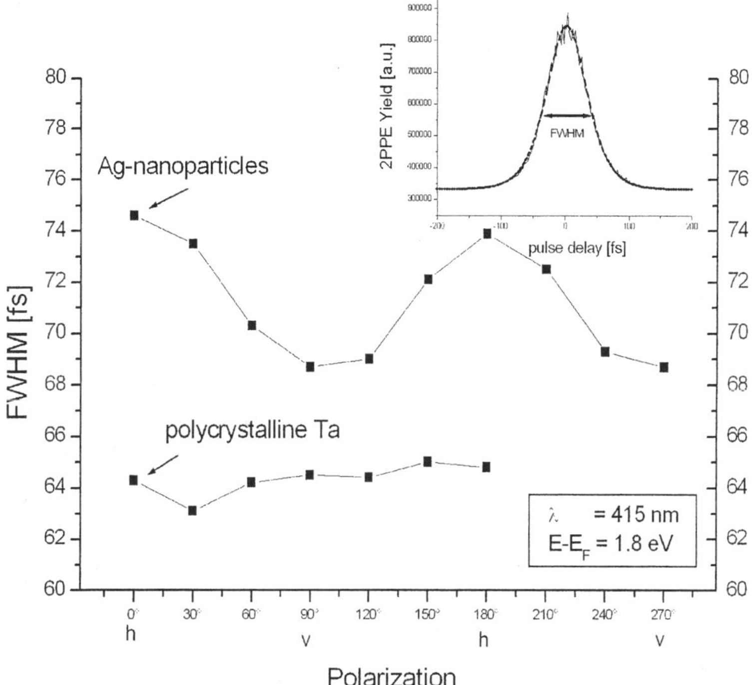

Figure 10. Nano-particles show a variation of the FWHM of the autocorrelation over rotating states of light polarization while tantalum shows no effect. In the inset, a typical autocorrelation measurement can be seen from which the data points have been derived.

As one can see the FWHM is reduced from almost 75 fs in the 'h' direction (short-axis mode, resonant plasmon excitation) to 69 fs in the 'v' direction (long-axis mode, off- resonant plasmon excitation). Further rotation of the polarization restores the long lifetime of the resonant Mie-plasmon excitation case. Please note that these FWHM values still include the autocorrelation of the laser pulse width. One can clearly see a difference in the FWHMs between resonance and off-resonance. This observation points unambiguously to a different electron dynamics.

It might be tempting to conclude that on-resonance collective excitation in general have a longer lifetime than excitation far from resonance. According to Liebsch *et al.* [19], the different damping rates obtained under on- and off-resonance conditions, as shown in Figure 10, can also be explained by different magnitudes of radiation damping along the short and long axis of the elliptical particles. This result is consistent with the fact that for spherical particles, radiation losses become more pronounced with increasing radius. Surface scattering plays a role only at radii less than about 5 nm. Moreover, Liebsch's theory predicts that collective plasma oscillations within the nano-particles are subject to the same microscopic Drude damping processes due to phonons, impurities, electron-electron interactions as off-resonance excitations [19]. None of these inelastic scattering mechanisms are reduced or absent if the electrons oscillate near the resonance excitation. However, the theory also points out that for an excitation, which matches the low frequency peaks in the extinction spectrum (Mie plasmon excited along the long axes) the radiation damping leads to a longer lifetime at off-resonance than at resonance.

In summary, it was demonstrated that the specific structuring of the material on a nanometer scale can govern the effective electron dynamics. In particular for defined silver ellipsoids the off-resonance excitation lifetime can be shorter or longer than the lifetime at resonance, depending on the incidence conditions of the laser light and on the dimensions of the metallic nanoparticles.

6. Acknowledgements

We are grateful to B. Lamprecht, H. Ditlbacher, J. R. Krenn and F. R. Aussenegg for providing us with many Ag-nanoparticle samples. We would like to thank M. Bauer and A. Liebsch for valuable discussions. This work was supported by the Deutsche Forschungsgemeinschaft (DFG).

7. References

1. Haight, R. (1995) Electron dynamics at surfaces, *Surf. Sci. Rep.* **21**, 275-325.
2. Schmuttenmaer, C. A., Aeschlimann, M., Elsayed-Ali, H.E., Miller, R.J.D., Mantell, D., Cao, J., and Gao, Y. (1994) Time resolved two photon photoemission from Cu(100): Energy dependence of electron relaxation, *Phys. Rev. B* **50**, 8957-8960.
3. Hertel, T., Knoesel, E., Wolf, M., and Ertl, G. (1996) Ultrafast electron dynamics at Cu(111): Response of an electron gas to optical excitation, *Phy. Rev. Lett.* **76**, 535-538.
4. Aeschlimann, M., Bauer, M., Pawlik, S., Weber, W., Burgermeister, R., Oberli, D., and Siegmann, H.C. (1997) Ultrafast spin-dependent electron dynamics in fcc Co, *Phys. Rev. Lett.* **79**, 5158-5161.
5. Petek, H. and Ogawa, S. (1997) Femtosecond time-resolved two-photon photoemission studies of electron dynamics in metals, *Progr. Surf. Sci.* **56**, 239-257.
6. Aeschlimann, M., Bauer, M., and Pawlik, S. (1996) Competing nonradiative channels for hot electron induced surface photochemistry, *Chem. Phys.* **205**, 127-141.
7. Ohms, T., Porath, R., Scharte, M., Beesley, J., and Aeschlimann, M. (to be published) Electron spin dynamics in a GaAs/metal interface.
8. Ferell, R.A. (1956), Angular dependence of the characteristic energy loss of electrons passing through metal foils, *Phys. Rev.* **101,** 554-563.
9. Quinn, J. (1962) Range of excited electrons in metals, *Phys. Rev.* **126**, 1453-1457.
10. Giuliani, G.F. and Quinn, J. (1982) Lifetime of a quasiparticle in a two-dimensional electron gas, *Phys. Rev. B* **26**, 442-446.
11. Kaveh, M. and Wiser, N. (1984), Electron-electron scattering in conduction materials, *Adv. Phys.* **33**, 257-372.
12. Abrahams, E., Anderson, P.W., Lee, P.A., and Ramakrishnan, T.V. (1981) Quasiparticle lifetime in disordered two-dimensional metals, *Phys. Rev. B* **24**, 6783-6786.
13. Luttinger, J.M. (1961) Analytic properties of single-particle propagators for many-fermion systems, *Phys. Rev.* **121**, 942-945.
14. Lamprecht, B., Leitner, A., and Aussenegg, F.R. (1999), SHG studies of plasmon dephasing in nanoparticles, *Appl. Phys. B* **68**, 419-423.
15. Vartanyan, T., Simon, M., and Träger, F. (1999), Femtosecond optical second harmonic generation by metal clusters: The influence of inhomogeneous line broadening on the dephasing time of surface plasmon excitation, *Appl. Phys. B* **68**, 425-431.

16. Klar, T., Perner, M., Grosse, S., von Plessen, G., Spirkl, W., and Feldmann, J. (1998), Surface-plasmon resonances in single metallic nanoparticles, *Phys. Rev. Lett.* **80**, 4249-4252.
17. Ashcroft, N.W. and Mermin, N.D. (1976) *Solid State Physics*, p. 2-27, Holt, Philadelphia.
18. Gotschy, W., Vonmetz, K., Leitner, A., and Aussenegg, F.R. (1996), Optical dichroism of lithographically designed silver nanoparticle films, *Opt. Lett.* **21**,1099-1101.
19. Scharte, M., Porath, R., Ohms, T., Aeschlimann, M., Krenn, J.R., Ditlbacher, H., Aussenegg, F.R., and Liebsch, A. (2001) Do Mie plasmons have a longer lifetime on resonance than off resonance? *Appl. Phys. B* **73**, 305-310.
20. Aeschlimann, M., Bauer, M., Pawlik, S., Knorren, R., Bouzerar, G., and Bennemann, K. H. (2000) Transport and dynamics of optically excited electrons in metals, *Appl. Phys. A* **71,** 5, 485-491.
21. The lightening rod effect means that the field enhancement can occur simply due to a concentration of the electric field at positions of maximum curvature (tip effect). Wokaun A. (1984) *Solid State Physics, p.223*, Ehrenreich, H., Thurnbull, T., and Seitz, F. (Editors), Academic, New York

ANALYTICAL TEM CHARACTERISATION OF CATALYTIC MATERIALS

D. S. SU
Department of Inorganic Chemistry
Fritz Haber Institute of the Max Planck Society
Faradayweg 4-6, D-14195 Berlin, Germany

Abstract. The application of analytical transmission electron microscopy to the characterisation of catalytic materials is presented by studying mechanically activated vanadium pentoxide V_2O_5. Local morphology and lattice distortion are studied by TEM while the oxidation states are determined by EELS. The mechanical activation can be described as a two-stage process: crushing of large crystals into small ones (macroscopic process) followed by amorphisation and reagglomeration of the fragments (microscopic process). No milling equilibrium state can be found. Energy-loss spectra reveal the reduction of vanadium via oxygen loss. The formation and distribution of V^{4+} or V^{3+} species depends on the history of milling.

1. Introduction

It is well known that catalysis, the science and technology of the modification of the velocity of chemical transformations in selected reactions, is a surface phenomenon. A large amount of surface sensitive techniques, especially spectroscopic methods like XAS, XPS, IRS, UPS are applied in the research of catalytic materials. The question is therefore, how can (transmission) electron microscopy (TEM), which is in essence a bulk technique, be applied in catalysis to provide relevant information for the explanation of the catalytic mechanism. The first answer comes from the catalytic material itself: most catalysts are nanostructured as deposited cluster systems or as microstructured – mesostructured solids. Due to its high lateral resolution, a TEM provides an excellent access to the local morphology, crystallography and composition of individual clusters or particles. Deposited clusters as small as 2 nm in size can be analysed via high-resolution imaging at a point resolution of less than 0.2 nm. For solid-state oxide catalysts, the role of the bulk beneath the active centres of the surface is not very well understood. Possible diffusion of lattice oxygen towards the surface could play an important role. Also lattice defects, grain boundaries and shear structures - features which can very well be studied in a TEM - could be additional factors influencing the catalytic reaction.

The second answer comes from the electron microscope itself: an analytic electron microscope equipped with the facilities of EELS and EDXS is a powerful tool in elucidating the local electronic structure and chemical composition of the binary or ternary catalytic materials at the nanometer scale. Since all catalytic processes concern the exchange of electrons between the reactants, information about the chemical valence and the local electron density of unoccupied states provided by core-level EELS becomes

L.M. Liz-Marzán and M. Giersig (eds.),
Low-Dimensional Systems: Theory, Preparation, and Some Applications, 241–252.

relevant. Due to the new development of monochromators and optimised spectrometers, high-energy resolution ELNES is available [1]. A comparison of bulk-sensitive ELNES with surface-sensitive NEXAFS could be interesting in order to proof the role of the bulk structure in catalysis.

One of the common methods in the structural characterisation in materials science is X-ray diffraction. Figure 1 displays X-ray powder diffraction patterns of ball milled V_2O_5 samples, compared with those of untreated V_2O_5. Milling of V_2O_5 results in a decrease of integral intensities of individual Bragg peaks, an increase of the full width at half maximum, and an increase of the background of all milled samples. These features can be attributed to the reduction of particle sizes, the build-up of lattice defects and strain/stress in the bulk of the crystallites, and the increase of amorphous components in the milled powder. However, such valuable integral information needs to be complemented by experiments probing the nanoscopic structural range. Since the catalytic performance is strongly dependent on the microstructure and the electronic state of the surface of small particles, such investigations are also important for the development of more efficient catalysts. In the present work, we show, using mechanically modified V_2O_5 as an example, how TEM/EELS can be applied to the structural and chemical characterisation of catalytic materials.

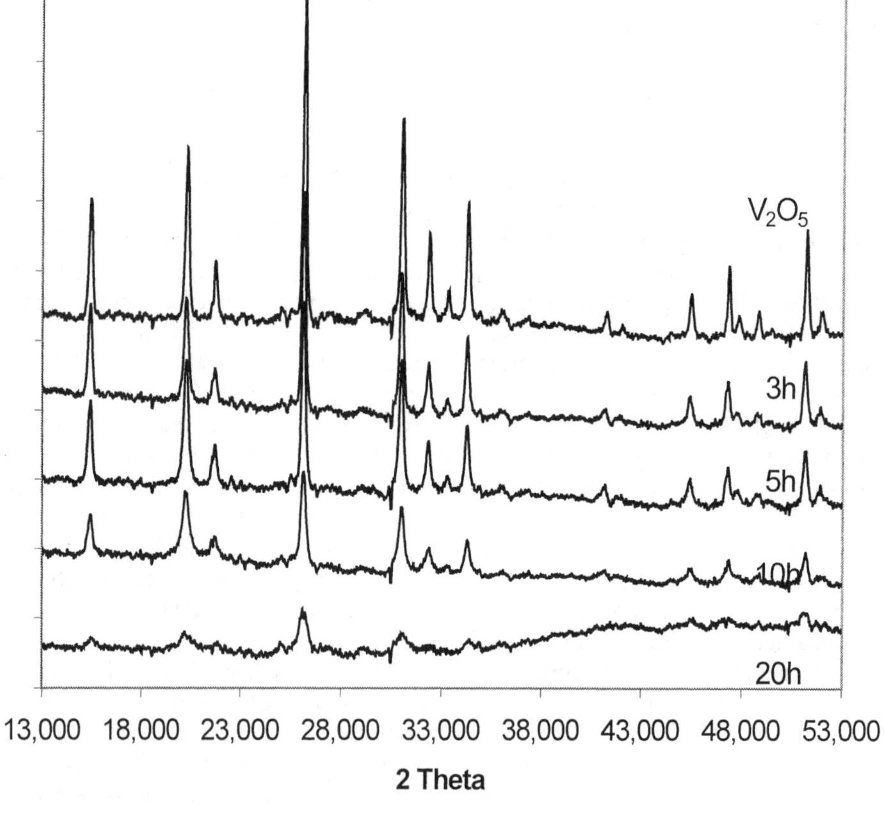

Figure 1. XRD of untreated and milled vanadium pentoxide

In recent years, mechanical activation (usually by means of ball milling) has been widely used in catalyst research [2-4] for the preparation of metal oxide catalysts. Ball milling preparation, which is technically simple and environmentally friendly, can be performed either by the simultaneous grinding of two or more oxide powders, or by the milling of only one prior to mixing and calcination. In addition, ball-milling is an effective method for the modification of the defect structure and the electronic properties of catalytic materials. Improvements in the catalytic performance of ball-milled catalysts have been reported. For instance, the yield for *n*-butane oxidation to maleic anhydride (MA) increases by about 5 % using a dry milled V_2O_5 precursor vs. the yield obtained with untreated V_2O_5, and yield increases even higher for wet milled V_2O_5 [3]. The catalytic performance of vanadyl pyrophosphate (VPP) catalysts is also influenced by ball-milling:. the MA selectivity and maximum MA yields increase noticeably [4].

2. Experimental

Vanadium oxide V_2O_5 was from J. T. Baker Chemicals B. V. and had a purity of > 99%. Ball- milling was carried out in a planetary ball-mill: 60 g of V_2O_5 together with six agate balls (1.5 cm diameter, 11 g) were placed into an agate vessel (250 cm^3 volume). Milling was performed for up to 20 h at approximately 150 rpm. Samples studied in the present work were taken after 3, 5, 10, and 20 h.

For TEM investigations, untreated and milled samples were dispersed onto copper mesh grids covered with holey carbon films. A Philips CM200 FEG transmission electron microscope, operating at 200 kV, and equipped with a GATAN image filter GIF100, was used. All selected area diffraction patterns were taken from an area of about 900 nm in diameter. The image filter, operated in the spectroscopy mode, was used to record EELS-spectra. All spectra were recorded from very fine particles and from the thin edge area of large particles to avoid any artefacts due to large thickness, and were corrected for backgrounds and multiple scattering [5]. In order to avoid any artefacts such as electron-beam induced structural changes, high-resolution images were taken using electron dose as low as possible. EELS-spectra were recorded with 2 A/cm^2 current density for 5 s. According to a previous study about stability of V_2O_5 under electron beam irradiation, no chemical changes under those conditions are visible [6].

3. Microstructure

Electron micrographs of untreated and milled samples are reproduced in Figures 2A – 2E. They reveal the morphological development of V_2O_5 and changes in the particle-size at various periods of milling in the planetary ball mill. Untreated V_2O_5 crystallite is shown in Figure 2A. Up to 3 h the main effect of milling is the crushing of large crystallites (Figure 2B). Further milling up to 10 h leads to the formation of fine particles whose size distribution becomes narrow and samples appear more homogenous in size (Figure 2D). In Figure 2D small particles of about 10 nm in size can be seen. Almost all particles exhibit a plate-like shape, as do those found in the untreated V_2O_5 sample. This confirms the finding by X-ray diffraction that fracture takes place in all directions [2]: the shape of particles up to 10 h milling time remains nearly the same, but their size decreases.

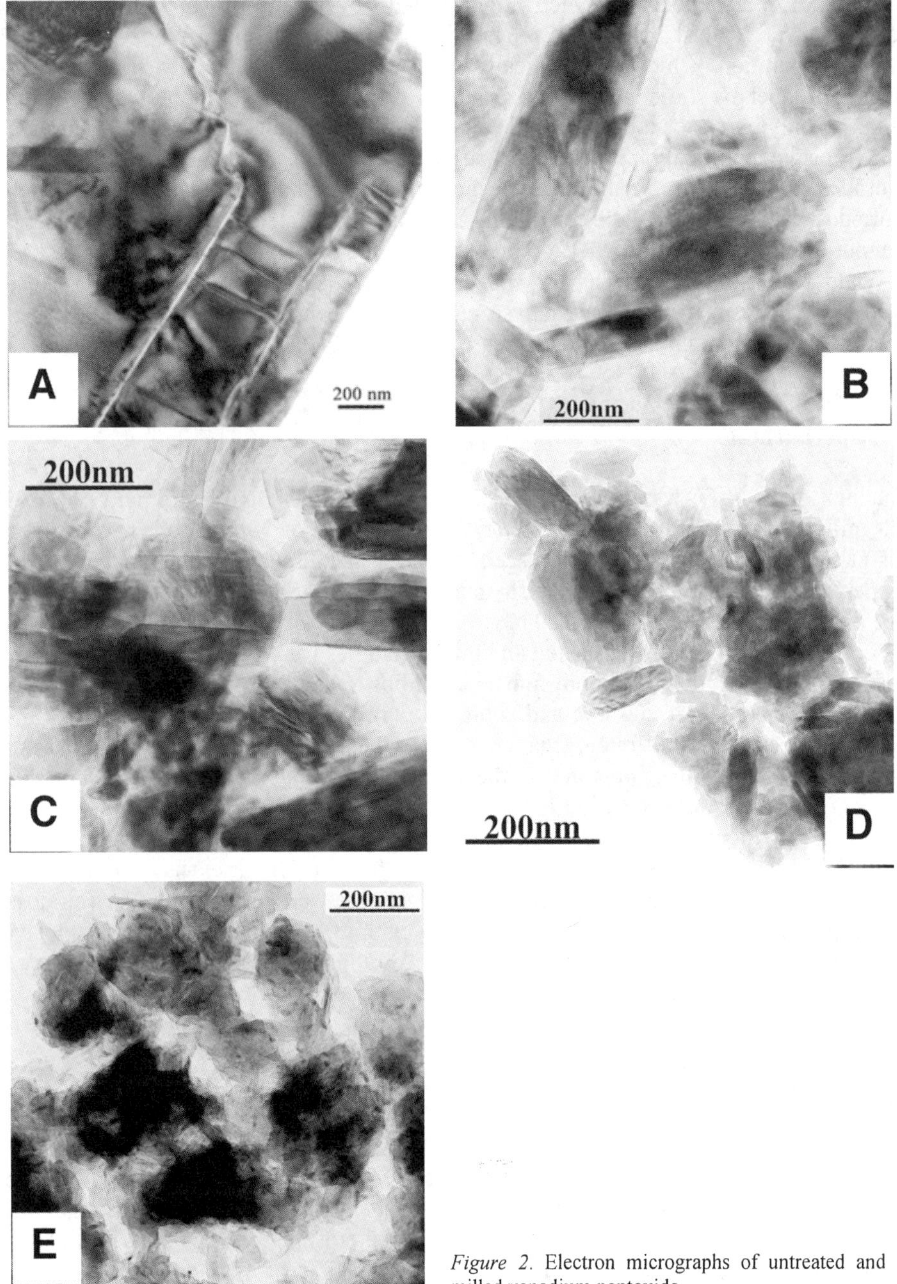

Figure 2. Electron micrographs of untreated and milled vanadium pentoxide

The increase of milling time up to 20 h leads to reagglomeration of small particles, the electron micrograph of the 20 h milled sample in Figure 2E does not show clear profiles of individual particles. Some scarf-like contrasts were obtained, which was never observed in the samples milled for up to 10 h.

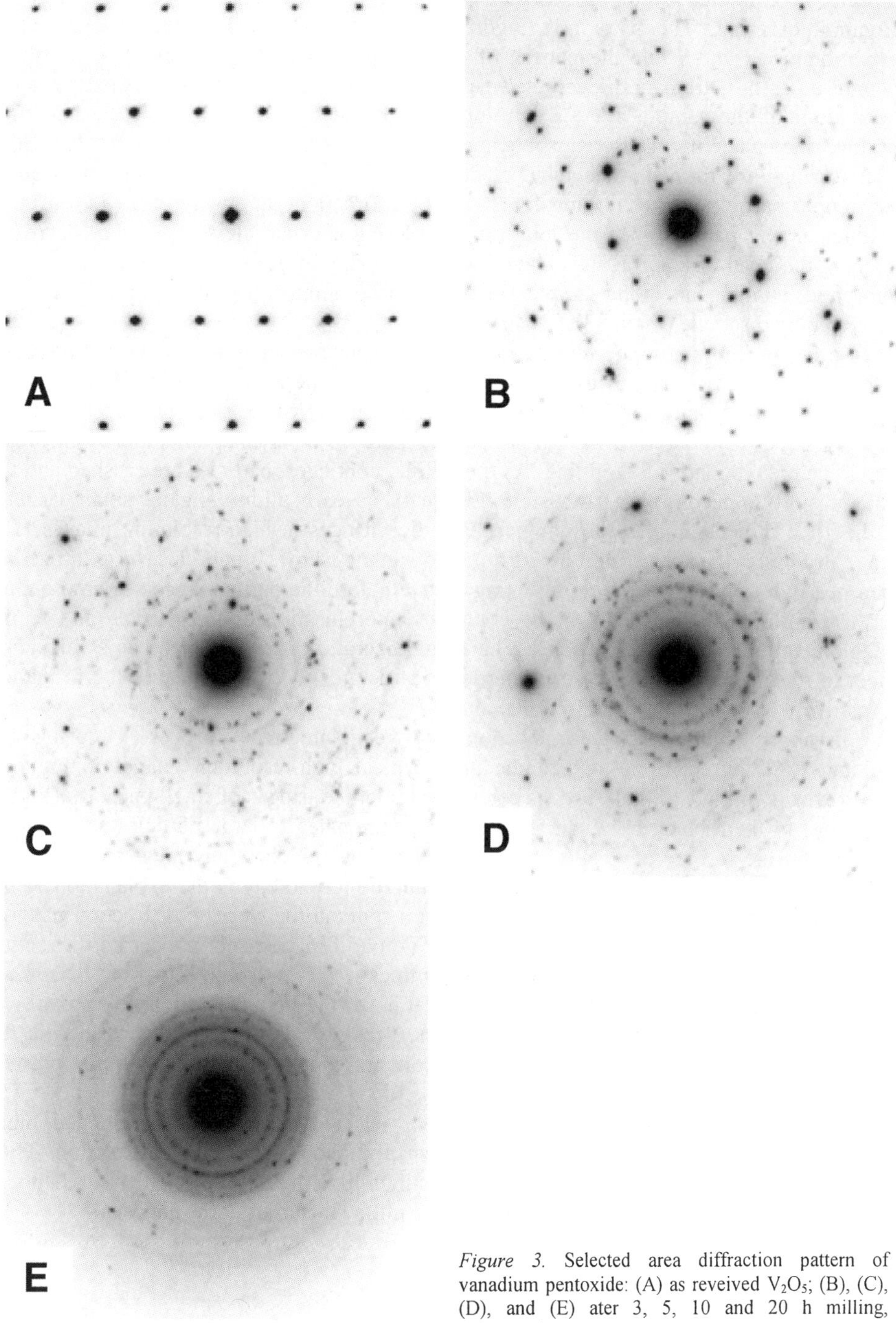

Figure 3. Selected area diffraction pattern of vanadium pentoxide: (A) as reveived V_2O_5; (B), (C), (D), and (E) ater 3, 5, 10 and 20 h milling, respectively.

Selected-area electron diffraction patterns in Figure 3 reveal the drastic loss of internal structural ordering. The pattern in Figure 3A can be identified as the 001-pattern of orthorhombic V_2O_5. While the pattern of the 2h milled sample (Figure 3B) can still be recognised as the 001-pattern decorated with additional spots due to the crushed particles, the pattern of the sample after 5 h milling (Figure 3C) becomes speckled. After 10 h milling, rings made up of discrete spots appear in the diffraction pattern (Figure 3D). Such discrete rings are formed when incident electrons are diffracted by randomly distributed nanoscopic particles. After milling for 20 h, the diffraction pattern develops into a more continuous ring pattern since the long range ordering and particle sizes further decrease, reaching the level of amorphous matter (Figure 3E). The diffuse background caused by the amorphous components in the samples becomes very prominent, especially in the pattern of the 20 h milled sample (Figure 3E).

In order to know how the milling induces the change at an atomic scale, we recorded high-resolution lattice images from samples after various periods of milling. The image in Figure 4A is from untreated V_2O_5, showing perfect lattice fringes without defects. Milling for 3 h induces disordered surface of particles (Figure 4B). The image in Figure 4C reveals the lattice breaking after 5 h milling. As the milling continues, particles are crushed and the lattice distortion becomes clearly visible (Figure 4D). Nearly all investigated small particles of the 10 h milled samples show imperfect lattice fringes in high-resolution images. The image in Figure 4E, taken from the 20 h milled sample, shows a particle whose lattice has been nearly totally destroyed, representing a state before the particle becomes amorphous. Further milling will then lead to the complete amorphisation of the particle and to the reagglomeration with other particles. The reagglomeration can take place as a relaxation effect to release the energy stored in the defect structures. It can even cause the formation of relatively large particles with irregular shape.

For understanding the amorphisation and reagglomeration processes during the prolonged milling time, we studied the three typical high-resolution images in Figure5. They reveal two kinds of contrast: (i) contrast with lattice fringes of small particles (Figure 5A), most of them being less than 5 nm in size, aggregated to large particles as shown in the TEM-image in Figure 3E and (ii) contrast that does not show any long-range order or periodic structures (Figure 5B). This two-fold micro-morphology is the explanation of the overlay of continuous diffraction rings with small spots in the electron diffraction patterns and of the strong background recorded both in X-ray diffraction and in electron diffraction patterns. An amorphous layer of about 5 nm in thickness can also be found on the surface of large particles (Figure 5C). Careful investigation of the high-resolution micrographs in Figures 5B and 5C reveals the existence of short-range order (marked by circles) in the apparently amorphous regions. The platelets contain an irregular array of oxide clusters with a well-defined internal structure.

The knowledge gained by means of TEM, selected electron diffraction and high-resolution imaging cannot be obtained by *integral* methods such X-ray diffraction. TEM allows the access to the information about ball milling microscopically: what happens during milling and which effects it induces on the milled materials at the nanoscale.

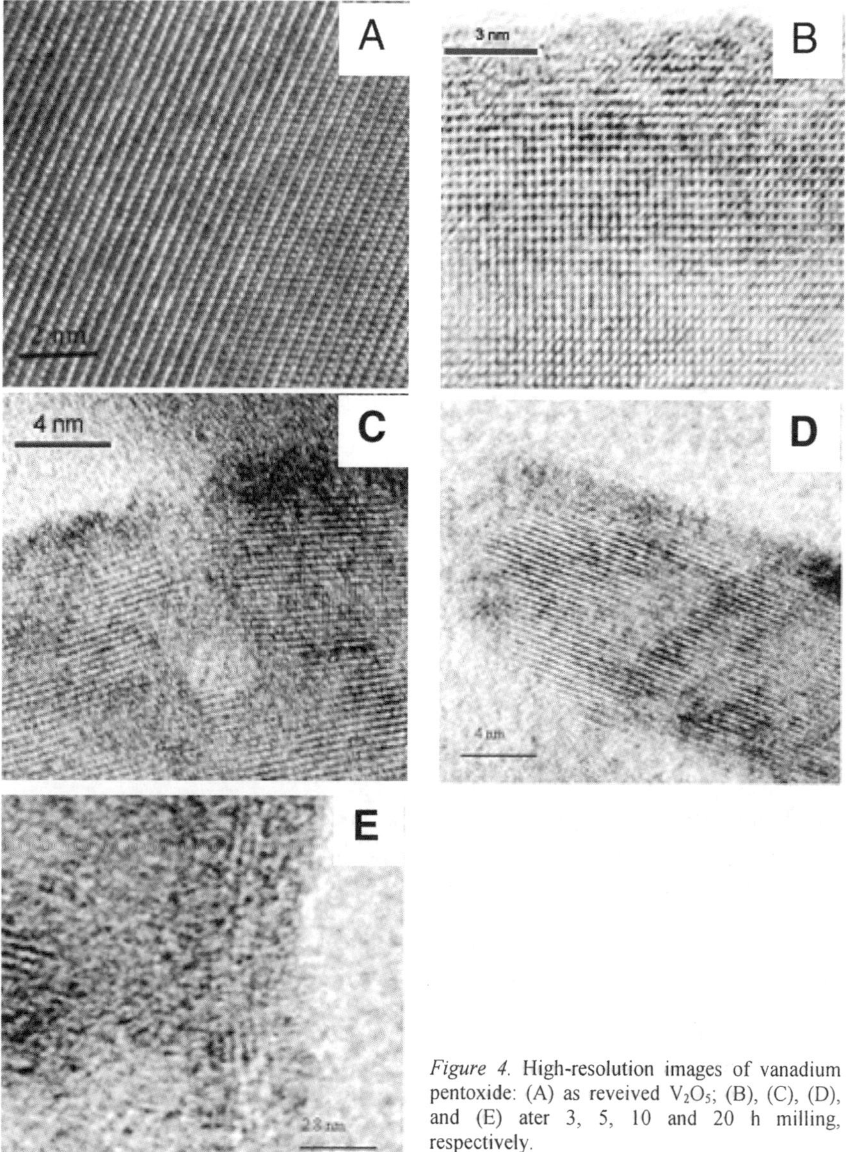

Figure 4. High-resolution images of vanadium pentoxide: (A) as reveived V_2O_5; (B), (C), (D), and (E) ater 3, 5, 10 and 20 h milling, respectively.

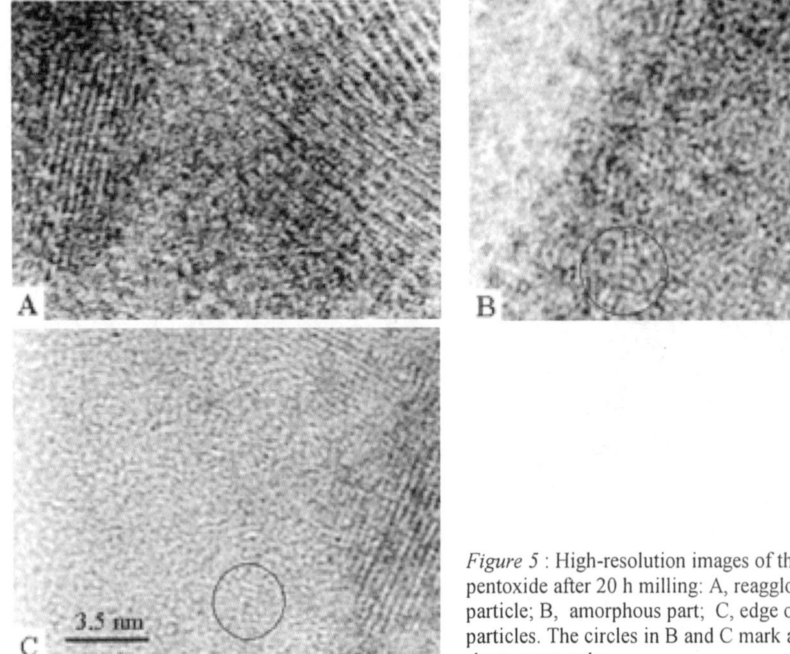

Figure 5 : High-resolution images of the vanadium pentoxide after 20 h milling: A, reagglomerated fine particle; B, amorphous part; C, edge of a large particles. The circles in B and C mark areas with short-range order.

4. Oxidation State

The observed structural changes must be accompanied by chemical changes. EELS, which is nowadays a standard facility in an analytic TEM, can be easily used to access to such information. Up to 10 energy-loss spectra of each sample at different positions were recorded. Vanadium 2p spectra were then retrieved from the measured data. Typical spectra of samples after 3 h and 20 h milling, taken at different positions in each sample, are shown in Figure 6, together with the spectrum of untreated V_2O_5. This spectrum is characterised by the vanadium $2p \rightarrow 3d$ transition (V L-edge): the two peaks, located at 519 and 525.7 eV, are the V L_3 and L_2 edges, attributed to the transitions from the V $2p_{3/2}$ and V $2p_{1/2}$ core levels to the unoccupied V $3d$ bands, respectively .

In comparison with the spectrum of untreated V_2O_5, significant changes in the spectra of milled samples are shifts of the V L_2- and L_3-edges to lower energy loss. These changes evidence that milling induces the chemical reduction of vanadium ions. This reduction is caused by the loss of oxygen atoms caused by milling. A decrease of the intensity of O K edge was observed as milling time increased. The chemical shift depends on milling time. For samples milled for 3 h and 5 h, spectra that are identical with the spectrum of untreated V_2O_5 can still be recorded. In the spectrum of the sample after 20 h milling the chemical shift of the V L_2- and L_3-edges to lower energy becomes significant. The chemical shifts measured for one sample at different positions vary, reflecting the chemical inhomogeneity of the samples introduced by milling. The measured V L_3 edge positions of all milled samples are summarised in Table 1. Using a correlation curve between the oxidation state of vanadium and the energy position of V L_3 edge [7], the oxidation states of vanadium ions after milling can be estimated (see Table 1).

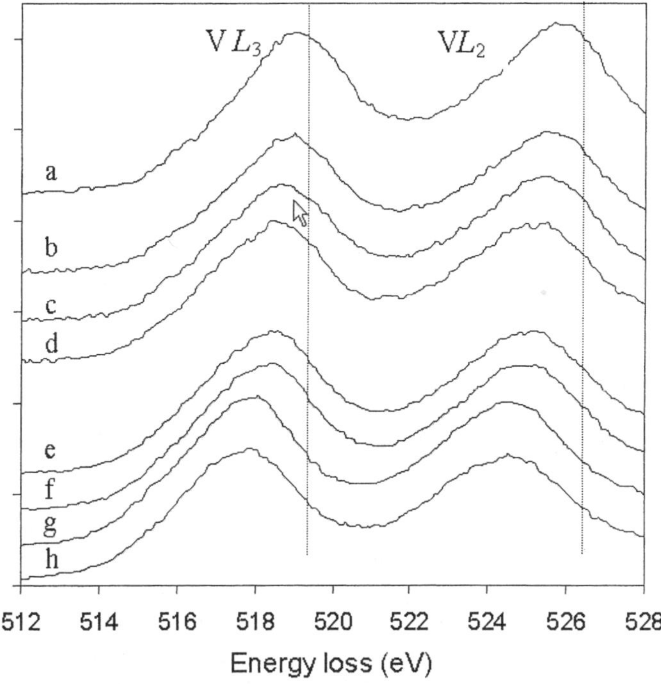

Figure 6. EELS-spectra of vanadium pentoxide: a, untreated; b-d, after 3 h milling; e-h, after 20 h milling.

In Table 1 we can see that the sample milled for 3h contains unreduced small particles (oxidation state 5) and reduced small particles (oxidation state 4.3). Milling up to 5 h induces further reduction of vanadium (down to 4.1), but unreduced particles can still be found. It is supposed that the unreduced particles in these two samples are formed during the milling by crushing of large particles. In the sample milled for 10 h, however, all the analysed particles are reduced; the highest oxidation state is 4.6 and the lowest oxidation state is 3.7. Milling for 20 h gives rise to the total reduction of the V^{5+} state and only V^{4+} and V^{3+} states can be detected. The main effect of milling up to 5 h is therefore the crushing of crystallites into small ones, with a minor loss of oxygen.

TABLE 1. V L_3 edge positions and oxidation states of untreated and milled vanadium pentoxide.

Sample	V_2O_5	3h	5h	10h	20h
V L3-peak positions	519.0	519.0-518.7	519.0-518.6	518.8-518.2	518.5-517.8
Oxidation state	5	5-4.3	5-4.1	4.8-3.7	4-3.4

For milling up to 20 h, the mechanical deformation and chemical reduction are equally effective. The inhomogeneity of oxidation states in all four milled samples can be explained as a dependence of the chemical state on the history of milling: the mechanical distortion of long range order is followed by a chemical reduction to a vanadium oxidation state below 5+ with two time constants in the order of several hours.

5. Discussion

Our investigation, performed by a combination of TEM and EELS, reveals that mechanical modification of V_2O_5 in a planetary agate ball mill can be divided into two stages, as illustrated in Figure 7. The first stage ranges from initial milling to about 5 h where large particles are crushed, with a minor loss of oxygen. This is a macroscopic process of mechanical deformation and fracture. The second stage describes the transformations induced by extended milling times up to 20 h when particles undergo fine grinding and reduction. Lattice defects become severe and the size distribution tends to narrow. Finally, amorphisation and reagglomeration take place leading to scarf-like particles, accompanied by the reduction of V^{5+} in small particles to V^{4+} and V^{3+}. This is a microscopic process that may have been ignored in earlier investigations. As revealed by high-resolution imaging, amorphisation and reagglomeration occur in a nanoscopic size range. Physically, in the first period the energy supplied by ball milling is partly exhausted by crushing the crystallites. In the second period, the mechanical energy is partly stored as strain energy in the particles. Most particles studied for this period show heavily deformed lattices. Therefore it seems that amorphisation and reagglomeration are consequences of extensive mechanical modification of V_2O_5.

By studying the chemical changes of milled sample using EELS, another interesting phenomenon was observed in the present study: the inhomogeneity in all milled samples after various durations of milling. In particular, the formation of V^{4+} and V^{3+} species depends on the history of milling: the mechanical distortion of long range order takes place after several hours and the reduction starts at all milling times, although it becomes strong in the second stage mentioned above. Obviously mechanical modification of V-based materials may not produce catalysts with homogenous chemical properties.

Milling equilibrium is reported for some vanadium oxide systems [4] where further milling causes only little macroscopic changes. Our results show, however, a continuous deformation and reduction of vanadium oxides at various periods of milling. Amorphisation and reagglomeration of particles take place. Therefore, the identification of a relevant parameter for controlling the milling becomes inevitable in the preparation of high performance catalysts.

One criterion can be the abundance of V^{4+} centres or of phases containing vanadium in the V^{4+} state, since it was found that vanadium oxide systems containing both V^{5+} and V^{4+} in a suitable ratio and abundance exhibit the best catalytic performance [3].

The oxidation state of vanadium can also be determined by other methods such as ^{51}V-NMR, EPR, and magnetic susceptibility measurements. However, all these methods, like X-ray diffraction, are integral methods and provide values averaged over a large number of particles. For instance, the average oxidation state of the sample after 20 h milling, determined by means of ^{51}V-NMR was found to be 4.8 [2]. This value is rather higher than the values summarised in Table 1.

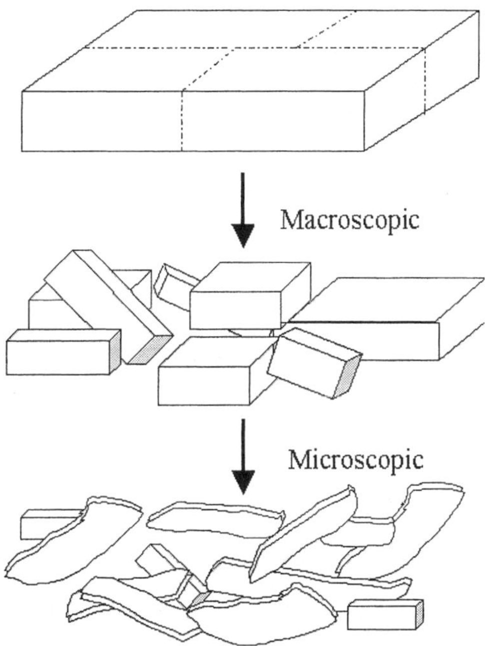

Figure 7. An illustration of the mechanical modification of V_2O_5. In the macroscopic process large particles are crushed into small ones. In the microscopic process amorphorisation and reagglomeration take place.

This is, however, no conflict to our results since the mentioned techniques do detect vanadium ions in the entire sample and do not provide information on the chemical state of vanadium within sub-micrometer regions. On the other hand, the present study does not analyse thick particles of the specimen. From the fact that our measurements were performed on very thin and small particles (less than 50 nm), it can be concluded that vanadium ions in the bulk of large particles could still be in a high oxidation state close to 5. However, the increase of specific surface area, an expected effect of ball-milling and important for the catalytic performance [2, 3], is achieved mainly because of the existence of a large number of small particles, the oxidation states of which are changed in a microscopic process (Figure 7).

6. Summary

Using mechanically activated V_2O_5, we show the application of analytic TEM in evaluating the microstructure and local chemistry in catalysis. Analytical TEM, with its excellent lateral resolution, provides not only a unique technique to access the information of catalytic materials at nanoscale. It delivers complete geometric structure information (SAED, HREM) that all the other analytic methods cannot provide. For the studied samples, the process of the activation of vanadium pentoxide V_2O_5 can be described as mechanical deformation (macroscopic process) followed by fine grinding, amorphisation and reagglomeration (microscopic process). No milling equilibrium state can be found. In the macroscopic process, the particle size decreases through fracture that takes place in all directions. Lattice defects are induced and the outer shell of most particles becomes amorphous. In the microscopic process small particles become completely amorphous and scarf-like particles grow through reagglomeration. Energy-loss spectra reveal the reduction of vanadium via oxygen loss. The formation of V^{4+} and V^{3+} species depends on the history of milling. In samples after 20 h milling, all studied small particles contain vanadium mostly in the V^{4+} and V^{3+} states.

7. Acknowledgements

The work at FHI is supported by SFB 546 of the Deutsche Forschungsgemeinschaft (DFG). The author thanks Prof. Knönziger for providing the samples and Prof. Schlögl for instructive discussion.

8. References

1. Tiemeijer, P. C., van Lin, J. J. A., and de Jong, A. F. (2001) Monochromized 200 kV (S)TEM, Microsc. Microanal. 7, 1130-1131.
2. Shubin, A. A.,. Lapina, O. B. N, Bosch, E., Spengler, J., and Knözinger, H. (1999) Effect of milling of V_2O_5 on the local environment of vanadium as studied by solid-state ^{51}V NMR and complementary methods, *J. Phys. Chem. B* **103**, 3138-3144.
3. Zazhigalov, V. A., Haber, J., Stoch, J., Bogutskaya, L. V., and Bacherikova, I. V. (1996) Mechanochemistry as activation method of the V-P-O catalysts for n-butane partical oxidation, *Appl. Catal. A: General* **135**, 155-161.
4. Fait, M., Kubias, B., Eberle, H.-J., Estenfelder, M., Steinike, U., and Schneider, M. (2000) Tribomechanical pretreatment of vanadium phosphates: structural and catalytic effects, *Catal. Lett.* **68**, 13-18.
5. Egerton, R. (1996) *Electron Energy-Loss Spectroscopy in the Electron Microscope,* Plenum Press, New York. 262-269.
6. Su, D. S., Wieske, M., Beckmann, E., Blume, A., Mestl, G., and Schlögl, R. (2001) Electron-beam-induced reduction of V_2O_5 studied by analytical electron microscopy , *Catal. Lett.* **75**, 81-86.
7. Su, D. S., Roddatis, V., Wilinger, M., Weinberg, G., Kitzelmann, E., Schlögl, R., and Knözinger, H. (2001) Tribochemical modification of the microstructure of V_2O_5, *Catal. Lett.* **74**, 169-175.

SCANNING PROBE MICROSCOPY CHARACTERIZATION OF CLUSTER SYSTEMS

WANDA POLEWSKA[1], RYSZARD CZAJKA[1], JAROSŁAW GUTEK[1], T. HIHARA[2] AND A. KASUYA[3]
[1]*Institute of Physics, Faculty of Technical Physics, Poznan University of Technology, ul. Nieszawska 13a, 60-965 Poznań, Poland*
[2] *Nagoya Institute of Technology (NIT), Department of Materials Science and Engineering, Gokisho-cho, Showa-ku, Nagoya 465-0097, Japan*
[3] *Center for Interdisciplinary Research, Tohoku University, Aramaki-Aza Aoba, Aoba-ku, Sendai 980-8578, Japan*

Abstract
Scanning Probe Microscopy (SPM) is one of the most powerful techniques allowing us to image, manipulate and modify matter at the nanometer scale. However, when imaging small cluster systems, the simple topographic interpretation could be misleading. We show STM images of small Ni clusters deposited on graphite (HOPG) substrate where they appear once as protrusions and secondly as hollows due to the nonlinear dependence on the density of states (DOS) vs. energy. Our results also show that the apparent height of imaged clusters is much closer to the real value of cluster diameter than an apparent diameter as it is seen in the STM images.

Electrochemical STM (ECSTM) allows "in situ" investigations of the processes taking place at the electrode-electrolyte interface. It is shown how the desired surface reconstruction can be induced on surfaces via potential control in the case of Tl monolayer deposited on Au(111) substrate. It is also shown that the observation of the building up and dissolution processes of Cu nanostructures can be traced in situ and in real time using ECSTM with the atomic resolution.

Atomic Force Microscopy (AFM) is suitable for investigation of clusters deposited on solid surfaces independently of their electronic properties. We present typical AFM images of Cr clusters deposited on Au/mica as well as AFM images which show the contrast between the Co metallic core and the surrounding ligand layer when using so called phase imaging option.

L.M. Liz-Marzán and M. Giersig (eds.),
Low-Dimensional Systems: Theory, Preparation, and Some Applications, 253–265.

1. Introduction

Twenty years have just passed since the first Scanning Tunneling Microscope (STM) was created by Gerd Binnig and Heinrich Rohrer [1] at IBM Lab in Ruschlikőn. This instrument enabled imaging of the solid surfaces with an atomic resolution in real space for the first time. Binnig with some other co-workers [2] developed this idea and they constructed an instrument, which allows imaging of any material, including insulators, called Atomic Force Microscope (AFM). Nowadays there are many mutations of these predecessors, which may work in different environments (vacuum, liquids, and gases) and investigate different physical properties of materials in the nanoscale. The whole technique based on these instruments is called Scanning Probe Microscopy on the basis of this simple fact that the image of a given surface property is collected via probe scanned above the investigated sample's surface.

Contrary to some other techniques as X-ray, SEM or TEM - SPMs enable the measurements in real space with a real atomic resolution. That means that individual atomic defects or adsorbents can be resolved at the investigated surface. This is a very important point because the present day material and electron technology is based on or creates structures of a few nanometers dimensions. The presence of a countable number of defects or dopants may strongly influence their properties and/or their operation.

These new materials and devices more and more often are composed of the nano "bricks" - clusters. Clusters belong to the mesoscopic systems, which means systems composed of a countable, from few to a few thousand, number of atoms or molecules. The physical properties of clusters are usually different from those of both bulk materials and individual atoms. These properties depend strongly on clusters dimensions. For example, small metal clusters do not behave as metals. Some of these properties are straightforwardly related to the density of electron states and Scanning Tunneling Microscopy/Spectroscopy (STM/STS) methods again are one of the best for their investigation.

In this article we would like to present a brief description of the SPM technique with some focus on electrochemical version of STM, and SPM's application to investigations of clusters deposited on different substrates. Electrochemical STM allows "in situ" investigations of the processes occurring at the electrode-electrolyte interface. The special construction of the EC STM gives the possibility to control and stimulate the electrochemical processes at the surfaces immersed in different electrolytes, also via tracing the movement of individual atoms or clusters. It makes the ECSTM a powerful instrument for investigations of the surface reconstruction, crystallization and dissolution of monocrystal electrodes, and investigations of corrosion phenomena.

2. Scanning Tunneling Microscope (STM)

Scanning Tunneling Microscope operates using the quantum-mechanical phenomenon of electron transfer ability through classically forbidden insulating barrier [3] - so called tunnel effect. If STM operates in vacuum, electrons may be transferred from the STM tip to the sample (or in the opposite direction) only when there are allowed and non-occupied electron states on the other electrode. Applying the bias voltage between the STM tip and sample as it is presented in Figure 1 can create such a situation.

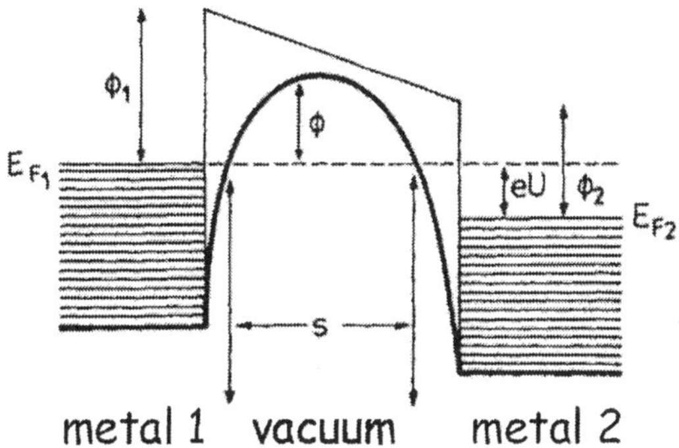

Figure 1. Diagram of energy levels for a tunnel junction biased with U voltage; **s**- the width of the potential barrier, φ_1 and φ_2 - work functions, for the left and right electrode, respectively.

Quantum calculations [4], lead to the general dependence of the tunnel resistance R vs. the tunnel barrier width s, as follows:

$$R \propto e^{Cs\sqrt{\overline{\varphi}}} \tag{1}$$

where $C = 4\pi \frac{\sqrt{2m}}{h} = 10.25\ (\text{eV})^{-1/2}\ \text{nm}^{-1}$

m - electron mass,
h - Planck's constant,
$\overline{\varphi}$ - average work function for the left and right electrodes.

According to this dependence I_t decreases exponentially with the probe-sample distance. Assuming typical values, $s\approx1$ nm and $\overline{\varphi}\approx5$ eV, one can evaluate that 1% change of s will cause 50% change of I_t value. This is the main reason why STM resolution in perpendicular to the surface direction is so extremely high.

Tunnel current is only one of the four main factors necessary for the STM to image the surface topography. The others are: a probe in the form of a sharp tip with curvature radius of several nanometers, a piezo-ceramic scanner which enables the probe scanning over the surface with sub-nanometer accuracy, and an electronic circuit which controls and regulates, via feedback control, the constant probe-sample distance. The schematic drawing is shown in Figure 2.

Figures 3a and 3b show well known images of the Si(111) crystal surface reconstruction (7×7) taken for opposite bias polarization. The reconstruction is developed through the top four atomic layers as it was proposed by Takayanagi et al. [5]. The two halves of the surface elementary cell are not equivalent and we have structure of *p3m1* symmetry instead of the expected one - hexagonal *p6mm*.

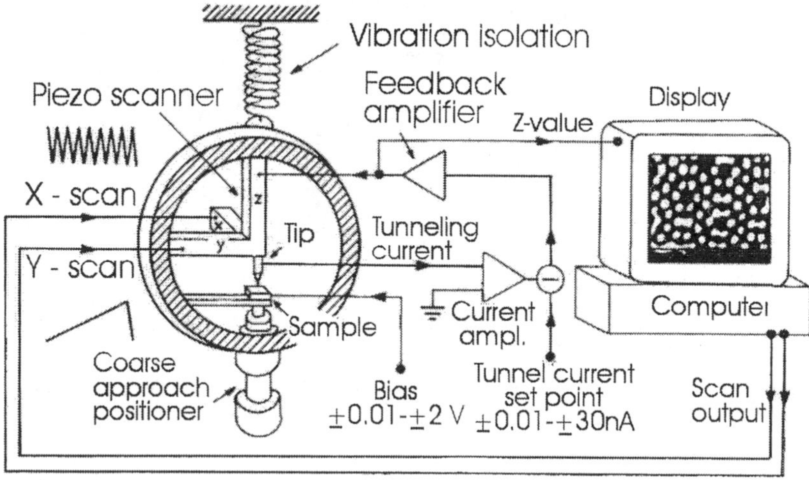

Figure 2. Schematic diagram of the STM: (a) the STM tip of a conical shape is attached to a piezo-ceramic scanner enabling the tip movement in XYZ. STM tip is moving at the constant distance from the investigated surface and reflects topography of the surface with atomic resolution.; (b) schematic diagram of the STM apparatus: STM stage consist of a scanner and the sample and is isolated from the external vibration via spring system; after approaching the tip towards the sample surface within 1 nm distance and registering the tunnel current, the tunnel current value is compared to the reference current. The differences appearing on the feedback loop output due to the change in tip-sample distance activate so called negative feedback loop. That means that detected increase of the tunnel current value leads to the decrease of the voltage applied to the Z-piezo. The changes in Z-piezo give the data for the STM image.

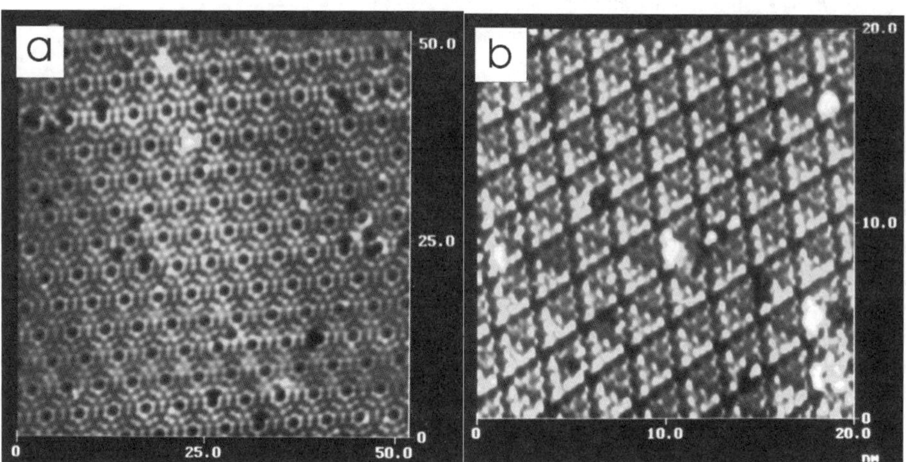

Figure 3. STM image of the Si(111) surface: (a) for the positive bias of the sample against the tip (+2V), 50x50 nm^2; (b) for negative bias (-2V), 20x20 nm^2.

One can realize it looking at Figure 3b, which is different from 3a. The conclusion is that STM images represent contours of the constant electron charge density taken at given space position and at given energy level (determined by the bias polarization) against the Fermi level. Therefore, the STM actually gives spectroscopic information and the simple topographic interpretation can be misleading. For example, the image of small Ni clusters deposited on HOPG exhibits them as protrusions for one bias polarization and as hollows for the opposite one - see Figures 4a and 4b.

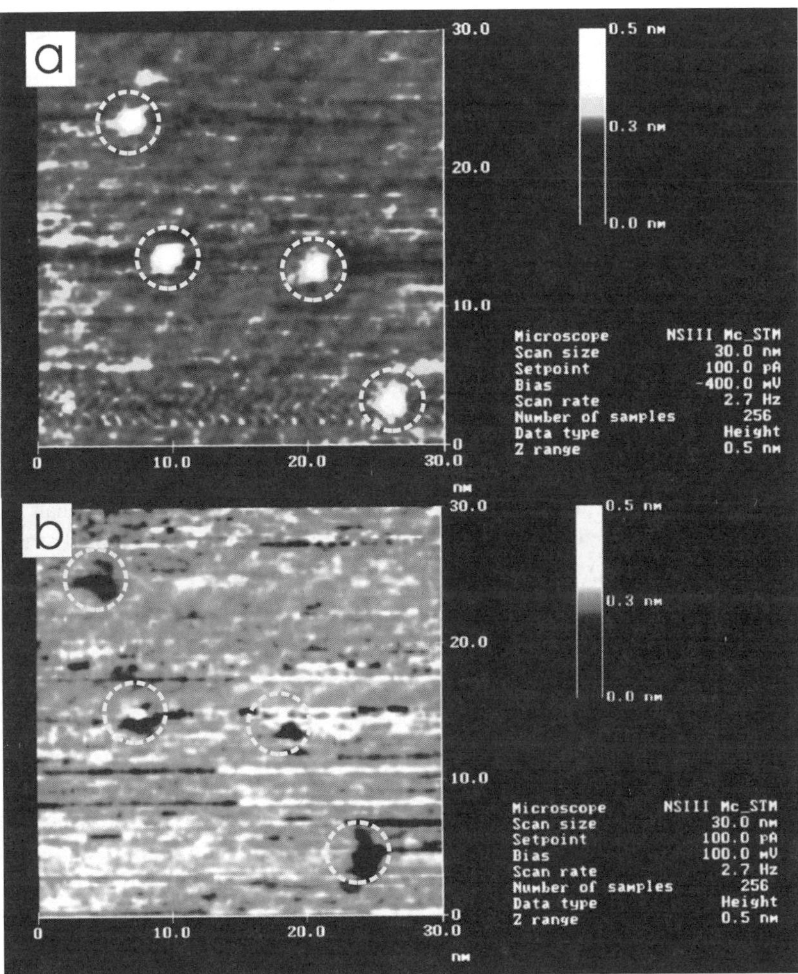

Figure 4. STM image of the same Ni clusters deposited on the graphite (HOPG): (a) for the negative bias of a sample against the tip (-0.4V); (b) for the positive bias (+0.1V). Cluster positions are marked by the dashed circles. The scanned area dimension is 30 nm x 30 nm; z-scale (black to white color) 0 to 0.5 nm.

The electron charge density is explicitly present in the tunnel current formula (2)

$$I \propto \int_0^{eV} \rho_s(r,E)\rho_t(r,-eV+E)T(E,eV,r)dE \qquad (2)$$

where: $\rho_s(r,E)$ and $\rho_t(r,E)$ represent the density of states for the sample and probe respectively evaluated at position r and for energy E measured against their individual Fermi's levels.
$T(E,eV,r)$ is the tunnel transmission probability for electrons with energy E at a given bias polarization:

$$T(E,eV) = \exp\left(-\frac{2s\sqrt{2m}}{\hbar}\sqrt{\frac{\phi_s+\phi_t}{2}+\frac{eV}{2}-E} \right) \qquad (3)$$

where: s - sample-probe distance,
ϕ_s *and* ϕ_t - work functions for the sample and probe respectively.
$\hbar$ - Planck's constant/2π.

Usually DOS for metal tips (W, Pt/Ir, and Au) are a smooth and continuous function of energy (or bias voltage) therefore any peculiarities seen in STM images appear due to the sample properties. In the case of Figure 4a the Ni clusters are seen as protrusions due to higher DOS than DOS for HOPG substrate at −0.4 V bias and as hollows in Figure 4b at +0.1 V bias due to opposite cluster/substrate DOS ratio. One can expect such peculiarities when imaging small clusters or any other low-dimensional structures (nanowires, quantum dots). The DOS is not a smooth function of energy in the latter case.

3. Electrochemical STM

The invention of STM enabled an arousal of the new branch of electrochemistry which might be termed “electrochemical microscopy” (ECSTM). Without this tool electrochemists would be still only guessing (based on I-V characteristics called voltammograms and UHV techniques) what is really going on at the solid – liquid interface. When working in an electrolyte we must take into consideration that both the substrate and the tip are susceptible to electrochemical charge transfer reactions. Without proper control, these processes can cause changes in surface morphology and in the chemical composition of the tip and the substrate. Also, the electrochemical current components can disturb the STM control system through increased noise and fluctuations in the tip current. These problems imply the necessity of the application of the potentiostatic STM, where we have a possibility to independently adjust the potential differences E_T and E_S between the reference electrode and the tip and the substrate respectively. To achieve the optimized reduction of the electrochemical tip current, very often referred to as “faradaic current”, the tip should be composed of such a material whose electrochemical behavior in a given electrolyte shows purely capacitive properties over a wide potential window. An additional and widely applied possibility to minimize the faradaic current is to cover the tip, except its very end, with an insulating layer – for example apiezon.

The most important advantage of the ECSTM is that it enables observations "*in situ*" of the processes taking place at the electrochemical interface such as for example: surface reconstruction, anion adsorption or metal electrode dissolution/deposition. Moreover, by changing the potential of the metal substrate it can be instantaneously seen how the shift in surface potential can affect these processes.

ECSTM seems to be a very useful tool in studying the structure of foreign metal adlayers, obtained via so called underpotential deposition (UPD) on single crystal electrode surfaces. At potentials relatively positive of the reversible Nernst potential of the metal electrode surface (underpotential) the substrate will bond with the foreign metal from the electrolyte, forming its monolayer on the top of the electrode. UPD monolayers are of interest to a broad range of researchers. Solid-state physicists can compare them to monolayers formed by UHV techniques. Those, working in bulk deposition, electroplating and deposition in microelectronics could use UPD, because the texture of bulk is strongly influenced by the UPD monolayer. Foreign metal adlattices, created by UPD may have different structures, depending on the electrolyte. This can be explained by the differing degrees of anion participation in the adlattice.

One of the first in situ ECSTM studies of UPD was made for copper on gold [6]. Here we present the UPD of Tl on Au(111) in 0.1M NaOH + 5mM TlF. The crucial factor taken into consideration while choosing the electrode potential window for UPD studies of this system were the data obtained from I-V characteristics [7] which have indicated two interesting regions. First of them was between –0.67 V and –0.38 V vs. Ag/AgCl reference electrode, where a Tl monolayer was expected atop the Au (111) electrode surface. Figure 5a [8] shows a typical high-resolution image of this monolayer, taken at –0.44 V. We can see the modulated, hexagonal lattice with the nearest-neighbor distance of (0.34 ± 0.02) nm. In addition, the hexagonal lattice is rotated by 6° ± 1° in comparison to the principal Au axes. The modulated pattern (called moiré pattern) is consistent with the hexagonal superstructure that arises from the mismatch between Tl and Au lattices spacing (see Figure 5b). The distance between the maxima of the superstructure is (1.5 ± 0.1) nm and within this there are 20 Tl atoms over 27 underlying gold atoms which gives the estimated coverage of $\Theta = 0.74$.

An increase in the surface potential to the value of –0.33 V (which means the transition to the second interesting potential region) results in partial desorption of Tl atoms from the hexagonal monolayer and co-adsorption of OH^-, which leads to the arousal of a new structure in the Tl monolayer. This structure is shown in Figure 5(c) and might be described as an aligned, hexagonal phase with the Tl – Tl spacing of (0.38 ± 0.02) nm (see Figure 5d). Now, the distance between maxima in moiré pattern is (1.15 ± 0.1) nm. Within this superstructure there are about 9 Tl atoms over 16 underlying Au atoms which gives the estimated coverage of $\theta = 0.56$. The existence of such type of phase, where interatomic distance in the deposited monolayer of Tl is larger than in the bulk, is possible due to the co-adsorption of partially charged OH^- anions which stabilize such an "open" structure.

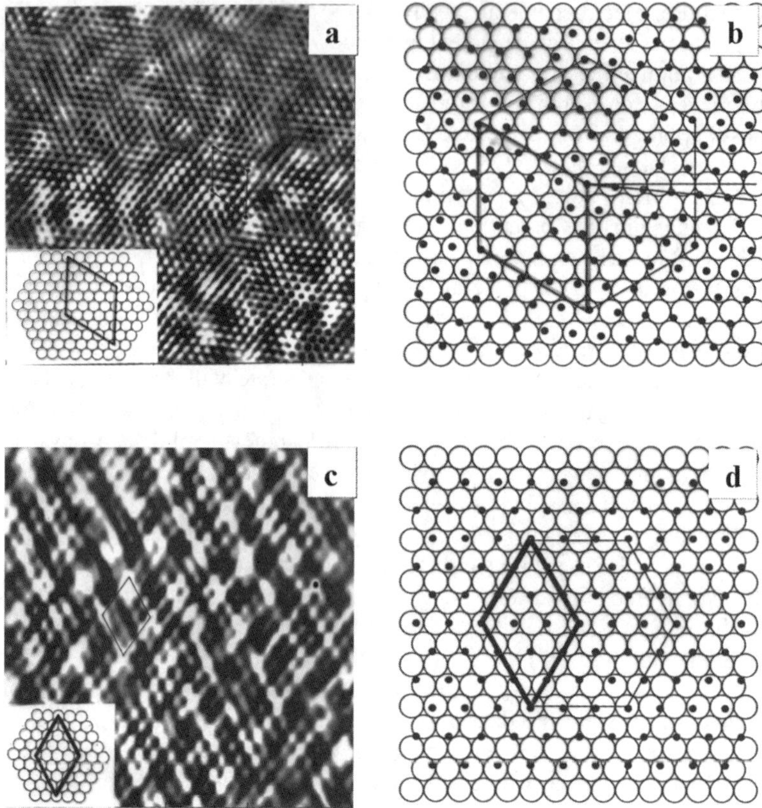

Figure 5. (a) Atomic resolution STM image (12 x 12 nm) of the rotated, hexagonal phase of Tl obtained at –0.49 V (vs. Ag/AgCl/3 M KCl) in 0.1 M NaOH + 5mM Tl^+. The Tl – Tl distance is 0.34±0.02 nm. The rotation angle of Tl adlattice versus Au(111) substrate is 6^0+1^0. A moiré pattern is seen with a repeat distance of 1.5±0.1 nm. The inset shows a model of the Au substrate with the schematic representation of the superstructure. (b) Model of the Au(111) surface covered with the rotated, hexagonal Tl adlayer, based on [8]. White circles denote Au atoms, black circles Tl atoms. (c) Atomic resolution STM image (8.7 x 8.7 nm) of the aligned – hexagonal phase of Tl obtained at –0.32 V. The Tl – Tl distance is 0.38±0.02 nm. The rotation angle is zero. A moiré pattern is seen with the repeat distance of 1.15±0.01 nm. The inset shows the model of the Au substrate with the schematic representation of the superstructure. (d) Model of the Au(111) surface covered with the aligned – hexagonal Tl adlayer, based on [8].

As it was mentioned above, ECSTM plays an important role in studying *in situ* electrochemical dissolution and deposition of metals. These processes are important in metal corrosion, technological etching, coating and metal refinement. Their possible application (especially for copper) in fabrication of modern ultra-large scale integrated circuits has raised new interest also in fundamental aspects of these processes.

Detailed studies of copper growth or dissolution on the atomic scale require significant advances in the time resolution of the STM due to high local rates of these processes. It might be done by reducing the scanning scheme (as it was shown, for

Cu(100) in HCl, time of walk "TOW" experiment described in ref. [9]) or by applying high – speed electrochemical STM, so called video-STM, described in ref. [10]. It allows obtaining atomic resolution images at acquisition rates up to 25 images per second. Figure 6 shows the direct observation of equilibrium fluctuations of Cu atoms (for Cu(100) in HCl) at Nernst potential. The presented sequence was chosen from the "video" taken with the rate of 5 images per second [11]. The copper terraces are covered by the ordered c(2×2) Cl adlayer (visible atomic lattice) and the steps exhibit the characteristic faceting along the {100} directions of copper structure, induced by this adlayer [12].

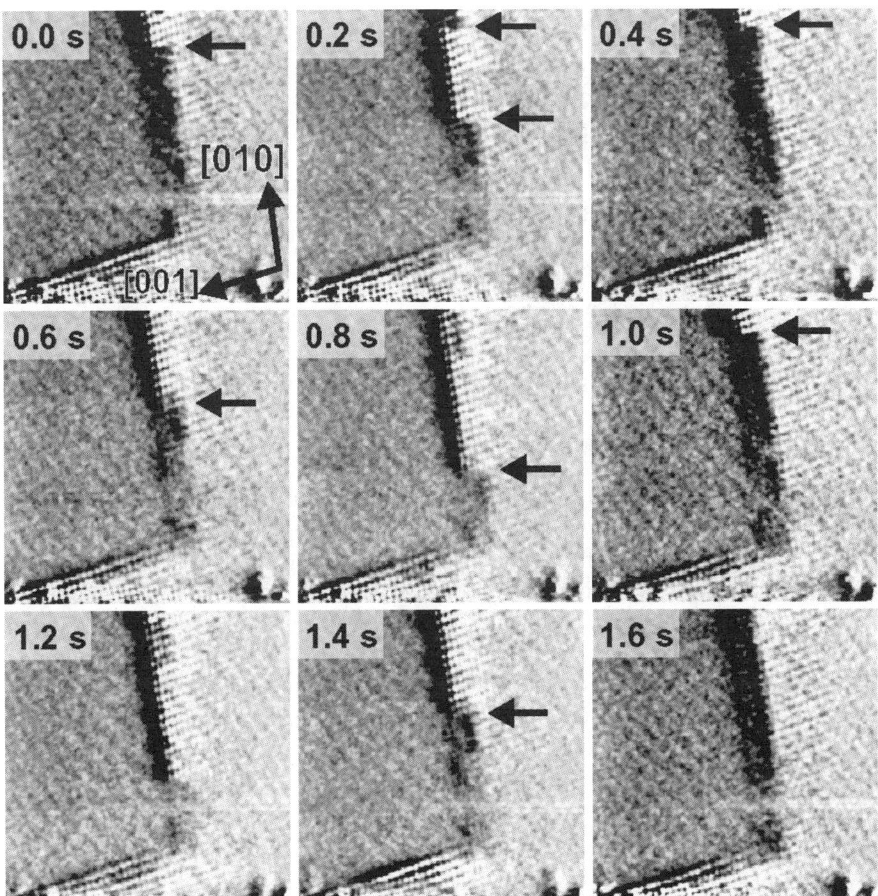

Figure 6. Series of in-situ video-STM images recorded on Cu(100) in 0.01 M HCl + $2{\cdot}10^{-6}$ M CuSO4 at –9 $mV_{Cu/Cu(I)}$ (190×115 Å^2), illustrating fluctuations of a ($\sqrt{2}\times\sqrt{2}$)R45° row along the step in [010] direction (kink position indicated by arrows)[12].

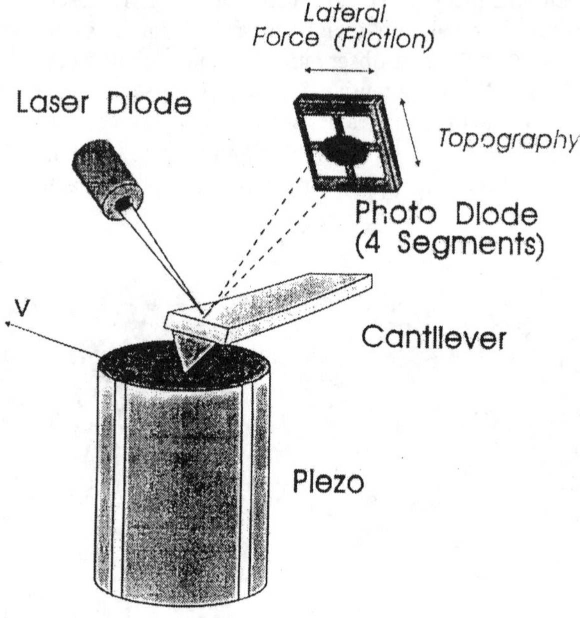

Figure 7. The diagram of the most used AFM apparatus with the laser beam deflection method for the detection of cantilever position.

Along the Cu steps the deposition or removal of rows is observed. The microscopic mechanism was studied by conventional STM, showing, that Cu(100) dissolution in HCl proceeds via removal of the primitive surface unit cell, i.e., ($\sqrt{2}$ x$\sqrt{2}$) rotated 45° units, containing two Cu and an adsorbed Cl atom, at kinks in the {001}–oriented steps. Arrows in Figure 6 indicate terminating kinks at the ends of these rows. A strong asymmetry can be observed between the two step directions of the single terrace. Growth and removal of rows occurs only along [010] direction, and not at the [001] direction. This different reactivity, resulting in local anisotropy, induced by c(2×2) Cl adlayer is the same as described in [12].

4. Atomic Force Microscope (AFM)

The STM principle excludes the possibility to image insulators. Therefore in 1986 Binnig et al. [2] have constructed the Atomic Force Microscope, which makes use of inter-atomic forces acting between atoms for imaging purposes of solid state surfaces. The general idea of AFM is shown in Figure 7.

Nowadays there are many derivatives of the original AFM construction. Their names usually are usually related to the kind of force used to create image of the given property of the investigated surface.

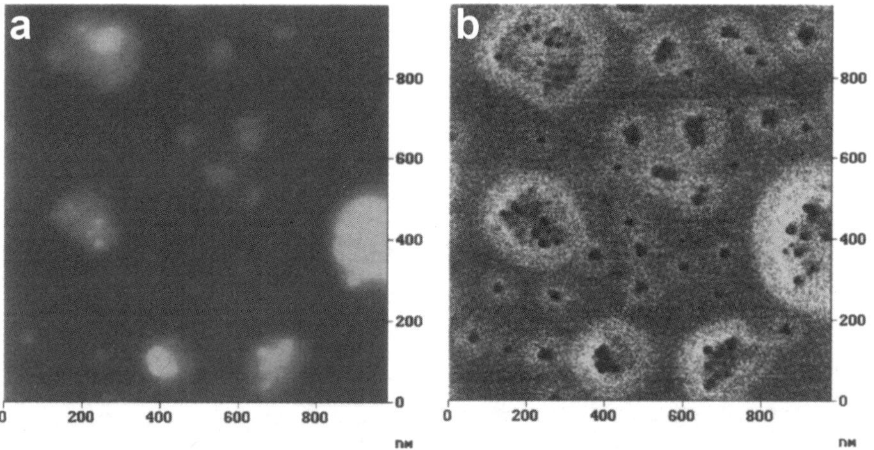

Figure 8. "Tapping mode" AFM images of Co clusters, 20 nm in diameter, deposited on graphite (HOPG) surface from toluol solution (a) topography, z-scale (black to white) 0 to 40 nm , and (b) phase-contrast image – Co clusters are seen as black dots inside the organic surrounding layer – bright spots. Scanned area dimension is 1000 nm x 1000 nm.

For example, we have Lateral (or Friction) Force Microscope LFM, Magnetic Force Microscope (MFM), Electrostatic Force Microscope (EFM) etc. [13]. All of these AFM derivatives present some advantages and disadvantages over other techniques, probing similar properties of solid surfaces, however description of this point is going far beyond the scope of this article. Let us concentrate on some examples of clusters' imaging by means of AFM. AFM may work in contact-mode, non-contact mode and intermittent modes. The latter one is using an oscillating cantilever which once per oscillation period is coming into contact with the sample. The data collected during the contact serve as data for surface topography; however the changes in the oscillation frequency or in oscillation phase, due to the interaction between the cantilever tip and the sample, may serve as a source of additional information as material contrast. Figures 8a and 8b show AFM images of Co clusters prepared by chemical methods [14], and deposited onto HOPG surface from the toluol solution. Using Nanoscope III working in "tapping mode" we can observe topography of the sample (a) and the phase image (b) where the contrast between the metallic core and organic surrounding layer may differentiate. The original Co clusters diameter was 20 nm on average and they are seen as black dots in Figure 8b. The original Co clusters diameter was 20 nm on average and they are seen as black dots in Figure 8b.

Clusters can also be created in Plasma Gas Condensation systems as it was described by Hihara et al. [15]. Figure 9 shows an AFM image of Cr clusters. AFM was working in non-contact mode using Needle Sensor probe (a sharp tip is vibrating at 1 MHz). We can observe the clusters as protrusions of about 15 nm in diameter (measured at the half of maximum height) distributed randomly on the Si(111) surface, which was not in accordance to the TEM results of the clusters deposited in identical process onto graphite membrane.

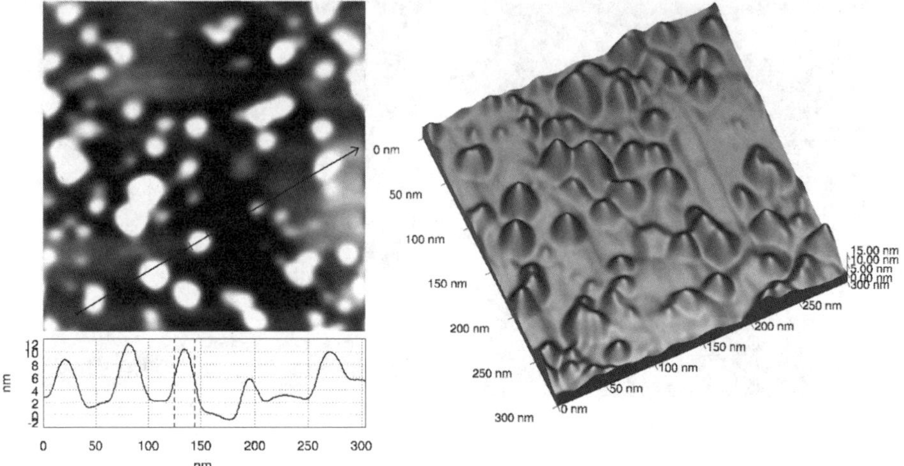

Figure 9. AFM non-contact (Needle-Sensor) image of Cr clusters (nominally 7.5 nm in diameter) deposited on Au/mica substrate (a) top view with cross-line section taken along black line; (b) quasi 3-dimensional light-shadowed image of the same sample area.

In general, in the case of imaging the objects of the same or smaller dimensions than the SPM probe, the lateral dimensions can differ from the real ones, e.g., they can be much bigger due to known tip artifact. Therefore the apparent height of imaged clusters is much closer to the real value of cluster diameter –7.5 nm in this case.

5. Summary

Scanning Probe Microscopy is the only analytical technique, which allows for imaging the surfaces and nanostructures with the true atomic resolution in the real space. SPMs also enable investigation of space resolved distribution of other physical properties, as electron density of states, friction, magnetic and electrostatic properties.

Surface reconstruction can be induced in a controlled way and imaged by means of Electrochemical STM. Dynamic processes of deposition and dissolution of metals, as it was shown in the case of Cu in HCl solution, can be imaged *in situ* and in real time with atomic resolution.

Phase imaging using AFM working in dynamic mode gives the contrast between the metallic cluster and the organic shell surrounding the metal core. SPMs resolution during imaging of objects of few nm in diameters is limited by the probe radius and may lead to false estimation of clusters' diameter. The apparent height of imaged clusters is close to the real value of cluster diameter, only.

6. Acknowledgements

The work was partially (as reg. Polish authors) done under Project KBN No. 5PO3B10021.

Authors are grateful to: Prof. R.J. Behm i Prof. O.M. Magnussen from Ulm University (Germany), Prof. R.R. Adzic and Dr. B. Ocko from Brookhaven National Lab (USA) for enabling the measurements on EC STM and valuable discussion.

Finally we would like to thank Prof. M. Giersig from Hahn Meitner Institute (Berlin, Germany) for supporting with the Co clusters in toluol solution and Prof. R. Wiesendanger for enabling us measurements on Nanoscope III at Micro-Research Center at University of Hamburg.

7. References

1. Binnig G., Rohrer H. (1982) Scanning tunnelling microscopy, *Helv. Phys. Acta* **55**, 726-735.
2. Binnig G., Quate C.F., Gerber C. (1986) Atomic force microscopy, *Phys. Rev. Lett.* **56**, 933-936.
3. Fowler R.H., Nordheim L. (1928) Electron emission in intense electric fields, *Proc. R. Soc., London A* **119**, 173-181.
4. Simmons J.G. (1963) Electric tunnel effect between dissimilar electrodes separated by a thin insulating film, *J. of Appl. Phys.* **34**, 2581-2590.
5. Takayanagi K., Tanishiro Y., Takahashi M., Takahashi S. (1985) Structural analysis of Si(111)-7x7 by UHV-transmission electron diffraction and microscopy, *J. Vac. Sci. Technol. A* **3**(3), 1502-1506.
6. Magnussen O.M., Hotlos J., Nichols R.J., Kolb D.M., Behm R.J. (1990) Atomic structure of Cu adlayers on Au(100) and Au(111) electrodes observed by in situ STM, *Phys.Rev.Lett.* **63**, 2929-2933.
7. Wang J.X., Adzic R.R., Ocko B.M. (1993) X-ray scattering study of Tl adlayers on Au(111) electrode in alkaline solutions – metal monolayer, OH^- coadsorption and oxide formation, *J.Phys.Chem.* **97**, 9754-9759.
8. Polewska W.,.Wang J.X, Ocko B.M., Adzic R.R. (1994) Scanning tunneling microscopy of electrodeposited Tl monolayers on Au(111) in alkaline solution, *J. of Electroanal. Chem.* **376**, 41-47.
9. Magnussen O.M., Vogt M.R. (2000) Dynamics of individual kinks during crystal dissolution, *Phys. Rev. Lett.* **82**, 357-360.
10. Magnussen O.M., Zitzler, L. Gleich B., Vogt M.R. and Behm R.J. (2001) In-situ atomic scale studies of the mechanism and dynamics of metal dissolution by high-speed STM", *Electrochim. Acta* **46**, 3725-3733.
11. Magnussen O.M., Polewska W., Zitzler L. and Behm R.J. (2002) In-situ atomic scale Studies of the mechanism and dynamics of metal dissolution by high-speed STM, submitted to *Faraday Disscussion*.
12. Vogt M.R., Lachenwitzer A., Magnussen O.M. and Behm R.J. (1998) In-situ STM study of the initial stages of corrosion of Cu(100) electrodes in sulfuric and hydrochloric acid solution, *Surf. Sci* **399**, 49-69.
13. Wiesendanger R. (1994) Scanning Probe Microscopy and Spectroscopy; Methods and Applications. Cambridge: University Press.
14. Giersig M., Hilgendorff M., (1999) The preparation of ordered colloidal magnetic particles by magnetophoretic deposition, *J. Phys. D: Appl. Phys.* **32**, L111-L113.
15. Hihara T., Sumiyama K., (1998) Formation and size control of a Ni cluster by plasma gas condensation, *J. Appl. Phys.* **84**, 5270-5276.

QUANTUM MECHANICAL DESIGN OF ELEMENTS OF MOLECULAR QUANTUM COMPUTERS BASED ON BILIVERDIN AND AZA–FULLERENE COMPOUNDS

A. TAMULIS, J. TAMULIENE, V. TAMULIS,
A. GRAJA*
Institute of Theoretical Physics and Astronomy, A. Gostauto 12, 2600 Vilnius, Lithuania
**Institute of Molecular Physics, Polish Acad. Sciences, Smoluchowskiego 17, 60–179 Poznan, Poland*

ABSTRACT: Quantum mechanical investigations of hydrogen and nitrogen atom Nuclear Magnetic Resonance (NMR) values of Cu, Co, Zn, Mn and Fe biliverdin derivatives and their dimers and aza–fullerene $C_{48}N_{12}$ adducts using *ab initio* and first principle methods indicate that these modified derivatives should generate from one to seven and eleven, twelve, eighteen, nineteen Quantum Bits (QuBits). It is developed analysis of localized orbitals that gives additional information concerning the nature of proton NMR of the investigated six QuBits generating Cu biliverdin derivative in different electronic states that allow to apply this analysis to other proton NMR quantum computing devices.

1. Introduction

Molecular nano–science and nano–technology devices working according quantum mechanics laws will bring us new features and possibilities due to dual particle–wave origin of matter. For example, valence electrons of molecules (which mainly determine features of nano–structures) possess strict quantum states, discrete quantum electronic, vibrational, NMR, EPR, etc. spectra.

Photoinduced electron charge transfer is going by quantum delocalized particle–wave trace (not like electrical current in present silicon–based PC computer chips)[1]. The study applied quantum mechanical *ab initio* and density functional theory–time dependent methods allow us to design light–driven, single supermolecular devices based on fullerene, biliverdin and photoactive molecules and supermolecules that could form basis for logically controlled organic molecular machines and molecular classical–digital and quantum computers[2–11]. Organic and organo–metallic molecular computers have advances in nano–size and pico– or even femto–second electron transfer as well as in principle new quantum computation[12]. Quantum computing logic elements possess not only sixteen possible two variable logic functions (as it is in our regular classical–digital computers)[1] but significantly more ones due to accounting of the electron spin direction.

L.M. Liz-Marzán and M. Giersig (eds.),
Low-Dimensional Systems: Theory, Preparation, and Some Applications, 267–280.

Experimentally observed NMR spectra of the Cu octaethylbilindione (CuOEB), CoOEB, ZnOEBOMe, Mn octaethylbilindione dimer (Mn2OEB) and Fe2OEBOMe[13–16] are in the large diapason that is useful for the design and construction the elements of the molecular quantum computers.

The aim of our work is the quantum mechanical investigation of the spectra of the proton and nitrogen atom NMR of several our modified biliverdin and aza–fullerene molecules in order to predict elements of the quantum computers that could generate the quantum bits. Our modification of various OEB and 2OEB is based on cutting octaethyl groups from biological biliverdin derivatives in order to make more clear the proton NMR spectra with certain well defined lines.

2. Used Theoretical Methods

The full geometry optimization of magnetically active molecules in the ground state was performed using density functional theory (DFT) in the framework of Becke's three parameter hybrid method and Perdew/Wang 91 gradient–corrected correlation functional (B3PW91) model in the 6–311G** basis sets using Gaussian 98 A.7 package (G98)[17–19].

The chemical shifts are obtained as the difference of the values of the tetramethylsilane (Si(CH3)4) and ammonia (NH3) molecules Gauge–Independent Atomic Orbital (GIAO) nuclear magnetic shielding tensor[20] on the hydrogen and nitrogen atoms and that of the magnetically active molecules.

In order to verify of above mentioned quantum mechanical methods and programs it was calculated relatively simple hydrogen bonding complex: pyridine–N–oxide (PyO) + hydrochloride (HCl). To perform a calculation of PyO+HCl in the presence of a solvent it was applied the Polarized Continuum (overlapping spheres) model of J. Tomasi *et al.* using the integral equation formalism[21].

Important coupling between solvent reaction field, proton transfer and chemical shift on C atom in para–position of PyO has been detected comparing the experimental and calculated NMR spectra[22]. Also note that the including of solvent effect improves the matching of the values of Cl^{-}–H^{+}O proton (^{1}H NMR) and oxygen (^{17}O NMR in R.E. Wasylishen *et al.* scale) shifts very efficiently (see Table 5 in[22]). Based on these found proton transfer and NMR shifts on atoms in complex PyO + HCl + aza–fullerene it was modeled magnetical switch and quantum computing gate[23].

3. Results and Discussions

3.1. DESIGN OF MOLECULAR DEVICES GENERATING QUANTUM BITS BASED ON PROTON NMR OF BILIVERDIN DERIVATIVES

The geometry of the single Cu, Co, Zn, Mn and Fe modified biliverdin molecules were obtained applied the Hartree–Fock (HF) method within 6–311G basis set with geometry optimization that is installed in the package G98[19]. For these investigations we used the initial experimental geometry of the various OEB and 2OEB molecules in solutions that are presented in above papers[13–16]. The modified Cu, Co and Zn biliverdin molecules possess C_2 symmetry and Mn, Fe biliverdin dimmers possess the inversion symmetry center. Then, the single point calculations were applied in order to obtain the proton NMR values of our modified biliverdin molecules by using methods HF/STO–3G and DFT/6–311G**/6–31G/STO–3G[19].

The quantum mechanical HF/STO–3G design of the proton NMR based on the quantum computer elements generating from one to six QuBits was done based on the Co biliverdin derivatives where different number of the symmetrical hydrogen atoms are replaced by the Cl atoms (see six QuBits example without Cl atoms in Figure 1).

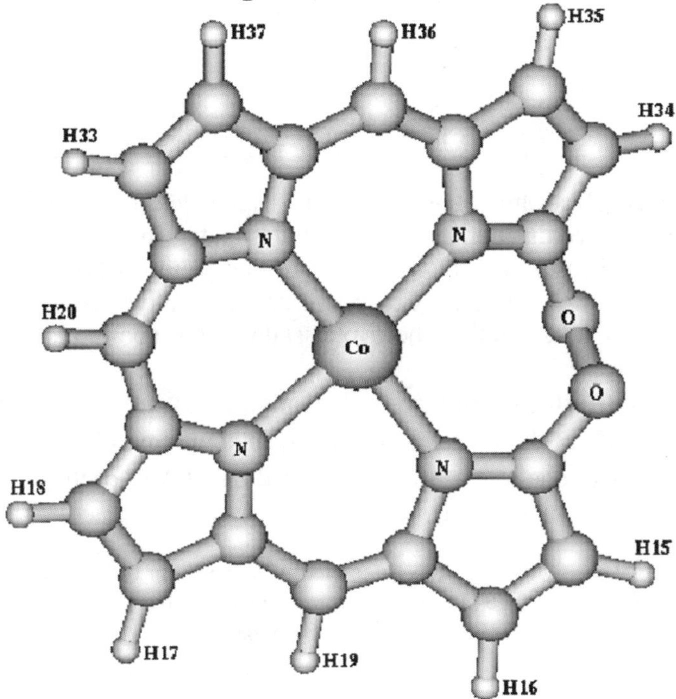

Figure 1. The quantum computing element generating six QuBits based on the proton NMR spectrum of the Co biliverdin derivative.

The proton chemical shifts are obtained as the difference of the values of the tetramethylsilane ($Si(CH_3)_4$) molecule GIAO nuclear magnetic shielding tensor on the hydrogen atoms and that of the Co biliverdin derivative (see Table I). The chemical shift values of the symmetrical H atoms in the Co biliverdin derivative are equal (Table I). Thus, in these investigated Co biliverdin derivatives the six different values of the proton NMR spectra are present that allow us to obtain up to six QuBits in such kind of the molecular quantum computing devices.

Spins of the Co and other biliverdin derivatives might be oriented by permanent magnetical field and then one by one flipped to opposite side by the magnetic field resonance frequencies.

TABLE I. The proton chemical shift (ppm) values in the various Co biliverdin derivatives obtained by HF/STO–3G. In the table the numbers of H atoms correspond to the Figure 1.

Metal	*QuBits*	*15, 34*	*16, 35*	*19, 36*	*17, 37*	*18, 33*	*20*
Co	1	–	–	–	–	–	7.75
Co		–	–	6.88	–	–	8.20
	2			6.88			
Co		–	7.95	6.50	–	–	7.99
	3		7.95	6.50			
Co		7.39	8.00	6.10	–	–	7.97
	4	7.39	8.00	6.10			
Co		6.33	7.54	5.75	6.72	–	6.40
	5	6.33	7.54	5.75	6.72	–	
Co		7.70	8.55	9.54	8.78	8.54	9.44
	6	7.72	8.55	9.54	8.78	8.54	

The various electronic states (singlet, triplet, etc.) of the six QuBits generating Cu biliverdin derivative were investigated in order to obtain more preferable one. Thus, the total energy and the proton NMR spectra of the various multiplicities of this Cu biliverdin derivative were investigated. According to our investigation the triplet state of the above molecule has the lowest energy (Table II).

It was found that the Cu biliverdin derivative LUMO of the β electron set is lower than HOMO of the α electrons. Thus, in some investigated cases one of the HOMO α electrons was artificially located on LUMO of the β orbital (see Triplet* and Pentet* in Table II).

TABLE II. Total energy of six QuBits Cu biliverdin derivative in various electronic states calculated by DFT B3PW91/6–311G. In the case marked by * the α electron is located on LUMO of the β orbital.

State	*Energy, a.u.*
Singlet	–2739.90719843;
Triplet	–2739.93651283;
Triplet (*)	–2739.93651471;
Pentet	–2739.87274311;
Pentet (*)	–2739.87274393;
Septet	–2739.74692076;

Additionally, the localization of the electrons were investigated that serves the G98 calculation of the localized orbitals[19]. It is important to underline that the localization of orbitals in the six QuBits Cu biliverdin device is very strict because the coefficients of the below mentioned atom orbitals are very close to 1.

The triplet state possesses the electrons that are located on the α $2p_Z$ orbital localized on all N atoms and β ones of the C(13), C(14) atoms (Fig. 2).

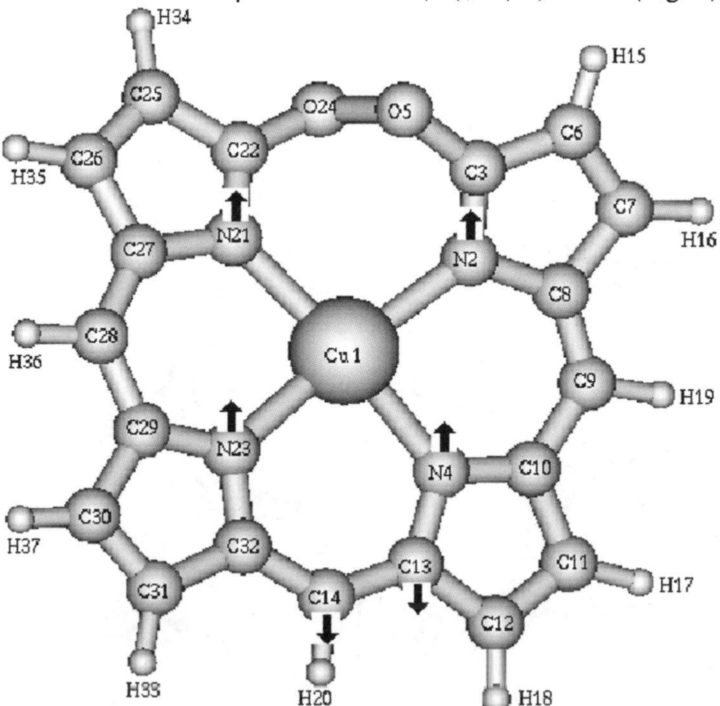

Figure 2. Six QuBits Cu biliverdin derivative when multiplicity is three (triplet). Arrows show the direction of spins of the electrons that are on the localized orbitals of the marked atoms.

The p_Z orbital is perpendicular to the plane of the above molecule. It implies, that chemical shift of H(20) atom of the above molecule should be smaller than that of other hydrogen atoms. The results of our calculation prove this assumption (see Table III).

TABLE III. The values of proton chemical shift (ppm) of six QuBits Cu biliverdin derivative in various electronic states obtained applied B3PW91/6–311G method. In the case marked by * the α electron is located on LUMO of the β orbital

State	*15, 34*	*16, 35*	*19, 36*	*17, 37*	*18, 33*	*20*
Singlet	6.44	6.42	6.19	6.4	6.75	5.79
Triplet	6.11	6.52	6.19	6.56	6.51	6
Triplet (*)	6.11	6.52	6.19	6.56	6.51	6.02
Pentet	5.03	5.06	3.88	5.2	5.32	4.16
Septet	6.71	6.2	6.34	6.59	6.45	6.72

In the case of septet, the electrons are located on the localized orbitals of the all four N atoms and C(3), C(10), C(27), C(29) atoms (Fig. 3). Thus, the values of the H(36) and H(19), H(35) and H(16) chemical shifts should be slightly smaller than that of other atoms, but not very much because the localized electrons are located enough far from these hydrogen atoms that is proved by results of calculations also (see Table III).

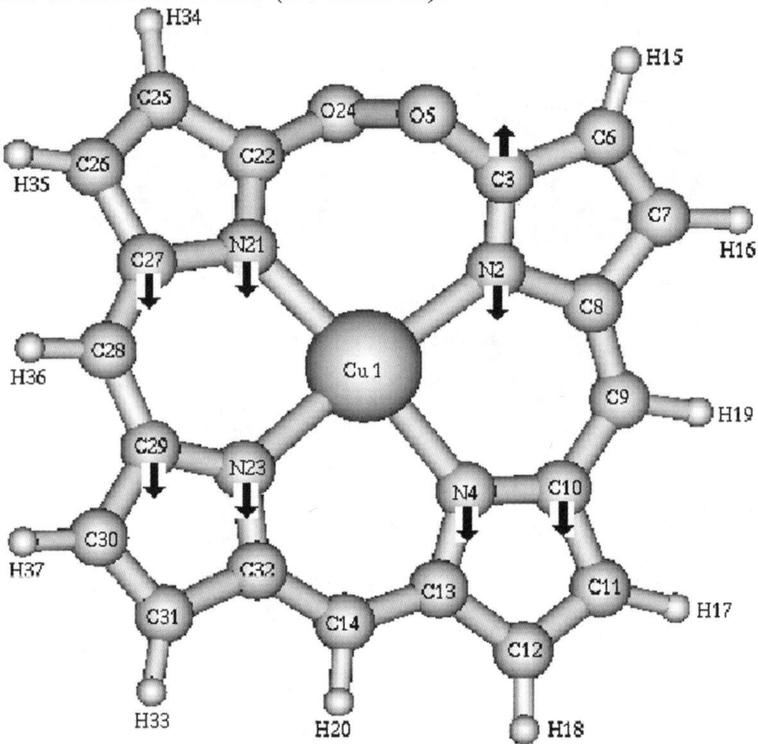

Figure 3. Six QuBits Cu biliverdin derivative when multiplicity is seven (septet). Arrows show the direction of spins of the electrons that are on the localized orbitals of the marked atoms.

It is necessary draw attention to the pentet state. In this case the unpaired electrons are located on the α ($2p_Z$) orbitals of the all four N atoms and the C(8), C(9), C(13) atoms, and the β ($2p_Z$) orbitals on C(14), C(28), C(27) atoms (Fig. 4). The presence of several localized electrons with opposite oriented spins leads to the larger range of chemical shift in comparison with the above–mentioned cases of triplet and septet (see Table III).

The analysis of localized orbitals applied for different electronic states of Cu biliverdin derivative gives additional information concerning the nature of proton NMR in the investigated six QuBits device that allow us to use this analysis in other proton NMR quantum computing devices. We are starting to create software for the evaluation of nuclear spin decoherence time due to coupling with localized orbitals applied for different electronic states. This methodics partially compensate not very good programs in the whole world calculating proton NMR spectra for transition metal species.

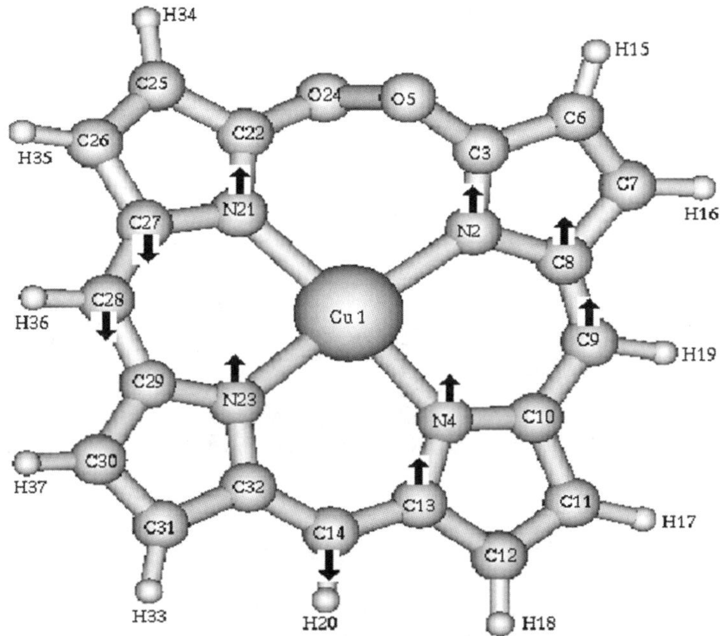

Figure 4. Six QuBits Cu biliverdin derivative when multiplicity is five (pentet). Arrows show the direction of spins of the electrons that are on the localized orbitals of the marked atoms.

The seven QuBits generating devices were designed from ZnOEB and CoOEB derivatives when the octaethyl groups are cut from OEB and the O atoms are replaced by H ones. (see Figures 5, 6). The values of the chemical shifts of that derivatives are presented in Tables IV and V.

TABLE IV. The seven values of the proton chemical shift (ppm) of the Zn biliverdin derivative calculated by B3PW91/ 6–311G in the doublet state. In the table the numbers of H atoms correspond to Figure 5.

H25, H33	*H26, H32*	*H27, H30*	*H28, H29*	*H35, H37*	*H38, H39*	*H36*
6.13	6.92	6.77	6.71	6.34	5.83	6.2

The small differences between values of NMR spectra of ZnOEB derivative obtained by G98 will be improved by using spin–orbit coupling that includes by using Pauli–Breit Hamiltonian[24]

$$H=H_0+H'$$

where H_0 means non–relativistic Hamiltonian using in G98 and H' includes spin related terms with second order relativistic corrections.

TABLE V. The seven values of the proton chemical shift (ppm) of the Co biliverdin derivative calculated by HF/6–311G in the singlet and triplet states

State	*H25, H28*	*H38, H39*	*H29, H34*	*H30,H 35*	*H31, H37*	*H32,H 36*	*H33*
Singlet	−3.68	−10.57	−0.27	−0.20	−1.22	−3.46	−7.66
	−3.68	−10.50	−0.26	−0.20	−1.22	−3.45	
Triplet	5.20	−0.80	5.60	7.44	6.98	7.41	0.76
	5.19	−0.93	5.60	7.45	6.97	7.41	

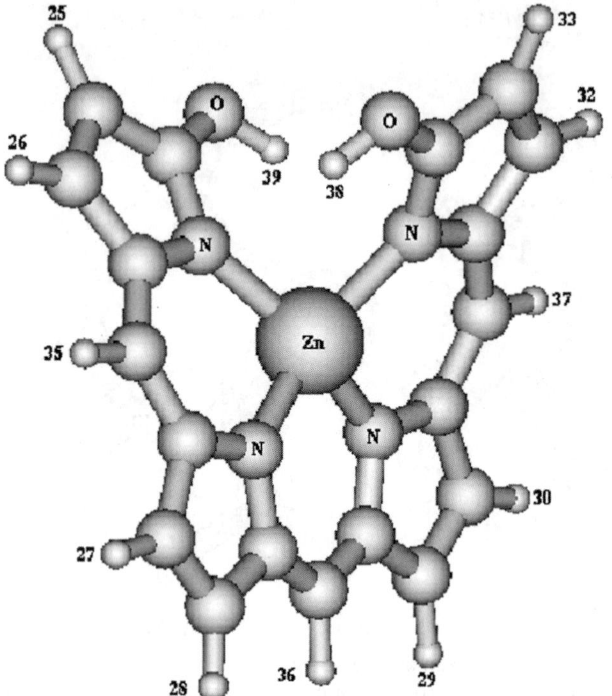

Figure 5. The seven QuBits generating device based on the Zn biliverdin derivative.

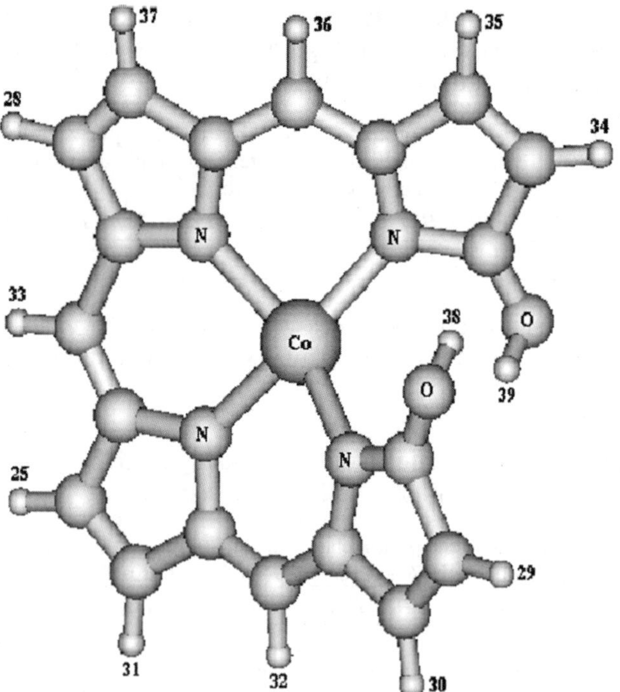

Figure 6. The seven QuBits generating device based on the Co biliverdin derivative

The eleven and twelve QuBits generating devices were designed from the Mn and Fe biliverdin dimmer derivatives when the octaethyl groups are cut from Mn2OEB and Fe2OEB (see Figures 7, 8 and Tables VI, VII).

TABLE VI. The eleven proton chemical shielding values (ppm) in the Mn biliverdin dimmer obtained by HF/STO–3G. In the table the numbers of the H atoms correspond that in Figure 7.

H71, H 74	*H50, H59*	*H64, H53*	*H51, H60*	*H63, H52*	*H69, H72*	*H70, H73*	*H57, H68*	*H65, H54*	*H56, H67*	*H66, H55*
−7.85	−1.72	5.16	2.94	5.52	3.86	−1.17	2.81	0.42	0.28	−1.6
−7.84	−1.71	5.17	2.94	5.52	3.85	−1.18	2.81	0.42	0.28	−1.6

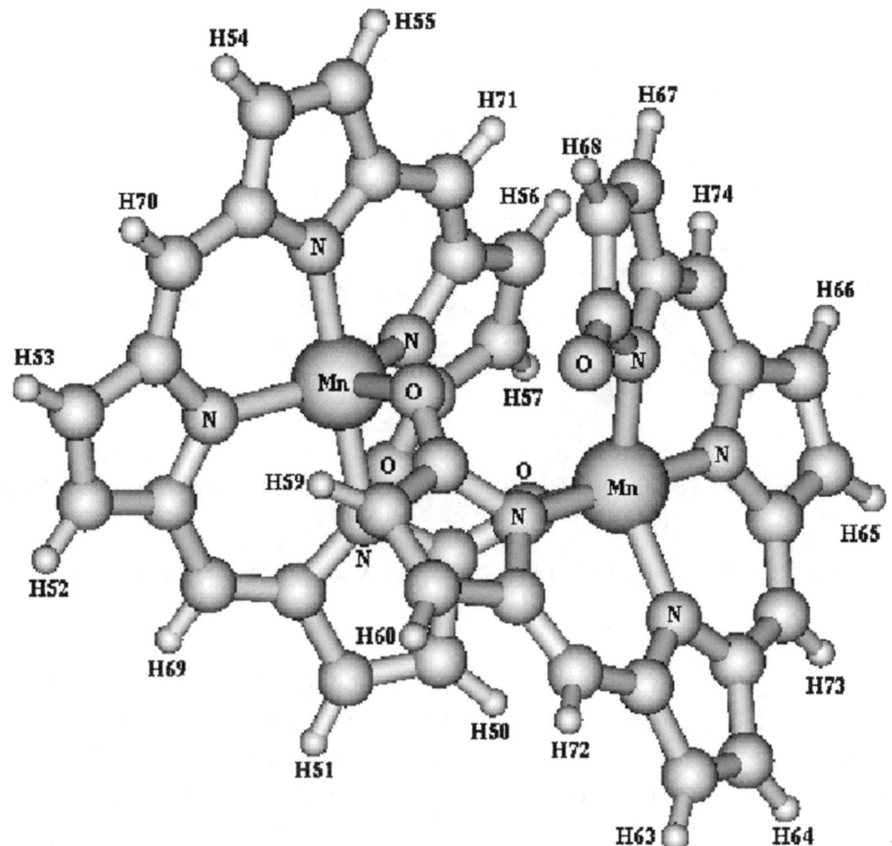

Figure 7. The quantum computing element generating eleven QuBits based on the proton NMR spectra of the Mn biliverdin derivative dimmer.

TABLE VII. The twelve proton chemical shift values in the Fe biliverdin dimmer obtained by HF/STO–3G. In the table the numbers of H atoms are in accordance with that in Figure 8.

H52, H54	*H55, H69*	*H62, H63*	*H56, H64*	*H61, H70*	*H71, H74*	*H73, H76*	*H57, H66*	*H60, H68*	*H58, H65*	*H59, H67*	*H72, H75*
14.33	13.1	9.2	4.78	0.92	−4.82	−4.9	2.63	5.92	6.83	12.3	20
14.31	13.1	9.19	4.79	0.92	−4.79	−4.92	2.63	5.94	6.84	12.3	20

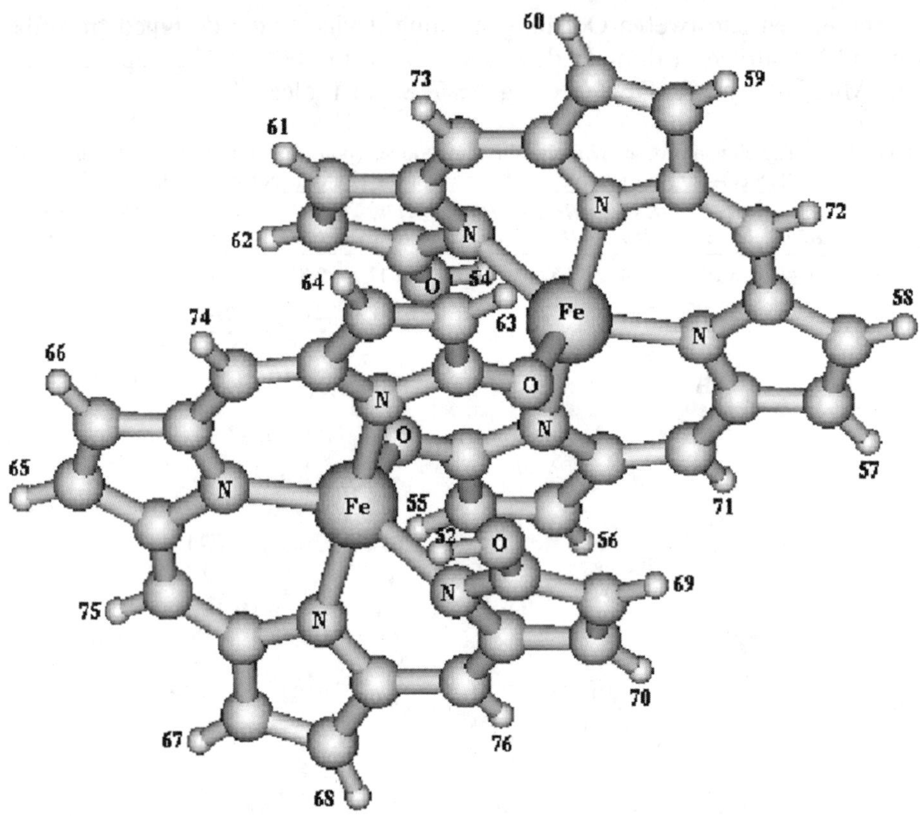

Figure 8. The quantum computing element generating twelve QuBits based on the proton NMR spectra of the Fe biliverdin derivative dimmer.

3.2. GENERATING QUBITS BASED ON AZA–FULLERENE $C_{48}N_{12}$ ADDUCTS

The design of the aza–fullerene $C_{48}N_{12}$ adducts was started from the symmetrical I_h C_{60} molecule replacing the twelve symmetrical C atoms in the cage by the N atoms and then the geometry of aza–fullerene $C_{48}N_{12}$ was optimized by HF/6–311G. The N atom NMR spectrum of this aza–fullerene possesses only two lines because of high symmetry of optimized $C_{48}N_{12}$ cage. In order to brake the symmetry the two adducts CH_2NHCH_2 and CCl_2NHCCl_2 at the 6–6 bonds of cage $C_{48}N_{12}$ are added (see Fig. 9, 10).

The N and H atoms chemical shifts (see Tables VIII, IX) are calculated as the difference of the GIAO nuclear magnetic shielding tensor values of the nitrogen and hydrogen atoms of the ammonia (NH_3) molecule and that of the aza–fullerene $C_{48}N_{12}$ adducts. For the obtaining right chemical shifts of H and N atom NMR spectrum was optimized the geometry of ammonia (NH_3) molecule by using HF/6–311G and then calculated H and N atom NMR spectrum.

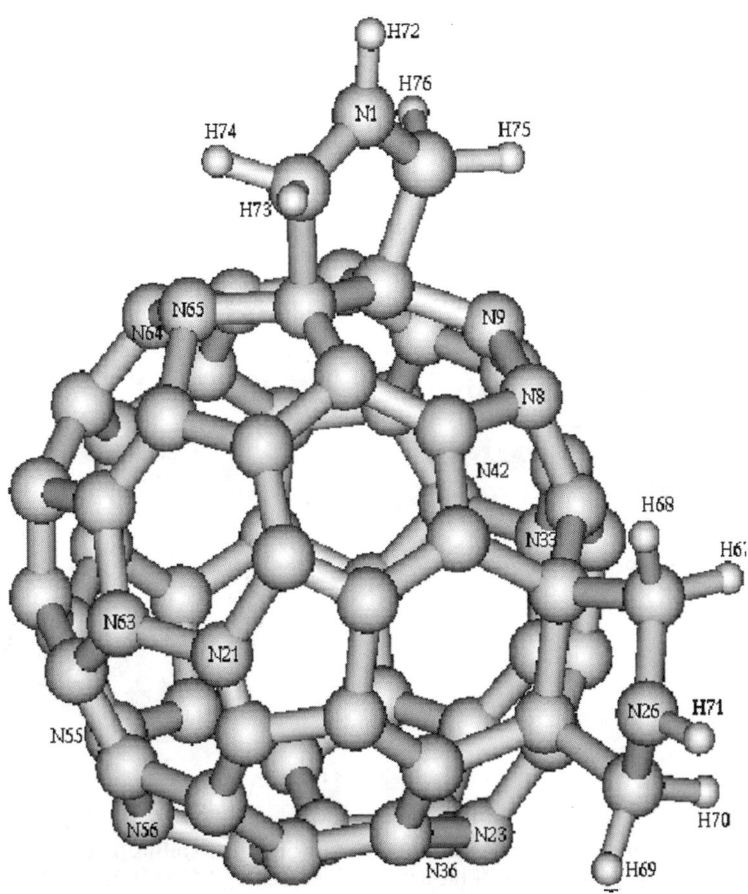

Figure 9. The quantum computing element generating nineteen QuBits based on the proton and nitrogen NMR spectra of the aza–fullerene $C_{48}N_{12}CH_2NHCH_2$ adduct.

TABLE VIII. The nineteen values of the N and H atoms chemical shifts in the aza–fullerene $C_{48}N_{12}CH_2NHCH_2$ adduct obtained by HF/6–31G. In the table the numbers of N and H atoms correspond to that in Figure 9.

N1	***N8***	***N9***	***N21***	***N23***	***N26***	***N33***
30.29;	119.10;	105.92;	128.60;	119.44;	28.02;	126.80;

N36	***N42***	***N55***	***N56***	***N63***	***N64***	***N65***
99.60;	141.40;	143.20;	126.01;	113.48;	144.24;	122.69

H67, H70	***H68, H76***	***H69, H73***	***H74, H75***	***H71, H72***
3.18; 3.19;	3.80; 3.57;	3.88; 3.86;	4.24 4.36;	0.14 0.40;

Analysis of calculation results show that the adduct CH_2NHCH_2 is splitting H and N NMR values not so strong in comparison with adduct CCl_2NHCCl_2.

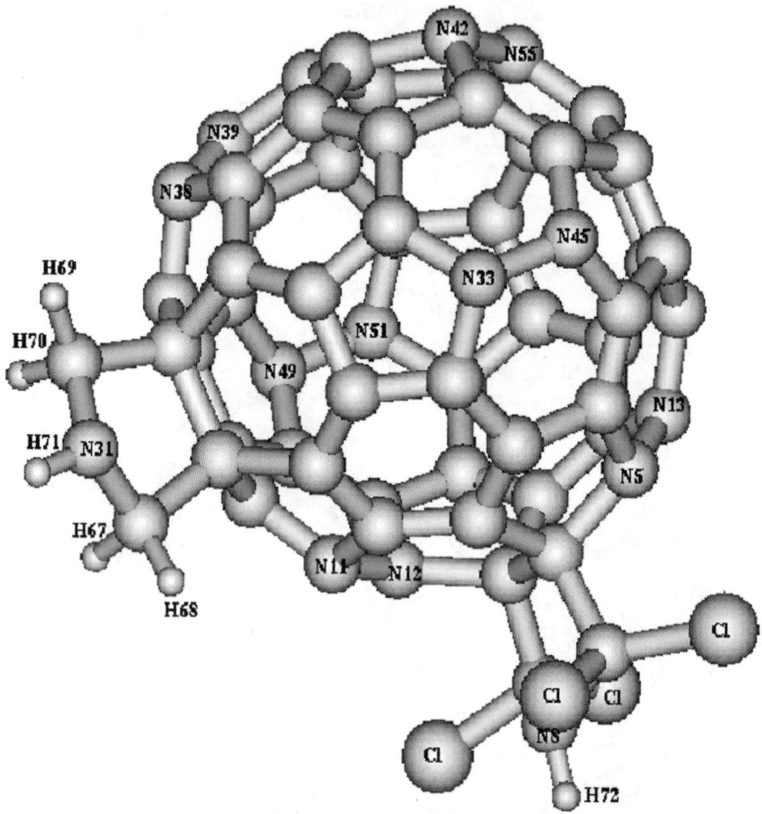

Figure 10. The quantum computing element generating eighteen QuBits based on the proton and nitrogen NMR spectra of the aza–fullerene $C_{48}N_{12}CCl_2NHCCl_2$ adduct.

TABLE IX. The eighteen values of the N and H atoms chemical shifts of the aza–fullerene $C_{48}N_{12}CCl_2NHCCl_2$ adduct obtained by HF/6–31G. In the table the numbers of N and H atoms correspond to that in Figure 10.

N31	*N5*	*N8*	*N11*	*N12*	*N13*	*N33*
28.00;	135.33;	128.11;	125.75;	120.77;	146.30;	128.58;

N39	*N38*	*N42*	*N45*	*N49*	*N51*	*N55*
99.51;	119.62;	126.34;	113.88;	127.36;	141.06;	143.16;

H67, H70	*H68, H69*	*H71*	*H72*
3.22; 3.22;	3.87; 3.89;	0.20;	3.98;

The chemical shift values of the symmetrical H atoms in the adducts $C_{48}N_{12}CH_2NHCH_2$ and $C_{48}N_{12}CCl_2NHCCl_2$ of aza–fullerene derivative are almost equal (see Tables VIII, IX). Thus, in the two different investigated $C_{48}N_{12}$ adducts the nineteen and eighteen different values of the N and H NMR spectra are presented that allow us to obtain up to nineteen QuBits in such a kind of molecular quantum computing devices.

CONCLUSIONS

1. The investigations of the H and N atom NMR values of the Cu, Co, Zn, Mn and Fe biliverdin derivatives and their dimmers and the aza–fullerene $C_{48}N_{12}$ adducts using *ab initio* and the first principles quantum mechanical methods indicate that modified derivatives should generate from one to seven and eleven, twelve, eighteen, nineteen QuBits.

2. Our developed analysis of localized orbitals gives additional information concerning the nature of proton NMR in the investigated six QuBits Cu biliverdin derivative that allow us to use this analysis in other proton NMR quantum computing devices. This methodic partially compensate not very good programs in the whole world calculating proton NMR spectra for transition metal species.

Acknowledgments

We are grateful to the Institute of Molecular Physics, Polish Academy of Sciences and Poznań Computer and Networking Centre (PCSS) for the possibility to use CRAY J916 and Gaussian 98 A.7.

References

1. Tamulis, A., Tamuliene, J., Tamulis, V. (2002) Quantum Mechanical Design of Photoactive Molecular Machines and Logical Devices, chapter in American Scientific Publishers *Handbook on Photochemistry and Photobiology*, accepted for publication.
2. Tamulis, A., Stumbrys, E., Tamulis V., Tamuliene, J. (1996) Quantum Mechanical Investigations of Photoactive Molecules, Supermolecules, Supramolecules and Design of Basic Elements Molecular Computers, in F. Kajzar, V.M. Agranovich, C. Y–C. Lee (eds.) NATO ASI series, High Technology, vol. 9, *Photoactive Organic Materials: Science and Applications,* Kluwer Academic Publishers, Dordrecht/Boston/London, p.p. 53–66.
3. Tamulis, A., Tamulis, V. (1998) Design of Basic Elements of Molecular Computers Based on Quantum Chemical Investigations of Photoactive Organic Molecules, *Proceedings of the SPIE Photonics WEST, Conference on Optoelectronic Integrated Circuits II* **3290**, 315–324.
4. Tamulis, A., Tamulis,V., Tamuliene, J. (1998) Quantum mechanical design of molecular implementation of two, three and four variable logic functions for electronically genome regulation, *Viva Origino* **26**, 127–146.
5. Tamulis, A., Tamuliene, J, Balevicius, M. L., Nunzi, J.M. (2000) Quantum Chemical Design of Multivariable Anisotropic Random–Walk Molecular Devices Based on Stilbene and Azo–Dyes, *Mol. Cryst. Liq. Cryst.*, **354**, 475–484.
6. Tamulis, A., Tamuliene, J., Balevicius, M.L., Nunzi, J.–M. (2000) Quantum mechanical investigations of photoactive molecules and design of molecular machine and logical devices in Bradley, D.D.C., Kippelen, B. (eds.) *Organic Photonics Materials and Devices II,* Proceedings of SPIE **3939**, 61–68.
7. Tamulis, A., Rinkevicius, Z., Tamuliene, J., Tamulis, V., Balevicius, M.L., Graja, A. (2001) *Ab initio* quantum chemical design of supermolecule logical devices, in Grote, J.G., Heyler, R.A. (eds.) *Optoelectronic Integrated Circuits and Packaging V,* Proceedings of SPIE, **4290** 82–93.
8. Tamulis, A., Rinkevicius, Z., Tamulis ,V., Tamuliene, J., Karna, S.P., Stickley, C.M. (2001) *Ab initio* quantum chemical design of photoactive molecular logical devices, *Nonlinear Optics,* **27**, 385–393.

9. Tamuliene, J., Tamulis, A. (2001) Quantum chemical a*b initio* design of molecular electronics tools for biotechnologies, *Biotech News International*, **6**, 12–14.
10. Tamulis, A., Rinkevicius, Z., Tamuliene, J. (2001) Molecular Classical Computer Basic Elements, in T. Gonis and P.E.A. Turchi, (eds.), NATO Science Series III Computer and Systems Sciences, *Decoherence and Implication in Quantum Computation and Information Transfer,* IOS Press, **182**, 358–365.
11. Tamulis, A., Rinkevicius, Z., Tamuliene, J., Tamulis, V., Graja, A., Gaigalas, A.K. (2002) Quantum Chemical Design of Light Driven Molecular Logical Machines, in A. Graja, B. R. Bulka, F. Kajzar (eds.) NATO Science Series II. Mathematics, Physics and Chemistry, *Molecular Low Dimensional and Nanostructured Materials for Advanced Applications",* Kluwer Academic Publishers, Dordrecht, **59**, 209–219.
12. Bouwmeester, D., Ekert A., Zeilinger A. (2000) *The Physics of Quantum Information*, Springer.
13. Balch, A.L., Mazzanti, M., Noll, B. C., Olmstead, M.M. (1993) Geometric and electronic structure and dioxygen sensitivity of the copper complex of octaethylbilindione, a biliverdin analog, *J. Am. Chem. Soc.*, **115**, 12206–12207.
14. Koerner, R., Olmstead, M.M., Van Calcar, P.M., Winkler, K., Balch, A.L. (1998) Reactivity of the verdoheme analogues, 5–oxaporhyrin complexes of Cobalt(II) and Zinc(II) with nucleophiles: opening of the planar macrocycle by alkoxide addition to form helical complexes, *Inorg. Chem.*, **37**, 982–988.
15. Balch, A.L., Mazzanti, M.,. Noll, B.C, Olmstead, M.M. (1994) Coordination patterns for Biliverdin–type ligands. Helical and linked helical units in four–coordinated Cobalt and five–coordinated Manganese(III) complex of octaethylbilindione, *J. Am. Chem. Soc.*, **116**, 9114–9122.
16. Koerner, R., Latos–Grazynski, L., Balch, A.L. (1998) Models for verdoheme hydrolysis. Paramagnetic products from the ring opening of verdohemes, 5–oxaporphyrin complexes of Iron(II), with methoxide ion, *J. Am. Chem. Soc.*, **120**, 9246–9255.
17. Parr, R.G., W. Yang, W., (1989) *Density–Functional Theory of Atoms and Molecules,* Oxford Univ. Press: Oxford.
18. Schlegel, H.B., (1994) in D. R. Yarkony (ed.), *Modern Electronic Structure Theory*, vol. 2., World Scientific Publishing: Singapore.
19. Frisch, M.J., Trucks, G.W., Schlegel, H.B., Scuseria, G.E., Robb, M.A., Cheeseman, J.R., Zakrzewski, V.G., Montgomery, J.A., Jr., Stratmann, R.E., Burant, J.C., Dapprich, S., Millam, J.M., Daniels, A.D., Kudin, K.D., Strain, M.C., Farkas, O., Tomasi, J., Barone, V., Cossi, M., Cammi, R., Mennucci, B., Pomelli, C., Adamo, C., Clifford, S., Ochterski, J., Petersson, G.A., Ayala, P.Y., Cui, Q., Morokuma, K., , Malick, D.K., Rabuck, A.D., Raghavachari, K., Foresman, J.B., Cioslowski, J., Ortiz, J.V., Stefanov, B.B., Liu, G., Liashenko, A., Piskorz, P., Komaromi, I., Gomperts, R., Martin, R.L., Fox, D.J., Keith, T., Al–Laham, M.A., Peng, C.Y., Nanayakkara, A., Gonzalez, C., Challacombe, M., Gill, P.M.W., Johnson, B., Chen, W., Wong, M.W., Andres, J.L., Gonzalez, C., Head–Gordon, M., Replogle, E.S., and Pople, J.A. (2000) *Gaussian 98, Revision A.7*, Gaussian, Inc., Pittsburgh PA, USA.
20. Wolinski, K., Hinton, J.F., Pulay, P., (1990) Efficient Implementation of the Gauge–Independent Atomic Orbital Method for NMR Chemical Shift Calculations, *J. Am. Chem. Soc.* **112**, 1164.
21. Cammi, R., Mennucci, B., Tomasi, J. (1999) Nuclear magnetic shieldings in solution: gauge invariant atomic orbital calculation using the polarizable continuum model, *J. Chem. Phys.* **110** 7627.
22. Balevicius, V., Tamuliene, J., Hadzi, D. (2002), ^{1}H, ^{13}C and ^{17}O NMR study of pyridine–N–oxide hydrogen bond complexes in liquid state, article in preparation.
23. Tamulis, A., Tamuliene, J., (2002) Quantum mechanical modeling of magnetical switch and quantum computing gate in complex: aza–fullerene pyridine–N–oxide hydrochloride,, article in preparation.
24. Mcweeny R., Sutcliffe, B.T. (1969) *Methods of Molecular Quantum Mechanics*, Academic Press, London.

QUANTUM INTERFERENCE AND CORRELATIONS IN ELECTRONIC TRANSPORT THROUGH NANODEVICES

B. R. BUŁKA, P. STEFAŃSKI AND S. LIPIŃSKI
Institute of Molecular Physics, Polish Academy of Sciences, ul. M. Smoluchowskiego 17, 60-179 Poznań, Poland

1. Introduction

Coherent electronic transport in nanostructures with strong electron-electron correlations has recently became a topic of broad interest, as it is fundamental to applications in quantum computing. The theoretical speed of quantum computers follows from the fact that they exploit the coherent superposition of wavefunctions. Coherence is achieved if the system size becomes smaller than the coherence length. In this case the typical quantum phenomena like interference or quantum many body effects as e.g. formation of many body resonances are observed. Quantum dot (QD) devices provide a well-controlled objects for studying these phenomena, as they offer the possibility of continuous tuning of the relevant parameters (see for example [1]). The gate voltage controls the number of the electrons at the dot and shifts the energy spectrum. Recent advances in technology allows to produce the QDs with tunable coupling between the electrodes and the dot. The energy spectrum of confined states of a nanostructure is discrete. A charging energy is, however, larger and Coulomb interactions are more important. Correlations between electrons lead to the Kondo resonance, which can be seen in transport through the QD [2]. Due to a specific quantum dot geometry the Kondo effect causes an enhancement of the conductance – in contrast to "classical" magnetic impurities in metals, where the resistance increases [3]. There are many experimental evidences of quantum interference and strong correlation effects in coherent transport through quantum dots [1, 2] as well as in carbon nanotubes [4].

In this paper we want to present modelling of the coherent transport through nanostructures, taking into account quantum interference as well as electronic correlations. First, the transport through the QD strongly coupled with the electrodes will be considered. In this case the Kondo resonance occurs together with the Fano resonance. Theoretical results will

L.M. Liz-Marzán and M. Giersig (eds.), Low-Dimensional Systems: Theory, Preparation, and Some Applications, 281–292.

be compared with the recent measurements of electronic transport through the QD [1]. Next, the transport through a metallic ring with the QD will be considered. We show how the Aharonov-Bohm effect can be seen in presence of the Kondo resonance.

Quite recently an increasing interest in spin-polarized transport is observed. This includes the giant magnetoresistance, spin-injection experiments and spin-polarized tunneling experiments, which have applications e.g. in nonvolatile magnetic random access memories MRAMs, magnetic field sensors and reader heads of magnetic hard disk drivers [5]. The importance of spin degrees of freedom in coherent transport results from unusually long spin-dephasing times approaching microseconds. A quantum dot coupled to ferromagnetic leads is an example of a ferromagnetic single electron transistor, where an additional control of a current is possible by changing of a magnetic polarization in the electrodes. Of special interest is the coherent low temperature regime, where the question of the stability of singlet Kondo state influenced by magnetic polarization of electrodes arises. These problems will be considered in the last part of our work.

2. Calculations of electronic current

Let us first present the method of calculation of the electronic transport. The current is determined from the time evolution of the occupation number $\hat{N}_L = \sum_{k,\sigma} c^\dagger_{kL,\sigma} c_{kL,\sigma}$ for electrons in the left electrode

$$J \equiv -e\langle \frac{d\hat{N}_L}{dt} \rangle = \frac{ie}{\hbar} \langle [\hat{N}_L, \hat{H}] \rangle \ , \tag{1}$$

where $\hat{H}$ denotes the Hamiltonian of the system, which is written as

$$\begin{aligned} \hat{H} = & \sum_{k,\alpha,\sigma} \epsilon_{k\alpha}\, c^\dagger_{k\alpha,\sigma} c_{k\alpha,\sigma} + \sum_{\sigma} \epsilon_0\, c^\dagger_{0\sigma} c_{0\sigma} + U n_{0\uparrow} n_{0\downarrow} \\ & + \sum_{k,\alpha,\sigma} t_\alpha\, (c^\dagger_{k\alpha,\sigma} c_{0\sigma} + h.c.) \ . \end{aligned} \tag{2}$$

The first term describes electrons in the left ($\alpha = L$) and the right ($\alpha = R$) electrode, the second and the third one correspond to electrons at the QD with the energy level ϵ_0 and the onsite Coulomb interaction U of two electrons with the opposite spins $\sigma = \uparrow$ and $\sigma = \downarrow$, the fourth term describes tunneling between the electrodes and the QD. Calculating the commutator in Eq.(1) one gets

$$J = \frac{ie}{\hbar} \sum_{k,\sigma} [t_L \langle c^\dagger_{kL,\sigma} c_{0\sigma} \rangle - c.c.] \ . \tag{3}$$

The thermal averages are expressed by the lesser Green function [6] as

$$\langle c^{\dagger}_{k\alpha,\sigma} c_{0\sigma} \rangle = \int \frac{d\omega}{2\pi i} G^{<}_{0\sigma,k\alpha\sigma}(\omega) \,. \tag{4}$$

3. Fano resonance in strongly coupled quantum dots

First experiment of the electronic transport through strongly coupled QD was performed by the MIT group [1]. The conductance $\mathcal{G}$, measured as function of a gate voltage, showed asymmetric peaks. Moreover, they found a reduction of $\mathcal{G}$ below a continuum level, which was strongly temperature dependent. These data indicated on the Fano resonance [7]. As it is well known, the Fano resonance is a quantum interference process between degenerate continuum of states and an evanescent (discrete) state and was observed in various systems. The scheme of the Fano resonance in transmission of electrons through the QD is presented in Fig.1. Apart from a travelling wave there is a standing wave localized at the QD, which results

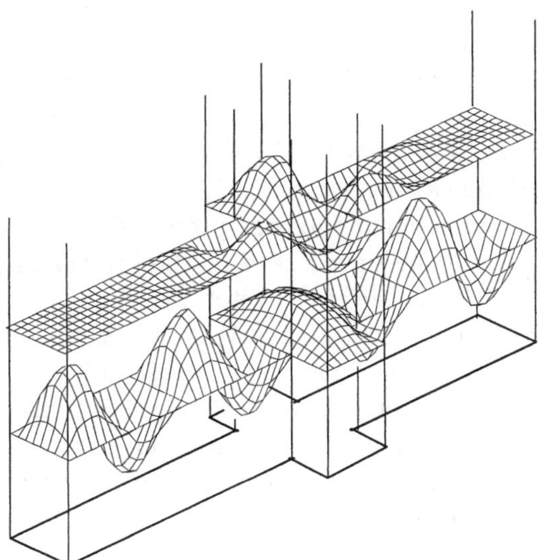

Figure 1. A schematic presentation of the Fano resonance in the electronic transport through a quantum dot. The QD is presented as a quantum well (in the central part) connected to two quantum wires. For a given energy of an incident electron, two waves occur in the system: a travelling wave scattered on the QD (bottom) and an evanescent state (top). Due to a different geometry of the QD (and its depth), both the waves can have different symmetry. Interference between both the waves leads to Fano resonance.

form different geometry of the QD and the leads. Both the waves interfere, what leads to the Fano resonance. For some values of the gate voltage one can find an odd number of electrons and an unpaired spin. Scattering of the travelling wave on the localized spin results in the Kondo resonance [2]. Due to a resonance peak in the density of states at the Fermi energy the conductance increases logarithmic with lowering of the temperature.

In order to describe the Fano resonance together with the Kondo resonance we proposed [8] to add to the Hamiltonian (2) the term

$$\hat{H}_{bridge} = \sum_{k,k',\sigma} [t_{LR} e^{ik+ik'} c^{\dagger}_{kL,\sigma} c_{k'R,\sigma} + h.c.] \,, \tag{5}$$

which describes a direct transmission of electrons between both the electrodes. Since the model includes an additional conducting channel, through which electrons pass without scattering on the QD, we called it as the bridge model. After some algebra one finds the expression for the current

$$J = \frac{2e}{h} \int d\omega [f_L(\omega) - f_R(\omega)] \{\alpha_{LR} \left| t_{LR} \right|^2 + \mathrm{Im}[\alpha_{00} G^r_{00}(\omega)\} \,, \tag{6}$$

where $\alpha_{LR} = 4\pi^2\rho^2/w^2$, $\alpha_{00} = -4\pi\rho z^-_L z^{+*}_L z^-_R z^{+*}_R / \, [w^2(|z_L|^2 + |z_R|^2)]$, $w = 1 + \pi^2\rho^2|t_{LR}|^2$, $z^{\pm}_L = t_L \pm i\pi\rho t_{LR} t^*_R$, $z^{\pm}_R = t^*_R \pm i\pi\rho t^*_{LR} t_L$. In derivations we assumed that the lesser, retarded and advanced Green functions in the electrodes are in the form $g^<_\alpha = 2i\pi\rho f_\alpha$ and $g^{r,a}_\alpha = \mp i\pi\rho$, where f_α denotes the Fermi distribution function for electrons in the α-electrode and ρ is the electronic density of states. The first term in (6) describes the background, whilst the second one is responsible for a resonate nature of the transport and quantum interference effects.

Independently the similar model was proposed by Heemeyer [9], which was studied by means of the slave boson method in the mean field approximation. This approach has several limitations, e.g. it is restricted to low temperatures $T \to 0$ and low voltages $V \to 0$. We used the equation of motion (EOM) method, which has no such restrictions and within which multiple scatterings and interference processes as well as the Kondo resonance were taken into account on the same level.

Although one could derive the current J for any value of the onsite Coulomb integral U, we used the procedure for $U \to \infty$, which is much simpler. The Green function is determined as

$$G^r_{00}(\omega) = \frac{1 - n/2 - a_{00}}{\omega - \epsilon_0 + i\Delta_0 - 2i\Delta_0 a_{00} - b_{00}} \,, \tag{7}$$

where n denotes the average number of electrons at the QD, $\Delta_0 = \pi\rho(|t_{L0}|^2 + |t_{R0}|^2)$, $a_{00} = (|z_L|^2 H_L(\omega) + |z_R|^2 H_R(\omega))/w$ and $b_{00} = (|z_L|^2 F_L(\omega) + |z_R|^2 F_R(\omega))/w$. Here, we use the functions

$$H_\alpha(\omega) = \frac{\rho}{w} \int_{-D}^{D} \frac{d\omega' f_\alpha(\omega') [G^r_{00}(\omega')]^*}{\omega - \omega'} \,, \tag{8}$$

$$F_\alpha(\omega) = \frac{\rho}{w}\int_{-D}^{D}\frac{d\omega' f_\alpha(\omega')}{\omega-\omega'}$$

$$= \frac{\rho}{w}\Big\{i\pi f_\alpha(\omega) + \ln\frac{2\pi k_B T}{D} + \mathrm{Re}\Psi\Big[\frac{1}{2} - i\frac{\omega-\epsilon_{F\alpha}}{2\pi k_B T}\Big]\Big\}, \tag{9}$$

where Ψ is the digamma function and $\epsilon_{F\alpha}$ denotes the position of the Fermi level in the α electrode.

At the temperature $T = 0$ one can find the Green function $G^r_{00}(\epsilon_F) = [1 - e^{2i\phi}]/(2i\Delta_0)$, where the phase ϕ is taken according to the Friedel sum rule [3] as $\phi = \pi n/2$. Putting G_{00} into (7) one can find the conductance in the limit $V \to 0$ as

$$\mathcal{G} = \frac{2e^2}{h}\{\alpha_{LR}|t_{LR}|^2 - \frac{1}{2\Delta_0}\mathrm{Re}[\alpha_{00}(1 - e^{i\pi n})]\}, \tag{10}$$

where n_0 can be associated with the relative position of the level $\Delta\epsilon = \epsilon_0 - \epsilon_F$ as $n = \frac{1}{2} - \frac{1}{\pi}\tan^{-1}[\Delta\epsilon/\Delta_0]$, which can be varied by the gate voltage applied to the QD.

For $T > 0$ the Green function and the conductance $\mathcal{G}$ were calculated numerically. In order to get the results for the case when the second electron is introduced onto the QD, the electron-hole transformation was applied together with a rescaling of the energy $\epsilon_0 \to -\epsilon_0 - U$. Fig.2 presents comparison of the experimental data [1] and our theoretical results for different temperatures. The conductance exhibits pairs of dips corresponding to an introduction of two additional electrons onto the QD. The transmitted electronic wave is scattered on the QD and changes its phase (according to the Friedel sum rule). The left dip in Fig.2 corresponds to the phase change by $\phi = \pi$ due to scattering on the QD with one additional electron, whilst the right peak – to $\phi = 2\pi$ when there are two electrons at the QD. The dips show that the Fano resonance occurs in the system, and there is a destructive interference between an electronic wave travelling through the bridge channel and that one scattered on the QD. An asymmetry between the left and the right dip is connected with the asymmetry between the left and the right junction. The curves in Fig.2 were obtained under the assumption of the large asymmetry of both junctions, with the hopping integrals $t_L = 0.6$ meV and $t_R = 4$ meV. For symmetric junctions ($t_L = t_R$) one gets a high peak apart from a dip and their shape is very different for the first and the second electron introduced onto the QD. If the ratio $t_L : t_R$ increases the peak in $\mathcal{G}$ decreases, but both the dips became more deeper and similar to each other (the asymmetry between them disappears). From the comparison of both the figures one can see that the bridge model captures all main features of the experimental data: the asymmetry between the left and the right dip as well as a small bump on the left hand side.

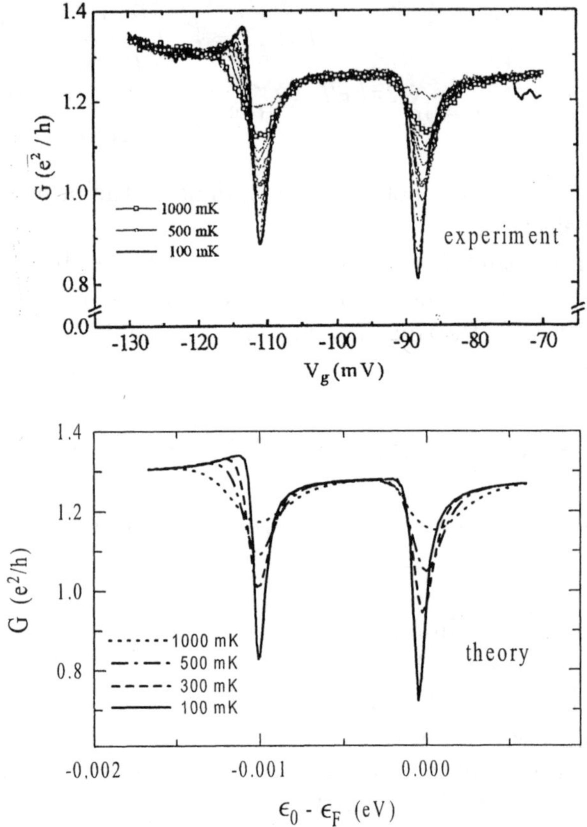

Figure 2. Comparison of the conductance measured for a strong coupled QD vs. the gate voltage for different temperatures (after Gores et al. [1]) and calculated for the bridge model within the EOM approach (the parameters were $t_L = 0.6$ meV, $t_R = 4$ meV, $t_{LR} = 320$ meV, $U = 1.1$ meV and the half-bandwidth $D = 1$ eV).

4. Kondo resonance and interference in metallic rings

In the fifties of the last century a question was raised: whether the vector potential $\mathbf{A}$ of the magnetic field $\mathbf{B} = \nabla \times \mathbf{A}$ is a measurable quantity in analogy to the potential ϕ of the electric field $\mathbf{E} = -\nabla\phi$. Aharonov and Bohm [10] proposed an experiment, which should show that $\mathbf{A}$ is really the measurable quantity. The idea was to take a beam of electrons, to split it and to send around a long solenoid. Far from the solenoid the intensity of the magnetic field B is very small, but the electronic waves are affected by the vector potential $\mathbf{A}$. There is a shift of the phase of the electronic wave $\phi_\Gamma = e/\hbar \int_\Gamma \mathbf{A} \cdot d\mathbf{s}$, which depends on a trajectory Γ. The electronic wave passing clock-wise around the solenoid has a different phase shift than that one passing counter clock-wise. The effect should be seen when both the waves interfere. The Aharonov-Bohm effect was observed in a mesoscopic metallic ring first by Webb *et al.* [11] in 1985. They measured the conductance, which was dependent on the magnetic flux Φ enclosed within the ring and oscillated with the period equal to the flux quantum $\Phi_0 = hc/e$. Nowadays one can produce metallic rings with a quantum dot in one of the arm [12]. Such a system allows to study quantum interference together with electronic correlations manifested by the Kondo resonance.

The bridge model is also suitable for studies of the electronic transport through the metallic ring with the Kondo impurity. In the Hamiltonian (2) and (5) the hopping integrals have to be taken as complex numbers $t_\nu = |t_\nu| e^{i\phi_\nu}$, where ϕ_ν corresponds to the shift of the phase of electronic wave passing through the ν arm of the ring ($\nu = LR$, L, R) in presence of the magnetic field. The shift of the phase is related to the magnetic flux enclosed in the ring $2\pi\Phi/\Phi_0 = \phi_{LR} - \phi_L - \phi_R$. At $T = 0$ one can derive the conductance from Eq.(10), which simplifies in the Kondo regime ($n \to 1$) to the form

$$\mathcal{G} = \frac{2e^2}{h} \frac{4|t_L|^2 |t_R|^2}{(|t_L|^2 + |t_R|^2)^2 w^2} \left[1 + \pi^4 \rho^4 |t_{LR}|^4 - 2\pi^2 \rho^2 |t_{LR}|^2 \cos(2\pi\Phi/\Phi_0)\right] . \quad (11)$$

As expected $\mathcal{G}$ oscillates with the period $\Phi_0 = hc/e$. Moreover, $\mathcal{G} \to 0$ for $t_L \to 0$, which means that the transport through the system becomes blocked. For the case $t_L = 0$ there is no electronic transport through the QD, although it is connected with the right electrode $t_R \neq 0$. Due to coherent electronic correlations between the wave travelling through the bridge channel and multiple scatterings on the QD, the transmission through the bridge channel is reduced and in the Kondo regime is completely blocked.

For finite temperature T and a source-drain voltage V we determined numerically the non-equilibrium Green function and the current. The results for the differential conductance dI/dV are presented in Fig.3 for various values of the flux Φ enclosed in the ring. It is seen that the curves show either a peak or a dip at $V = 0$. This is an evidence of either a constructive or a destructive interference process between electronic waves travelling through the ring in the presence of the Kondo resonance. We propose to perform an experiment, in which one can observe a continuous evolution of the shape of dI/dV with a change the flux Φ – the peak should be transformed into the dip. This effect can be seen only for the QD in the Kondo regime ($\epsilon_0 \ll \epsilon_F$). If the energy level ϵ_0 is shifted to the empty state regime ($\epsilon_0 \gg \epsilon_F$), and the electronic transport is uncorrelated, the differential conductance does not show any structure around $V = 0$.

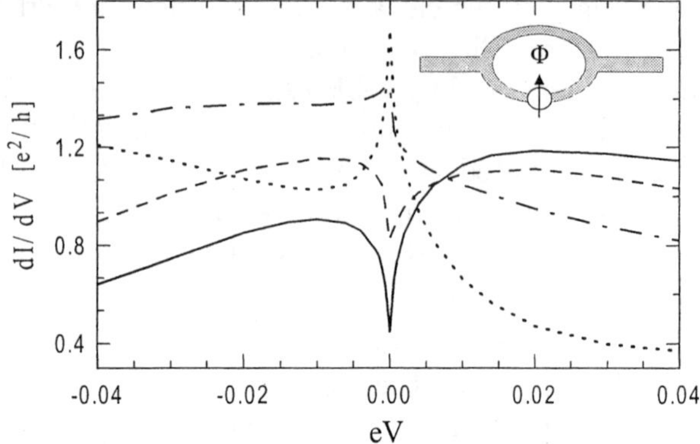

Figure 3. Differential conductance dI/dV as a function of the source-drain voltage V in the metallic ring with the quantum dot for various magnetic flux Φ enclosed within the ring $\Phi = 0.5hc/e$ (dotted), $0.25hc/e$ (dash-dotted), $0.125hc/e$ (dashed) and 0 (solid). In the calculations $t_L = t_R = 0.1$, $t_{LR} = 0.2$, $T = 2 \times 10^{-6}$ and $\epsilon_0 = -0.05$. The half-bandwidth D in the electrode is taken as the unity.

5. Correlations in coherent transport through magnetic nanostructures

Let us now analyze the transport between ferromagnetic electrodes in the presence of electronic correlations. Apart from interest in magnetoresistance we hope to understand better the Kondo resonance and the formation of the singlet state.

The procedure is similar as the described above, with a difference that there are two different conducting channels for electrons with the spin orientation $\sigma = \uparrow$ and $\downarrow$. Therefore, the calculations of the Green functions and the current take into account an explicit dependence on the spin σ. For $T = 0$ the derivations are simplified and one gets the relation

$$\sin(\pi m) = \frac{\Delta_\uparrow - \Delta_\downarrow}{\Delta_\uparrow + \Delta_\downarrow} \sin(\pi n) \qquad (12)$$

linking the spin accumulation m and the average number n of electrons at the QD. Here, $\Delta_\sigma = \Gamma_{L\sigma} + \Gamma_{R\sigma}$ denotes the broadening of the resonant level for the electrons with the spin σ and $\Gamma_{\alpha\sigma} = \pi\rho_{\alpha\sigma}t_\alpha^2$ is the spin-dependent tunneling rate. The relation (12) shows that in the Kondo regime (when $n \to 1$) the spin accumulation $m \to 0$ and the system achieves the unitary limit with the singlet state. Although the tunneling rates $\Gamma_{\alpha\sigma}$ are different for the opposite spin orientations, the electron with the spin $\sigma = \uparrow$ occupies the QD on average so long as the electron with $\sigma = \downarrow$.

The conductance determined at $T = 0$ can be expressed as

$$\mathcal{G} = \frac{e^2}{h} \sum_\sigma \frac{4\Gamma_{L\sigma}\Gamma_{R\sigma}}{\epsilon_0^2} \quad \text{in the empty state regime,} \tag{13}$$

$$\mathcal{G} = \frac{e^2}{h} \sum_\sigma \frac{4\Gamma_{L\sigma}\Gamma_{R\sigma}}{\Delta_\sigma^2} \quad \text{in the Kondo limit ,} \tag{14}$$

respectively. The magnetoresistance is an important quantity characterizing an efficiency of magnetic nanojunctions. It is defined as the relative difference between the current measured for the parallel (P) and the antiparallel (AP) orientation of magnetization in the electrodes $MR \equiv (I_P - I_{AP})/I_P = (\mathcal{G}_P - \mathcal{G}_{AP})/\mathcal{G}_P$. Using the relation $P_\alpha = (\rho_{\alpha\uparrow} - \rho_{\alpha\downarrow})/(\rho_{\alpha\uparrow} + \rho_{\alpha\downarrow})$ between the polarization P_α and the electronic density of states $\rho_{\alpha\sigma}$ in the α-electrode [13], one can determine MR

$$MR = \frac{2P_L P_R}{1 + P_L P_R} \,. \tag{15}$$

for the empty state regime. It is the Julliere formula [13], as one could expect for the uncorrelated transport of electrons. It is useful to define the coefficient $\alpha = (t_L^2 - t_R^2)/(t_L^2 + t_R^2)$ describing asymmetry between the left and the right junction. For the Kondo regime we present the formulae of MR calculated in two cases: i) for the system with equal polarization of the electrodes $P_L = P_R = P$

$$MR = \frac{P^2(1 - 3\alpha^2 + \alpha^2 P^2 + \alpha^4 P^2)}{(1 - \alpha^2 P^2)^2} \,; \tag{16}$$

and ii) for a large asymmetry between the left and the right junction ($\alpha \to 1$)

$$MR = -\frac{2P_L P_R}{1 - P_L P_R} \,. \tag{17}$$

It is seen that in the second case the magnetoresistance is negative and its absolute value is larger than that one (15) for the uncorrelated electronic transport.

Fig.4 presents the results for all range of energies ϵ_0 and for finite temperatures. Fig.4a shows the conductance peak, which increases with a decrease of T and is shifted toward the Kondo regime. Similar dependences were observed in measurements of $\mathcal{G}$ through nonmagnetic systems with the QD (see for example [1, 2]). The magnetoresistance is presented in Fig.4b. For the uncorrelated transport ($\epsilon_0 \gg \epsilon_F$) the magnetoresistance is given by the Julliere formula (15), which for our case is 0.2. In this range MR is weakly temperature dependent. In the mixed valence range ($\epsilon_0 < \epsilon_F$)

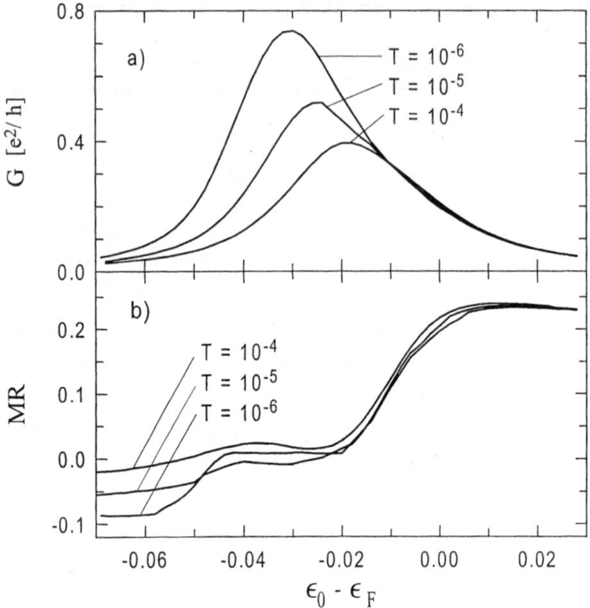

Figure 4. The conductance for the parallel configuration of the magnetization (a) and the magnetoresistance as a function of the relative position of the dot level $\epsilon_0 - \epsilon_F$ for the system with the asymmetric junctions $t_L = 0.02$, $t_R = 0.06$ and for different temperatures $T = 1 \times 10^{-6}$, 1×10^{-5} and 1×10^{-4}. The magnetic polarization of the electrodes is taken as $P_L - P_R = 1/3$.

one can see the drop of MR. This results from the correlations between electrons with the opposite spins passing through the QD. Since the considered system is with asymmetric junctions, MR achieves negative values in the Kondo regime ($\epsilon_0 \ll \epsilon_F$). From Eq.(16) the minimal value of MR is -0.103 at $T = 0$. Both $\mathcal{G}$ and MR are strongly temperature dependent in this regime. It is known that due to the Kondo resonance a sharp peak is formed at the Fermi level ϵ_F in the electronic density of states, which corresponds to spin fluctuations at the QD. The peak decreases logarithmic with T and disappears above the Kondo temperature T_K. Since at a low temperature $\mathcal{G}$ depends on the value of the density of states at ϵ_F, an increase of T reduces the peak and leads to a strong decrease of the conductance.

6. Summary

In this work we have shown that quantum interference and correlations are important for electronic transport through nanostructures. We presented modelling and studies of coherent transport phenomena: the Fano and Kondo resonances, recently observed in electronic transport through quantum dots. Due to a specific quantum dot geometry the Kondo effect greatly enhances conductance – in contrast to the "classical" case of magnetic impurities in metals. The Fano resonance in quantum dots manifests itself as a quantum interference between the bridge channel and a channel formed by the Kondo resonance in the vicinity of Fermi level. Thus, this effect is induced by strong electronic correlations seen in the conductance.

We also investigated interference processes seen in electronic transport through a metallic ring with a quantum dot. This geometry allows us to investigate the Aharonov-Bohm effect in presence of electronic correlations. Within our bridged Anderson model the sign of quantum interference can be changed by the value of magnetic flux enclosed in the ring. In the Kondo regime one can observe continuous evolution of the shape of the differential conductance with the magnetic flux, where the sharp peak at low voltages is transformed to a dip. We hope that our predictions based on the Aharonov-Bohm effect will stimulate further experimental studies.

Moreover, we presented studies of spin-dependent transport in the system with the ferromagnetic electrodes. Such the system is interested not only for its potential application in nanoelectronic, but also for fundamental research. We showed that the magnetoresistance effect is diminished when strong electron correlations come into play, its value can be even negative for a large junction asymmetry. Although transfer rates for electrons with the opposite spin orientation are different in the magnetic system, nevetheless due to the Kondo resonance the conductance reaches the unitary limit and the singlet state is formed.

Acknowledgement

The work was supported by the Committee for Scientific Research (KBN) under Grant No. 2 P03B 087 19.

References

1. Gores, J., Goldhaber-Gordon, D., Heemeyer, S., Kastner, M. A., Shtrikman, H., Mahalu, D. and Meirav, U. (2000) The Kondo effect in a single-electron transistor *Phys. Rev. B* **62**, 2188.
2. Kouwenhoven, L. and Glazman, L. I. (2001) Revival of the Kondo effect *Physics World* **14**, 33; and references therein.
3. Hewson, A. C. (1993) *The Kondo problem to heavy fermions*, Cambridge University Press.
4. Liang, W. J. , Bockrath, M., Bozovic, D., Hafner, J. H., Tinkham, M. and Park, H. (2001) Fabry-Perot interference in a nanotube electron waveguide *Nature* **411**, 665.
5. see for example: Mitani, S., Fujimori, H., Takanashi, K., Yakushiji, K., Ha, J.-G., Takahashi, S., Maekawa, Ohnuma, S., Kobayashi, N. Masumoto, T., Ohnuma, M. and Hono, K. (1999) Tunnel-MR and spin electronics in metal-nonmetal granular system *J. Magn. Magn. Mat.*

192, 179, and references therein.

6. For a technique of the nonequilibrium Green functions and its applications in electronic transport, see: Haug, H. and Jauho, A.-P. (1998) *Quantum Kinetics in Tranport and Optics of Semiconductors*, Springer Verlag.
7. Fano, U. (1961) Effects of configuration interaction on intensities and phase shifts *Phys. Rev.* **124**, 1866.
8. Bułka, B. R. and Stefański, P. (2001) Fano and Kondo resonance in electronic current through nanodevices *Phys. Rev. Lett.* **86**, 5128.
9. Heemeyer, S. (2000) Interference in resonant tunneling through a single-electron transistor, Ph.D. thesis, Massachusetts Institute of Technology.
10. Aharonov, Y. and Bohm, D. (1959) Significance of electromagnetic potentials in the quantum theory *Phys. Rev.* **115**, 485.
11. Webb, R., Washburn, S., Umbach, C. and Laibowitz, R. (1985) Observation of h/e Aharonov-Bohm oscillations in normal-metal rings *Phys. Rev. Lett.* **54**, 2696.
12. Yacoby, A., Heiblum, M., Mahalu, D. and Shtrikman, H. (1995) Coherence and phase sensitive measurements in a quantum dot *Phys. Rev. Lett.* **74**, 4047; Gerland, U., von Delft, J., Costi, T. A. and Oreg, Y. (2000) Transmission phase shift of a quantum dot with Kondo correlations *Phys. Rev. Lett.* **84**, 3710; van der Wiel, W.G., De Franceschi, S., Fujisawa, T., Elzerman, J. M., Tarucha, S. and Kouwenhoven, L. P. (2000) The Kondo effect in the unitary limit *Science* **289**, 2105.
13. Julliere, M. (1975) Tunneling between ferromagnetic films *Phys. Lett.* **54A**, 225.

ON CONSTRUCTION OF MOLECULAR MEMORY BASED ON FULLERENE ADDUCTS

P.BYSZEWSKI[1,2], Z.KLUSEK[3,4]
1 Institute of Physics PAS, al. Lotników 32/46, 02-668 Warsaw, Poland,
2 Institute of Vacuum Technology, ul. Długa 44/50, 00-241 Warsaw, Poland,
3 Advanced Materials Research Institute, University of Northumbria, Ellison Building, Ellison Place, Newcastle upon Tyne, NE1 8ST, United Kingdom,
4 Department of Solid State Physics, University of Łódź, ul. Pomorska 149/153, 90-236 Łódź, Poland

The $C_{60}ONCC_5H_4FeC_5H_5$ cycloadduct was prepared in the reaction between C_{60} and ferrocene oxime. The compound dissolved in dichloroethane was deposited on HOPG and observed by UHV STM/STS methods. The molecules of C_{60}ONCFn formed long straight chains extending for several microns, and in some of the chains the adducted groups were clearly visible. The *dI/dV(V)* plots exhibit pronounced maxima interpreted as resonance tunneling to molecular discrete electronic states. The STM/STS observations are discussed within the terms of semiempirical quantum chemical molecular modeling.

1. Introduction

With the development of microscopic scanning tunneling or atomic force microscopy (STM, AFM) techniques new methods of dense information storage are being searched. Possibility of recording and reading information by polarization of nanoscopic domains on ferroelectric P(VDF/TrFE) thin films using a conductive atomic force microscope was proven by Matsushige et al. [1]; Locquet [2] recorded information on ferroelectric $LaTiO_{3.5}$ with AFM. Kwon et al. [3] proposed a memory unit of molecular dimensions consisting of a carbon nanocapsule C_{480} containing endohedral fullerene $K@C_{60}^+$ that could be shifted between two ends of the capsule by an external electric field. It was shown that self-assembled monolayers might be utilized to built molecular memory devices [4] and Seminario et al. [5] considered π-electrons conjugated systems with NH_2 and NO_2 substitutes whose electrical behavior would depend on transferred charge and thus could be used for storing information.

L.M. Liz-Marzán and M. Giersig (eds.),
Low-Dimensional Systems: Theory, Preparation, and Some Applications, 293–303.

The investigations presented here intended to find molecules whose structure may be changed externally so that information might be stored using matrices of molecules that might adopt different stable conformations. For recording and reading information, STM would be used. Position of a functional group in complex molecules might be changed relatively to the rest of the molecule due to instantaneous charge tunneled from the STM tip or under the influence of the electric field generated by the STM tip. The two molecular states have to be separated by a potential barrier preventing spontaneous transition between them though low enough to allow changes of conformation under an external influence. The cyclic molecules with concentrated density of states are appropriate for these experiments, better suited than chain molecules because the high density of electronic states facilitates the STM observations.

The technique allows investigating molecules with open energy gap between highest occupied molecular orbital and the lowest unoccupied molecular orbital (HOMO-LUMO) like C_{60} and organic molecules that are insulating in their bulk structure, as well as self-assembled monolayers of molecules or dense layers of organic adsorbates. Pure C_{60} fullerenes weakly interact with the substrate predominantly by Van der Waals forces, however the interaction of functionalized fullerenes depends on adducted functional groups e.g. phenylated C_{60} on Si(111)-(7x7) was found to diffuse and rotate due to the presence of the phenyl groups until ordered layers were formed [6].

We investigated fullerene derivative C_{60}ONCFn deposited on highly oriented pyrolitic graphite (HOPG) by the STM method. The complex molecule consists of ferrocene (Fn, $C_5H_5FeC_5H_5$) bound to C_{60} at the 6-6 bond by a heterocyclic ring.

The STM topographic and tunneling spectroscopy measurements were carried out to find out what kind of structures the C_{60}ONCFn complexes form on HOPG and what sort of influence the molecule–molecule and molecule-substrate interactions have on the molecular energy spectrum. We expect that because the interaction with the graphite surface is weak, predominantly of the Van der Waals type, we study the inherent properties of C_{60}ONCFn complexes.

The structure of C_{60}ONCFn complex, a possibility of conformation change by external electric field of the complex and STM topographic and spectroscopic measurements were analyzed using semiempirical quantum chemistry calculations based on the Neglect of Diatomic Differential Overlap approximation with PM3 parameterization. In the model, the Slater type atomic orbitals are used and electron-electron interaction is accounted for by the self consistent field method. The model, parameterized by Stewart [7] reproduces well the structure of fullerenes and organic molecules also containing transition metals.

In this paper STM topographic observations of C_{60}ONCFn molecules deposited on graphite and interpretation of the images are presented.

2. Preparation and Properties of C_{60}ONCFn

The C_{60}ONCFn complex was prepared using C_{60} and ferrocenecarboxaldehyde (FnCHO) as the substrates [8]. The reaction proceeded through the following steps:

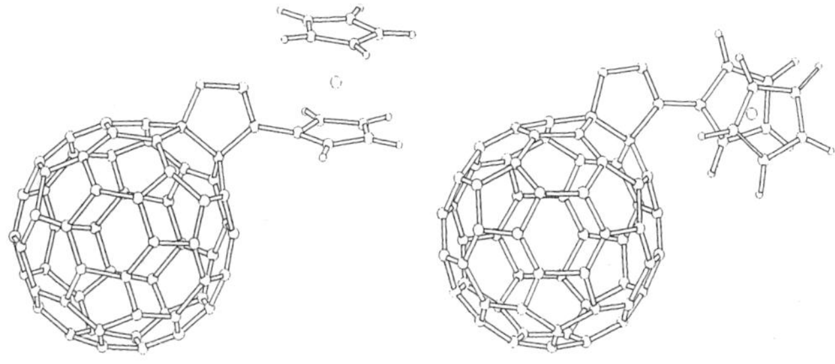

Figure 1. Two conformations of the C_{60}ONCFn complex.

$$FnCHO : NH_2OH:HCl \rightarrow FnCH{=}NOH \text{ in ethanol;} \quad (1)$$

$$FnCH{=}NOH : C_4H_2O_2NCl \rightarrow FnC{=}N^+O^- \text{ in } CH_2Cl_2; \quad (2)$$

$$FnC{=}N^+O^- : C_{60} \rightarrow C_{60}ONCFn \text{ in toluene;} \quad (3)$$

The mixture obtained from the reaction was concentrated and purified by chromatographic method using silica gel column with toluene as the eluent.

Two possible structures of the molecule optimized in the restricted Hartree-Fock approximation in a singlet state are shown in Figure 1. In order to verify if the C_{60}ONCFn complexes adopt anticipated structure, the ^{1}HNMR, ^{13}CNMR and absorption spectra in the 400-600 nm region were measured.; the observed spectra comply with both propositions. The binding energy calculated for both conformations differs by only ~2 kcal/mol that is within an accuracy of the calculation method; the potential barrier between the states evaluated from the transition state is <15 kcal/mol. It has to be assumed that the purified reaction product contains both types of complexes that might be recognizable by STM experiments. The largest distance between atoms in the molecules is 1.5 nm, thus together with its electron cloud the molecule extends for over 1.8 nm. The C_{60}ONCFn molecules have intrinsic electric dipole moment $d \approx 2.5$ Debye parallel to the heteroatom ring attached to C_{60}.

Looking for a possibility of changing externally the structure of molecules by an electric field or ionization, the calculations were performed for complexes in uniform external electric field of various intensities and orientations relative to the molecule and various charge states. The applied electric field simulates the experimental conditions during the STM measurements. Modeling of charged complexes or in the electric field was performed in the unrestricted Hartree-Fock approximation in an appropriate spin multiplicity state. This approximation allows different spatial and energy distribution of molecular orbitals (MO) populated by electrons of different spin polarization.

The calculations were repeated for gradually increasing electric field until the Fn group rotated by such an angle that it relaxed to conformation another than the initial one when the field was switched off. To induce transition the electric field of intensity ~3 V/nm oriented almost perpendicular to the C_5H_5 ring was required. The transitions between parallel and perpendicular positions of C_5H_5 and $ONCC_{60}$ rings are reversible. The change of conformation takes place because of anisotropic polarizability of the Fn group. It results from calculations that orientation of the Fn group is independent of the ionization state thus the instantaneous charge tunneled from the STM tip can not modify the conformation.

The STM images of molecules may differ due to their high polarizability because of occupied molecular orbitals close to HOMO may be shifted between C_{60} and adducted Fn by the electric field generated by STM tip depending on orientation and intensity of the scanning field. It results from the calculations that C_{60}ONCFn cycloadduct is suitable for planned STM experiments because of the intrinsic degree of freedom though the whole molecule deposited on the substrate may rotate in the electric field.

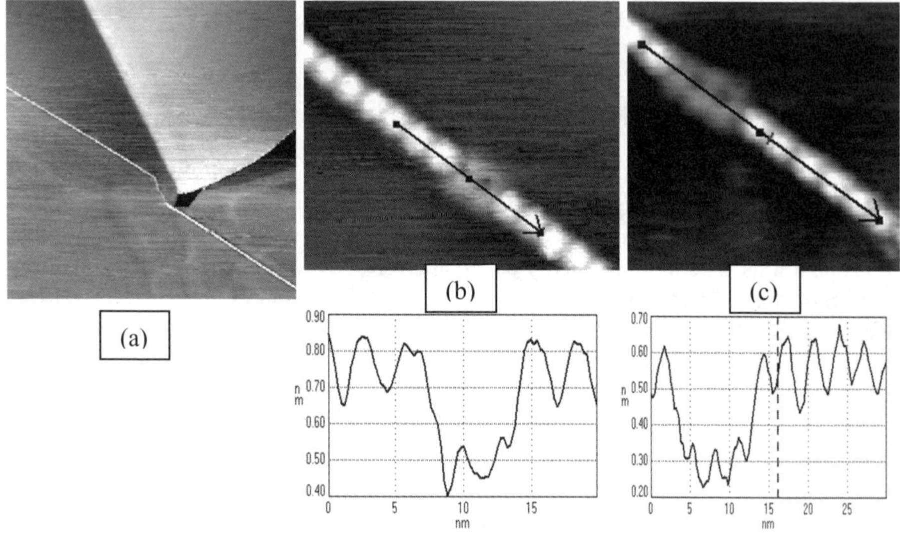

Figure 2. STM images of C_{60}ONCFn molecular chain. a) 2-D image of 300×300 nm^2 area of HOPG with deposited C_{60}ONCFn complexes scanned at: U=0.1V and i=0.1pA; b) enlarged defected fragment of a molecular row with z-profile along the arrow; c) another defected fragment of a molecular row with z-profile.

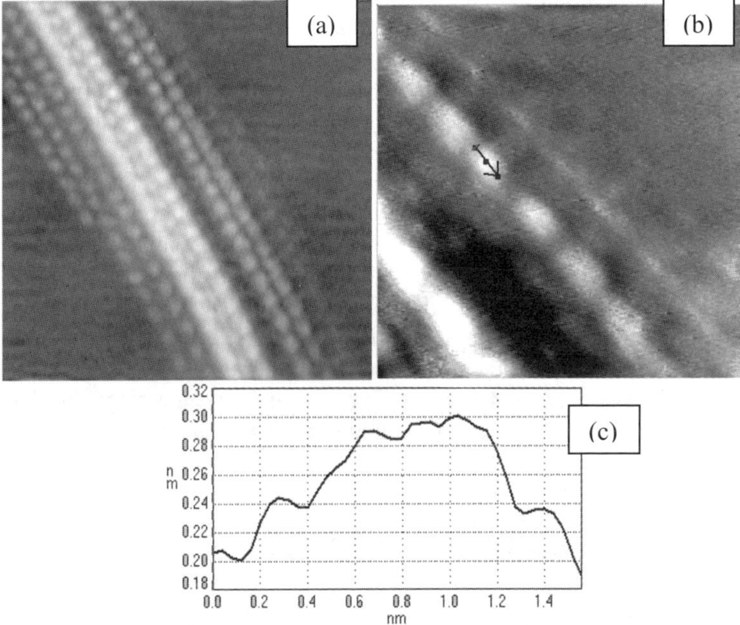

Figure 3. The STM topographic image of HOPG with deposited C_{60}ONCFn observed at scanning parameters: U=0.05V i=0.06nA. a) 2-D STM image of 42×42nm^2 area; b) enlarged image of 15×15 nm^2 fragment of single molecular row; c) z-profile along single molecular row at place shown by the arrow.

3. STM Observations of Fullerene Derivatives

A few μg of the C_{60}ONCFn dried compound were sonicated for 30 min. in 2 cm^3 of 1,2-dichloroethane, a droplet of the solution was deposited on freshly cleaved HOPG (0001) basal plane and the sample was loaded to the UHV STM chamber. The STM tips were prepared by mechanically cutting the Pt90%-Ir10% alloy wires.

We found several scattered clusters of C_{60}ONCFn complexes after depositing the solution on the HOPG substrate, however the main and most interesting features were long linear chains of the molecules, some of them are shown in Figures 2-4. The straight row shown in Figure 2a extends for over 300 nm, the average diameter of bright spots-"beads" is approx. 3 nm. It was determined from STM image registered at still higher resolution at a place where the atomic structure of the substrate was visible that the row was aligned with the <-110> direction of HOPG. It is interesting to notice that the distance between C atoms in graphite lattice along this direction is ~0.24 nm, twelve times smaller than characteristic spacing within the row thus the substrate may neither impose periodicity in the molecular row nor its direction. The periodicity in the row and its width do not correspond to the size of single molecule therefore one has to assume that all regular beads consist of the same number molecules.

Apart from the apparent parallel shift of the row, probably induced by the defect of the substrate (Figure 2a), there are two noticeable defects in the arrangement of molecules within the row shown in detail in Figures 2b and 2c, together with z-profiles scanned along the row. The defect presented in Figure 2c consists of two five-member rings of size ~3.5 nm with very similar brightness of the vertices. The presence of the defects helps to hypothesize on a structure of regular beads.

The size of the ring results from the presence of the adducted Fn groups, so that for pure C_{60} the ring would have a cross-section smaller than 2 nm. The pentagonal rings shown in Figures 2b and 2c with a cross-section of ~3.5 nm may be built of C_{60}ONCFn if the long axis of molecules are aligned with the edges of the pentagon and C_{60} at the vertices. The molecules have to be oriented with their electric dipole moment so as to compensate the total electric dipole of the cluster. The arrangement suggests that intermolecular electrostatic interactions dominate over Van der Waals interactions.

The regular beads are smaller by approx. 0.5 nm than the rings and have lateral dimensions of ~3 nm but their apparent height of ~0.65 nm is larger than that of the rings. Because there is no molecular structure visible, they are probably built of closely packed four C_{60}ONCFn molecules.

The apparent height of the molecules determined experimentally differs from the value of ~1.3 nm expected from the modeling. One of the possible reasons for this effect will be discussed later.

There were also other long parallel rows of C_{60}ONCFn molecules of the type shown in Figure 3. We suppose that the central, brightest i.e. highest feature shown in Figure 3a consists of two layers of molecules and the side rows of single linearly arranged molecules. The visible bright objects differ in length although their average size roughly corresponds to the calculated dimensions of the molecules. Since the objects in the side chains have elongated forms we assume that molecules are arranged with the ferrocene group along the chains, the z-profile measured along single molecule is shown in Figure 3c. The z-profile exhibits a variation in height of the measured molecule, so that the first part of the molecule along the arrow may be tentatively ascribed to the ferrocene adduct. The height of the molecule is again smaller than the prediction.

In another fragment of the molecular row, the structure of the deposited molecules could be easily observed; the recorded image is shown in Figure 4 and at higher resolution in Figure 5. Here many of the large bright objects have attached smaller fragments. We suppose that main parts of the large bright objects correspond to C_{60} and the small fragments, best seen in the chain at the right, correspond to the ferrocene group attached to C_{60}. The difference in the shape of small fragments (see Figure 5) may indicate different orientations of the Fn group.

The average distance between the ferrocene groups within the single molecular row is ~ 2.2 nm almost the same as the separation between the rows. If the distance between the fullerenes was determined by the size of C_{60} then it should be close to 1 nm as in the fullerite crystals. It means that Fn groups are not visible at each molecule constituting the rows. It is not probable that every other object is pure C_{60} because purity of the compound excludes presence of unreacted C_{60} at this proportion and the dichloroethane solvent molecule is too small to explain the distance.

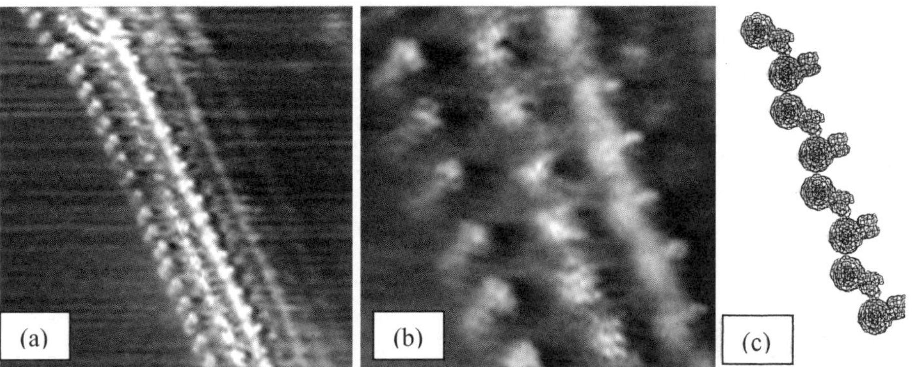

Figure 4. The STM topographic image of HOPG with deposited C_{60}ONCFn and possible arrangement of molecules in the chain. a) 2-D STM image of $45{\times}45 nm^2$ area observed at scanning parameters U=0.2 V, i=0.1 nA; b) 13.5*13.5 nm^2 fragment recorded at scanning parameters: U=0.05 V, i=0.06 nA; c) model of a chain.

Therefore, as "spacers" may be also C_{60}ONCFn complexes not aligned with the Fn group along the rows because then the separation between visible Fn would be ~2.8 nm but tilted relative to the direction of the rows.

Because of the dipole moment of C_{60}ONCFn molecules, the long range electrostatic interaction enforces arrangement of molecules to minimize energy and net electric field. It is possible if molecules adopt various orientations relative to the neighbors and the substrate e.g. in C_{60}ONCFn dimer with the molecules' electric dipoles oriented anti-parallel: ↑↓. One of the possible arrangements of complexes is shown in Figure 4c. For the construction of the model, the structure of the dimer was optimized using PM3 model then the charge distribution around the dimer was calculated to simulate an effective diameter of C_{60} equal to 1 nm and finally the dimers were placed at proper distances to reconstruct the STM image.

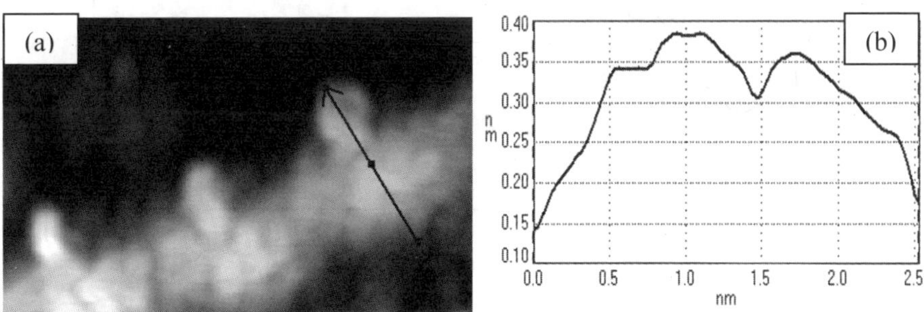

Figure 5. STM image of the fine structure of the C_{60}ONCFn molecule. a) topographic view; b) z-profile along the arrow.

Similar reconstructed STM images may be obtained for both possible conformations of the C_{60}ONCFn molecule. Although there is no experimental evidence of this particular orientation of the complexes within the dimers such arrangement explains the characteristic distances and fine steps along the chains shown in Figure 4b.

According to the z-profile shown in Figure 5b the complex extends horizontally for approximately ~2.3 nm. The size of the molecule determined by the chemical modeling agrees well with the lateral dimensions shown in Figure 5b, however, the height determined by the STM technique (z-profile in Figure 5b) is as previously lower than expected.

4. STS Measurements

The molecular chains are conducting at any voltage in the applied range of ±2 V allowed by the experimental setup. The current increases monotonically with the applied voltage observed in the *I(V)* characteristics even though the HOMO - LUMO energy gap of the isolated molecule is relatively large. The current at low voltages results from the tunneling of electrons to the density of states (DOS) of the conducting substrate continuously distributed in the energy scale.

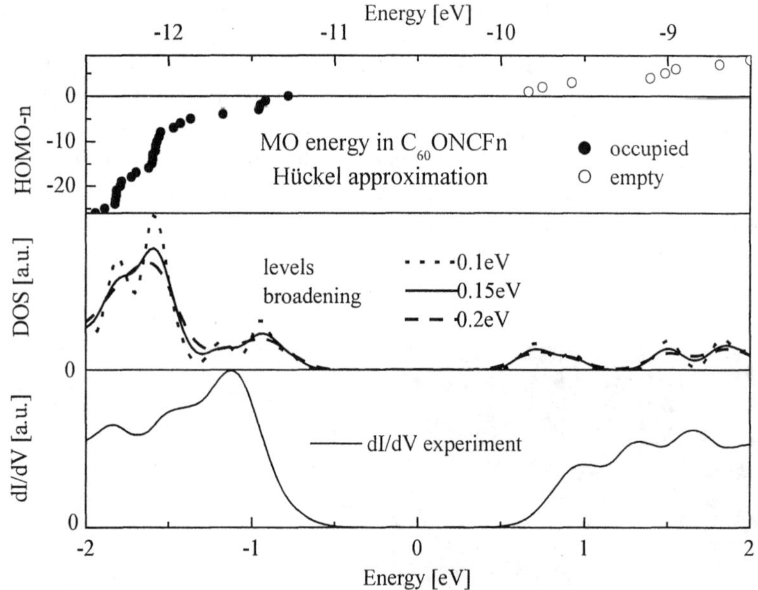

Figure 6. Energy of MO calculated in extended Hückel approximation (upper panel); DOS for various broadening (center panel) and measured *dI/dV(V)* (lower panel).

The current characteristics *I(V)* recorded at several sites across the molecular chains, revealed no substantial differences in conductance. Discrete electronic states of deposited molecules only slightly modify total DOS of the substrate-molecule-STM tip system and thus in most cases they can only be seen on the plots of *dI/dV(V)* derivative. The positions of peaks of *dI/dV(V)* also did not depend on position across the molecular chains although the amplitudes of the peaks varied. The *dI/dV(V)* derivative shown in Figure 6 is compared with the DOS of C_{60}ONCFn molecule calculated from the energy distribution of MO (open and solid dots in the upper panel of Figure 6) calculated using extended Hückel theory applied to molecules with the structure optimized with the PM3 model. All MO were broadened using Gauss function with the same width and number of states vs. energy was found summing contribution from all levels at each energy (lines in the central panel of Figure 6). The calculations were performed for different level broadening looking for a similarity with the experimental *dI/dV(V)* plot. The DOS line was shifted horizontally to adjust the Fermi energy with the center of the HOMO-LUMO gap. The plot calculated for 0.15 eV broadening agrees best with the experimental results. The broadening is much larger than the thermal energy, which suggests that tunneling does not occur from single isolated molecules but from dimers or larger clusters with energy spectrum weakly modified by the interaction. There is a difference in the behavior of the amplitudes of the observed *dI/dV* peaks and calculated DOS of the occupied states with lowering the energy, for the amplitude in experiment suddenly decreases at negative sample bias while the DOS amplitude increases. This is probably another indication that measurements were performed on clusters of C_{60}ONCFn molecules where occupied orbitals originating from the fivefold degenerate h_u and h_g orbitals in isolated C_{60} were close in energy (see Figure 6) and underwent further different splitting in each molecule due to the intermolecular interaction or interaction with the substrate. The interaction may produce almost continuous distribution of MO in energy thus broaden the DOS peaks and obliterate the singularities.

5. Imaging of the Open Gap Molecules

The C_{60}ONCFn has open HOMO - LUMO gap and thus are not conducting in topographic experiments or STS measurements in low voltages range. Different mechanisms were considered to explain the conductivity of the organic molecules observed by STM, like work function modulations [9] or a decrease of the HOMO–LUMO energy gap due to molecule–molecule or molecule–substrate interactions [10].

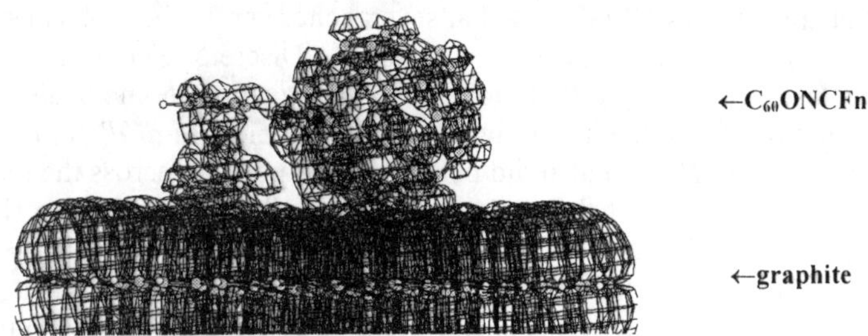

Figure 7. HOMO isodensity contour of C_{60}ONCFn:graphene ensemble. Apparent heights of the molecule measured from level of graphite MO, not position of carbon atoms, is ~0.7 nm.

We suppose that the electronic states of a conducting substrate participate in the tunneling process occurring in these experiments. Semiempirical quantum chemical calculations performed for an optimized ensemble consisting of C_{60}ONCFn placed 0.3 nm above graphene sheet showed that close to the HOMO-LUMO gap of the molecule the energy of the electronic levels was not affected by the interaction with the substrate. All MO of C_{60}ONCFn close to the gap of the molecule occur at the same energy as in the isolated molecule. However, the substrate states extend locally over the molecule making electrons tunneling possible.

The apparent height of any molecule would be diminished by this effect because the STM tip maintains the same overlap with wave function of the imaged object at the constant current mode. Thus, the height of the object is measured relative to a given spatial density of the wave function not from the graphite level. The isodensity contour of C_{60}ONCFn HOMO is displayed in Figure 7. The height of ~0.7 nm, rather than 1.3 nm would be observed at appropriate voltage and distance of STM tip from the sample.

6. Conclusions

The C_{60}ONCFn complexes produce well arranged linear forms on HOPG substrates that may be imaged by the STM method. In order to explain the observed molecular structures one has to take into account the electric dipole of single molecules. The interaction is insufficient to explain the formation of molecule chains, and therefore they probably grow while the solvent dries.

The molecular modeling indicates that the open gap molecules may be imaged at low scanning voltages because the states of conducting substrate envelop the molecules then the vertical dimensions shown by the experiments are significantly reduced. Because of admixing of the substrate states to the molecular states, any prominent singularities of conductivity ought to be searched in molecular systems extending high above the substrates like C_{60} multi-adducts. Until now we were unable to distinguish the parallel conformation of the adduct from the perpendicular one, and unable to modify the conformation with the STM tip.

7. Acknowledgements

This work was partially supported by the Polish Committee for Scientific Research under grant No 2 P03B 061 19.

8. References:

1. Matsushige, K., Yamada, H., Tanaka, H., Horiuchi, T., and Chen, X. Q. (1998) Nanoscale control and detection of electric dipoles in organic molecules, *Nanotechnology* **9**, 208-11.
2. http://www.omicroninstruments.com/products/afm_stm/r_afmst7.html.
3. Kwon, Y-K. Tomanek, D. and Iijima, S. (1999) "Bucky shuttle" memory device: synthetic approach and molecular dynamics simulations, *Physical Review Letters* **82**, 1470-1473.
4. Reed, M. A., Chen, J., Rawlett, A. M., Price, D. W., and Tour, J. M. (2001) Molecular random access memory cell, *Applied Physics Letters* **78**, 3735-3737.
5. Seminario, J. M., Zacarias, A. G., and Derosa, P. A. (2001) Theoretical analysis of complementary molecular memory devices, *J. Phys. Chem.* A **105**, 791-795.
6. Upward, M. D., Moriarty, P., Beton, P. H., Birkett, P. R., Kroto, H. W., Walton, D. R. M., and Taylor, R. (1998) Functionalized fullerenes on silicon surfaces, *Surf. Sci.* **405**, L526-531.
7. Stewart, J. J. P. (1989) Optimization of parameters for semiempirical methods. I. Method, *J. Computational Chem.* **10**, 209-220; Stewart, J. J. P. (1989) Optimization of parameters for semiempirical methods. II. Applications, *J. Computational Chem.* **10**, 221-229.
8. Popławska, M., Byszewski, P., Kowalska, E., Diduszko, R., and Radomska, R. (2000) Preparation and structure of ferrocene derivative C_{60} adduct, *Synthetic Metals* **109**, 239-245.
9. Spong, J. K., Mizes, H. A., LaComb, L. J., Dovek, M. M., Frommer, J. E., and Foster, J. S. (1990) Contrast mechanism for resolving organic molecules with tunnelling microscopy, *Nature* **338**, 137-139.
10. Fisher, A. J. and Blochl, P. E. (1993) Adsorption and scanning-tunneling-microscope imaging of benzene on graphite and MoS_2, *Phys. Rev. Lett.* **70**, 3263-3266.

ELECTROCHEMICAL APPLICATION OF CARBON NANOTUBES

E. FRACKOWIAK[a] K. JUREWICZ[a], S. DELPEUX[b], F. BEGUIN[b]
[a]*Institute of Chemistry and Technical Electrochemistry,*
Poznan University of Techology,
60-965 Poznan, ul. Poiotrowo 3,Poland
[b]*CRMD, CNRS-Universite, 1B rue de la Ferollerie,*
45071 Orleans,France

Abstract

Different types of nanotubes have been used for storage of energy, e.g. lithium insertion and electrochemical capacitors. In both cases the electrochemical properties of nanotubes are dominated by their mesoporous character mainly due to the entanglement and a presence of a central canal. The most promising electrochemical application of carbon nanotubes seems to be their usage for supercapacitors. Capacitance values of pristine nanotubes varied widely from 10 to 80 F/g depending mainly on their microtexture, presence of central canal, amorphous carbon and catalyst impurities. Special modification has been performed to increase the capacitance values. For instance, a coating of nanotubes by conducting polymers, e.g. polypyrrole (PPy) supply pseudocapacitive effects connected with quick faradaic reactions. The open entangled network of the nanotubular composite seems to form a volumetric electrochemical capacitor where the charge has a three-dimensional distribution. On the other hand, a significant increase of capacitance was reached by development of nanotubes specific surface area, i.e. micropore volume, through chemical KOH activation. Modified nanotubes and their nanocomposites with PPy supplied capacitance values of 80 to 170 F/g, hence, they are attractive electrode materials for supercapacitors.

1. Introduction

Carbon nanotubes, due to their unique morphology (e.g. helical or fishbone arrangement of graphitic layers, presence of central canal, entanglement, bundle formation) and interesting conducting and mechanical properties are of great interest for many applications such as nanowires, special capillaries, nanocomposites and components for nanoscale electronic devices [1-3]. The possibility of a large scale production of high purity carbon nanotubes permits to consider more carefully their usage in different areas [4]. Nanotubes have been taken into account for some electrochemical applications, for example lithium insertion [5-10], support for catalysts [11-12] and hydrogen storage [13-17] which could have potential use in storage of energy (lithium-ion batteries or fuel cells).

In this paper a critical discussion is presented on practical application of nanotubular materials for energy conversion in Li-ion accumulators and especially supercapacitors, i.e. energy devices delivering a high power due to the fast propagation of charges.

L.M. Liz-Marzán and M. Giersig (eds.),
Low-Dimensional Systems: Theory, Preparation, and Some Applications, 305–318.

2. Experimental

2.1. NANOTUBULAR MATERIALS

Different types of multiwalled carbon nanotubes (MWNTs) were used for electrochemical investigations. Catalytic MWNTs were synthesized by decomposition of acetylene using cobalt supported on silica at 700 °C and 900 °C (A/CoSi700, A/CoSi900) or on zeolite at 600 °C (A/CoNaY600) [18-21]. High purity MWNTs were prepared at 600 °C on Co particles from solid state solution ($A/Co_xMg_{(1-x)}O$) [4]. Chemical vapour deposition of propylene at 800 °C within an alumina membrane yieldied bamboo-like nanotubes (P/Al800) [22]. In all cases, after the preparation of the nanotubes, the catalyst support or the alumina template were dissolved with 72 % hydrofluoric acid, which also allowed to eliminate an important part of cobalt. The samples A/CoSi700 and A/CoSi900 were additionally treated in dilute nitric acid for further elimination of the free Co particles. After these treatments, the samples were filtered, washed several times with distilled water and dried at 150 °C under vacuum. For comparison, catalytic Graphite Fibrils™ from Hyperion Catalyst International Inc. (USA), referred to as Hyperion MWNTs, were also used. They contain 1.2 wt % Fe from the residual catalytic precursor.

The purified nanotubes have a high oxygen content or surface functionality, e.g. 5-10% for the samples A/CoSi700 and A/CoSi900, which is caused by the post-treatment in nitric acid. This is more remarkable in the case of A/CoSi700 for which transmission electron microscopy showed a surface coating of the nanotubes by pyrolytic carbon. Surface area and micro/mesopore volume were obtained from nitrogen adsorption/desorption isotherms at 77 K (Micromeritics ASAP 2010). Prior to the adsorption experiments, the samples were outgassed (10^{-6} mbar) at 350 °C during 12 hours.

From the nanostructural and microtextural characterisations, the catalytic MWNTs appear as a web of curved nanotubes forming often intertwined entanglements [18-21], while P/Al800 are stiff and with bamboo-like morphology. Mostly the nanotube tips are opened, except for A/CoNaY600, for which the aromatic carbon layers are remarkably continuous and straight. The material A/CoSi900 has a so-called fishbone morphology with an ill-defined central canal [20], therefore it is closer to nanofilaments rather than to nanotubes.

In some experiments single walled nanotubes (SWNTs) as bucky paper from Rice University (USA) were used.

2.2. ELECTROCHEMICAL EXPERIMENTS

2.2.1. *Lithium insertion*

Lithium insertion into the nanotubular carbon material has been performed in a two electrode Swagelok® cell where a lithium disk was the counter electrode as well as the reference electrode. The carbon electrodes were prepared in the form of pellets with 85 wt.% content of MWNTs, 5 wt.% of acetylene black and 10 wt.% of binding substance (Polyvinylidene fluoride, PVDF-Kynar flex 2801, Atochem, France). The electrolyte was 1M $LiPF_6$ dissolved in a mixture (1:1) of ethylene carbonate (EC) and diethylcarbonate (DEC) (Merck). Galvanostatic charge/discharge cycling with a current load of 20 mA/g of carbon was performed in order to estimate the degree of lithium insertion and extraction, using a multichannel potentiostat/galvanostat MacPile II (Biologic, France).

2.2.2. *Electrochemical capacitors*

The above described nanotubes have been used either in the form of pellets or bucky paper for the assembly of supercapacitors. Two electrodes (positive and negative) of comparable mass ranging from 1 to 20 mg were connected with golden or steel current collectors depending on the electrolytic medium. Swagelok® type teflon cells were used for the montage of capacitor, including a glassy fibrous paper separator and 6M KOH or 1M H_2SO_4 as an electrolytic solution. Voltammetry and galvanostatic charge/discharge cycling with potential limitation (VMP-Biologic, France and BT2000-Arbin, USA) were used for the evaluation of capacitance values. For a comparison, the electrochemical characteristics were also investigated by impedance spectroscopy with a SOLARTRON SI 1260 analyser (Schlumberger) in the frequency range from 100kHz to 1 mHz at open circuit voltage with 10 mV amplitude

3. Lithium storage

Lithium-batteries are based on intercalation and/or insertion materials between which lithium ions are transferred through electrolyte during charge and discharge processes. For the negative electrode the graphite based materials are the most often used, however there is still a demand for better electrochemical characteristics. Among many different carbons a nanotubular material has been also proposed as electrode for lithium storage in aprotic media [5-10]. A significant reversible capacity has been found from 400 mAh/g up to ca. 800 mAh/g, however, with some drawbacks. The high divergence between insertion and extraction (so called hysteresis) and a great irreversible capacity (800 mAh/g) during the first cycle are observed (Figure 1).

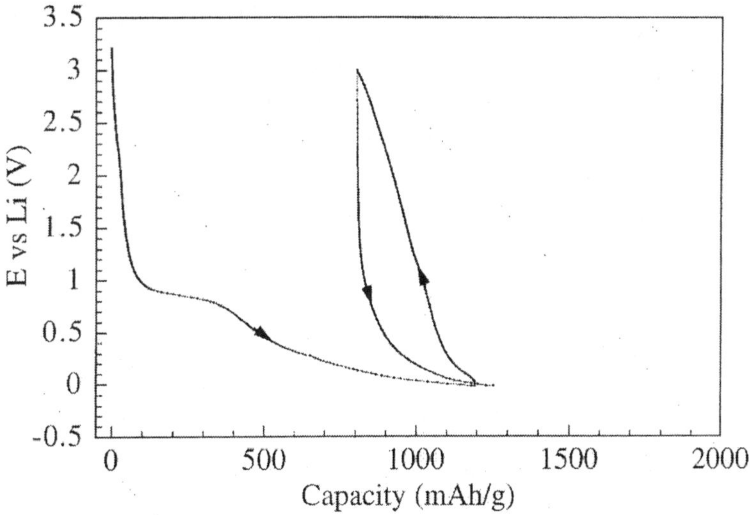

Figure 1. Lithium insertion/deinsertion in A/CoSi900 carbon nanotubes. Current load 20 mA/g.

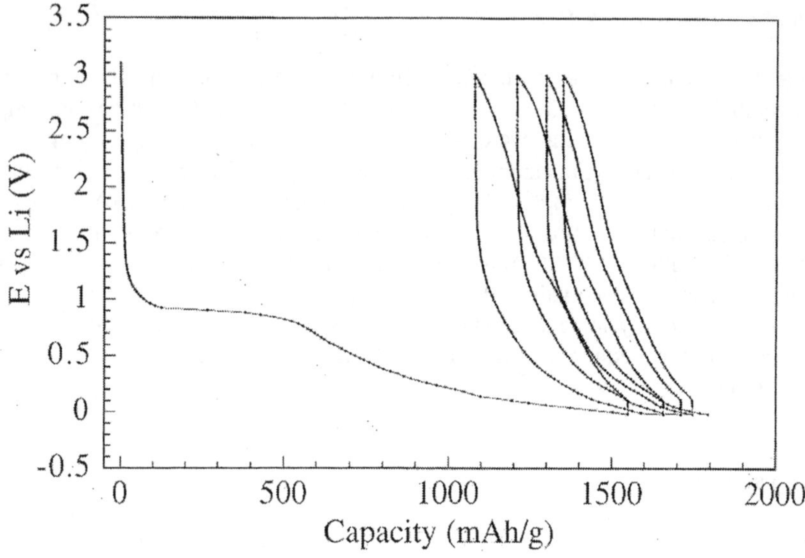

Figure 2. Electrochemical insertion of lithium into MWNTs (P/Al800) during subsequent cycles

Apart from hysteresis, the lack of plateau during lithium deinsertion excludes this material for lithium storage. Additionally, during subsequent cycling the continuous formation of solid electrolyte interphase (SEI) is observed (Figure 2). It can bc explained by penetration of the solvated lithium ions into mesopores which then decompose yielding a lithium carbonate. Simultaneously, the reversible capacity also diminishes with cycling and the capacity loss is estimated as about 30% after 10 cycles.

Finally, even if a storage capacity for lithium in nanotubular materials exceeds typical values for the graphite intercalation compound LiC_6 (372 mAh/g), all the mentioned drawbacks preclude this material in practical applications for Li-ion batteries. On the other hand the presence of a mesoporous network facilitates the diffusion of ions to the electrode/electrolyte interphase which suggests promising properties of this material for supercapacitor applications.

4. Supercapacitors

Energy storage in supercapacitors combines a pure electrostatic attraction of the ions in the electrical double layer and quick faradaic pseudo-capacitance reactions [23]. The stored energy is based on the separation of charged species across the electrode/solution interface. The electrochemical capacitor is built from a positive electrode with electron deficiency whereas the second electrode is negative, i.e. with electron excess. The capacitance values are strictly connected with the nature and surface of the electrode/electrolyte interface. Generally, the more developed surface area the higher the ability for charge accumulation. The presence of micropores (below 2 nm) are crucial for the formation of a double layer, while mesopores (from 2 to 50 nm) play a role during both adsorption and transport.

The main advantage of supercapacitors is the ability of a high dynamic charge propagation that allows a rapid recovery of energy. Due to the open network of nanotubes, the accessible electrode/electrolyte interface facilitates charging of double layer, hence, this material seems to be very promising. The electrochemical characteristics of supercapacitors built from MWNTs and SWNTs have been investigated and correlated with the microtexture and elemental composition of the materials [24-29].

The capacitance was calculated mainly from the galvanostatic discharge and voltammetry characteristics at 2 mV/s scan rate. In 6 M KOH aqueous electrolyte, most of the nanotube based capacitors give voltammograms with regular box-like shape. The higher the BET specific surface area and oxygen content of the nanotubes, the higher the values of capacitance. Capacitance values as high as 80 F/g of carbon nanotubes have been found, even if the specific surface area of the materials reaches only a moderate value of maximum 450 m^2/g [25-27]. The functionalisation of A/CoSi700 by hot nitric acid increases further the capacitance value up to 130 F/g, and a reversible, ill-defined redox peak due to an oxygenated surface functionality is noted [25]. However, the increase of capacitance is not stable with cycling typical for this type of pseudocapacitance due to the presence of surface groups. Another effect not strictly connected with the charging of an electrical double layer has been demonstrated on Hyperion™ catalytically grown nanotubes. Depending on the electrolytic solution, the capacitance reached different values (14 F/g in 6 M KOH and 78 F/g in 1 M H_2SO_4), clearly showing that redox pseudocapacitive reactions take part due to the dissolution of iron impurities (1.2 wt%) [27].

SWNTs (Rice University) have been also investigated for comparison and they supplied a value of 40 F/g. After their annealing at 1650 °C, the capacitance diminished to 18 F/g due to a better arrangement of the tubes in the bundles, that hinders the diffusion of solvated ions towards the active surface [27].

Generally for activated carbons there is almost a linear relationship between the specific surface area and the capacitance. In the case of MWNTs and SWNTs, the specific surface area is moderate, ca. 400 m^2/g, and microporosity is very limited. Hence, the ability for charge accumulation in the electrode/electrolyte interface strongly depends on the ions accessibility to the outer walls and central canal of nanotubes, and on the total number of defects. On the other hand, the presence of a dense pyrolytic carbon outer layer is able to aggravate ions storage. However, if the micropores of pyrolytic carbon covering the nanotubes are large enough to be penetrated by the solvated ions, they can play a great role in the charging of the double layer. Additionally, the values of capacitance are enhanced if the tips of nanotubes are open, which seems to be a proof of accumulation of charges in the central canal. However if the canal is too large, i.e. over the diameter of a few solvated ions, it does not play any positive role for charging the double layer. It was the case of template nanotubes P/Al800, obtained with different times of deposition, where the values of capacitance varied from 5 to 40 F/g depending on the diameter of central canal (100 nm and 10 nm, respectively). In the case of a very large canal and only a few concentric graphitic layers which form the nanotube wall, the electrochemically active electrode/electrolyte interface is very small. Hence, optimised carbon nanotubes for a supercapacitor should possess a great number of graphene layers, an open central canal with diameter below 5 nm (taking into account the size of a few solvated ions) and an interconnected entanglement. All the defects and roughness of walls are very favourable for charging the electrical double layer. Optimal would be additional stable pseudoeffects which could significantly enhance the values of capacitance, e.g. conducting polymers or other electroactive species deposited on the nanotubes.

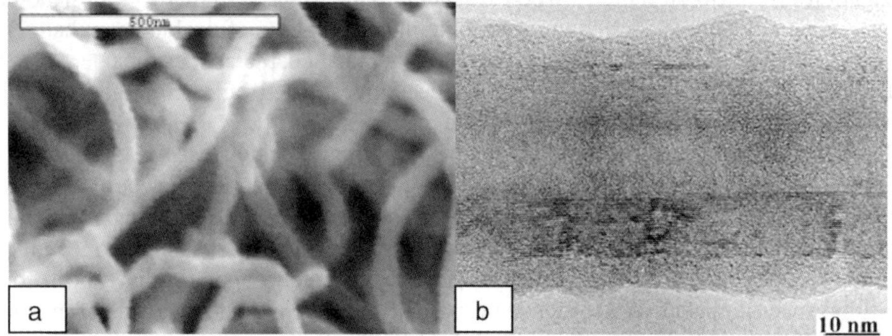

Figure 3. Multiwalled nanotubes A/CoNaY 600 with electrodeposited PPy:
(a) general population of coated nanotubes by SEM; (b) single MWNT/PPy (TEM)

4.1. NANOTUBE/POLYPYRROLE COMPOSITE

A novel type of composite electrodes based on MWNTs with deposited polypyrrole (PPy) has been used for the assembly of supercapacitors. Chemical and electrochemical polymerisation of pyrrole has been considered in order to get a homogenous layer of PPy on the nanotubular materials. The homogeneity and the thickness of the PPy layer in the composite material were estimated by Scanning Electron Microscopy. Chemical deposition of PPy on nanotubes or fibres supplied a non-homogenous type of deposit with a tendency to form aggregates. On the other hand, the application of the electrochemical method gave a very unique deposit of a homogeneous PPy film with a good conductivity. SEM and TEM micrographs of the nanotubes A/CoNaY600 with electrodeposited PPy (Figure 3) show a homogenous coating; by comparison with the diameter of the pristine material, the film thickness is equal to 5 nm.

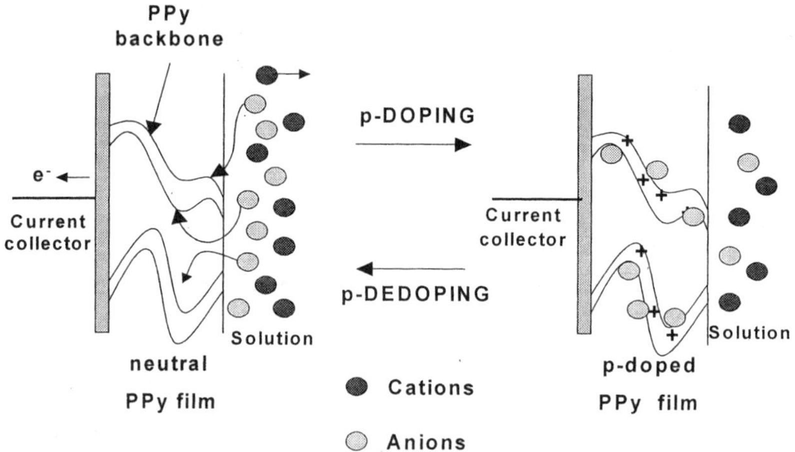

Figure 4. Schematic representation of the doping process of PPy. Positive charging of polymer chains is compensated by attraction of counterions.

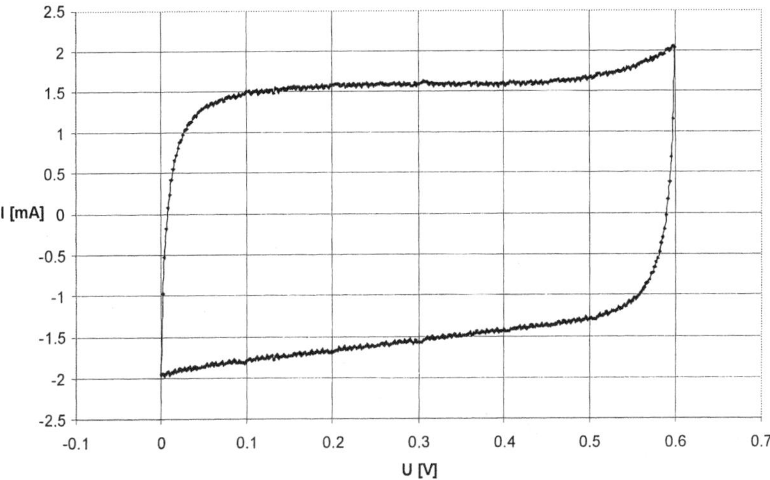

Figure 5. Voltammetry characteristic of a supercapacitor built from MWNT/PPy nanocomposite. Scan rate 2 mV/s

Cyclic voltammetry with a potential scan rate from 1 to 10 mV/s and galvanostatic charge/discharge cycling from 0 to 0.6 V or higher voltage limitation (0.8 V; 1.0 V; 1.2V) were performed to estimate capacitance values. The effect of PPy on the capacitance properties of A/Co700 measured by voltammetry is shown in Figure 5. The square shape of the voltammogram for this nanocomposite would suggest a pure electrostatic attraction, even if pure PPy usually gives more irregular characteristics, confirming a good synergy between PPy and MWNTs. This kind of capacitive behaviour is confirmed by the linear discharge on the galvanostatic curve (Figure 6).

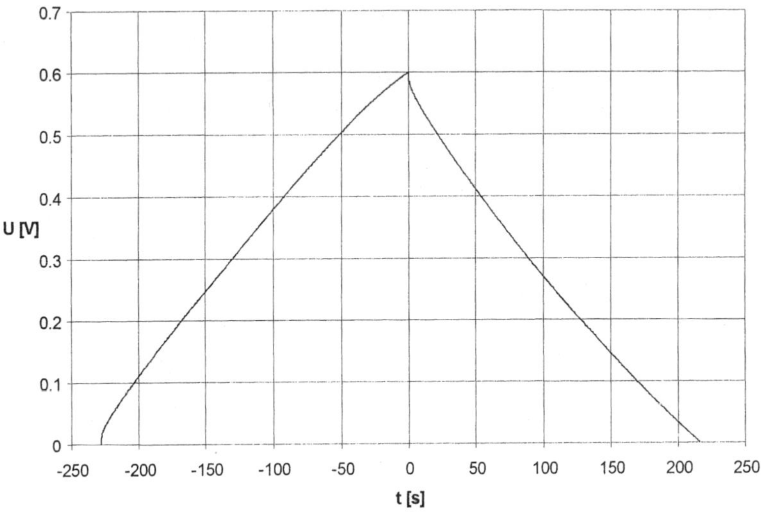

Figure 6. Galvanostatic charge/discharge of a supercapacitor built from MWNT/PPy composite.

TABLE 1. Specific capacitance (F/g) of nanotubes after doping with PPy

Sample	S_{BET} (m^2/g)	C (F/g)	C (F/g) with PPy
A/CoSi700	430	65	141
Hyperion	290	78	146
P/800Al	50	5	123
A/CoNaY600	130	50	165

Galvanostatic discharge seems to be the most reliable method for getting the values of capacitance in F/g of active nanotubular material for all the investigated samples (Table 1). From the analysis of these data, we can see that the nanotubular materials with electrochemically deposited polypyrrole yield significantly higher values of capacitance than the pristine nanotubes. A maximum value of ca. 170 F/g has been obtained for MWNTs prepared at 600 °C and modified by a PPy layer of 5 nm. This is about twice higher than the value obtained with the pristine nanotubes, cà. 80 F/g, thus proving a synergy effect between nanotubes and PPy.

It seems that the open entangled web of the nanotubular composite allows to form a volumetric electrochemical capacitor where the charge has a three-dimensional distribution. It is remarkable that pristine template nanotubes P/Al800 with a large central canal and an extremely low specific surface area (below 50 m^2/g) demonstrate negligible values of capacitance (5 F/g). In this case mainly electrodeposited PPy is responsible for the high value close to 130 F/g of nanotubular composite material. Taking into account the large diameter of the central canal for these tubes, ca. 100 – 200 nm, it might be expected that a thin PPy film also covers the inner core which still remains accessible for the electrolyte. The high value of capacitance compared to pristine P/Al800 originates from inner and outer coating of the nanotubes.

Due to the open network of mesopores in the nanotube/PPy composites, the bulk of the thin PPy layer is fully involved for quick pseudofaradaic processes, which is the reason for the high values of capacitance. Taking into account that some of the investigated capacitors have been cycled over 2000 cycles (Figure 7), with a charge loss which never exceeded 20%, nanotubes coated by a thin layer of conductive PPy seem to be an efficient model for the realisation of long durability materials. Hence, this new kind of nanocomposites, easily manufactured at reasonable costs, presents very promising perspectives for future capacitor application of nanotubes.

4.2. NANOTUBES ACTIVATED BY KOH

A great interest has been recently devoted to the further modification of carbon nanotubes to find a wide potential application of these attractive materials. For efficient charging of the electrical double layer, i.e. for high values of capacitance, a developed surface area is demanded and the role of micropores is essential. In the case of nanotubes, specific surface area is very moderate, i.e. from 200 to 400 m^2/g, hence, even if accessibility to electrode/electrolyte interface is perfect due to the presence of mesopores, the almost negligible microporosity limits the high values of specific capacitance.

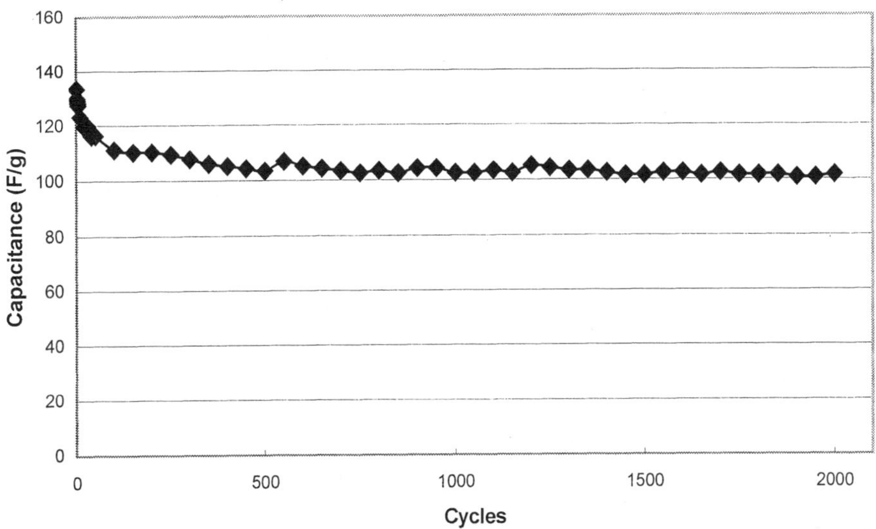

Figure 7. Cycling of supercapacitor built from MWNT doped by PPy. Current load 350 A/g.

Generally, the purified MWNTs can supply values of capacitance from 10 to ca. 100 F/g and values are strongly depending on their final microtexture. Essentially, the better graphitized carbon walls the lower values of capacitance and the defected surface of nanotubes enhances accumulation of charges. Hence, controlled development of nanotubes porosity especially in the range of micropores can significantly modify electrochemical properties of nanotubes and create a new attractive material.

A significant development of nanotubes microporosity has been obtained by chemical activation with KOH that gives a novel advanced materials with various potential uses, e.g. electrodes for supercapacitors, supports for catalysts or absorbents for gas storage. Development of surface area of carbon by KOH activation is well known, however, the mechanism of this process is still under discussion. During KOH activation considerable amounts of potassium carbonate and hydrogen are formed but a crucial role is played by metallic K formed above 700 °C according to the reactions:

$$KOH + H_2 \rightarrow 2K + H_2O$$

$$K_2O + C \rightarrow 2K + CO$$

Mobility of the free vapour of potassium formed at 774 °C and its intercalation and/or insertion into the nanotubes through all the outer defects on the walls and through tips are responsible for the development of microporosity. Insertion of K is at the origin of graphitic layer separation, in turn, formation of internal microporosity quite often with a cage-like structure. Careful TEM observation reveals that in some cases the creation of porosity takes place also inside of nanotubes that proves the assumption of K insertion into the central canal. The optimal weight ratio of activation agent, i.e. KOH to C (4:1) has been found to get at least two-fold increase of surface area during activation performed at 800 °C.

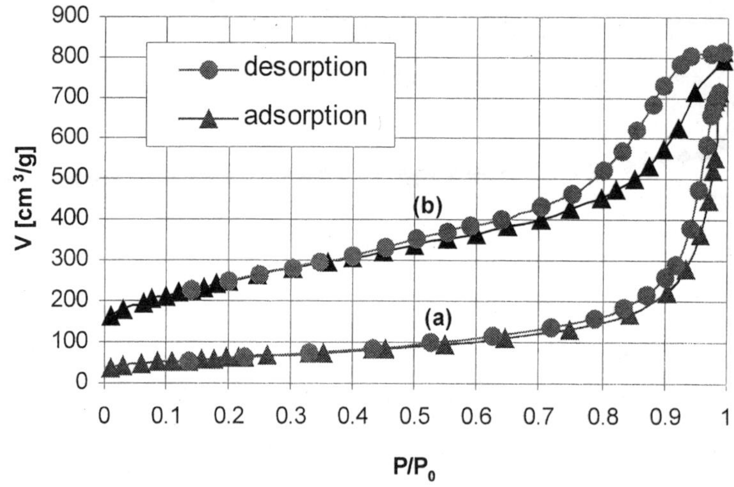

Figure 8. Nitrogen adsorption/desorption isotherm for A/$Co_xMg_{(1-x)}O$ nanotubes: a) before; b) after KOH activation

Two types of MWNTs have been used for KOH activation: A/CoSi700 and A/$Co_xMg_{(1-x)}O$. At the surface of A/CoSi700 sample, a thin 1-2 nm pyrolytic layer of carbon is observed, nanotubes have outer diameters ranging from 15 to 25 nm and central canals ranging from 5 to 10 nm. A/$Co_xMg_{(1-x)}O$ nanotubes have better graphitised and thinner walls where the outer diameter varies from 9 to 14 nm and the central canal from 3 to 6 nm. Depending on the nanotubular material, a weight loss of 20 to 45 % has been found after the activation process. For the sample A/CoSi700, the weight loss was over 40% due to the presence of the disordered pyrolytic coating which undergoes activation more easily, whereas for A/$Co_xMg_{(1-x)}O$ nanotubes with strongly graphitised walls, a burn-off varied from 20 to 30%. The surface area increased from 430 m^2/g to 1035 m^2/g for A/CoSi700 and from 220 to 885 m^2/g for A/$Co_xMg_{(1-x)}O$ sample after KOH activation.

It is noteworthy that the activated material still possesses a nanotubular character with many defects at the outer walls that provide a significant increase in microporosity. The lack of nanotubes destruction is confirmed by nitrogen adsorption at 77 K (Figure 8). The typical mesoporous character for nanotubes A/$Co_xMg_{(1-x)}O$ before and after activation is documented by a IV type isotherm, however, hysteresis is definitely more pronounced in the case of activated nanotubular material. It confirms a wide distribution of pores in such a material. Nitrogen adsorption of the activated samples shows an important uptake at low relative pressure that reveals a significant development of microporosity. Micropore volume for both activated samples varies from 0.4 to 0.45 cm^3/g whereas before activation it ranges from 0.1 to 0.2 cm^3/g with a higher value for A/CoSi700. Nitrogen adsorption data correlate very well with microscopy observations. A formation of micropores due to cracks, defects, and surface irregularities was clearly seen.

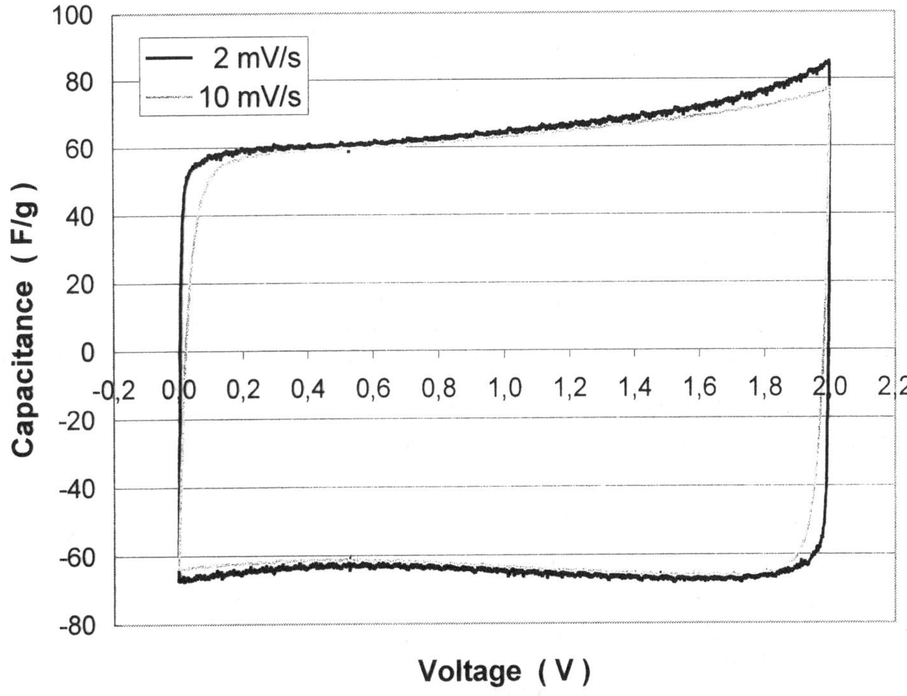

Figure 9. Charge/discharge of a supercapoacitor built from A/CoSi700 activated nanotubes at two different scan rates (2 and 10 mV/s). Electrolyte: 1.4 M $TEABF_4$ in acetonitrile

KOH activation seems to be quite a simple method for the enhancement of electrochemically active surface area of nanotubes. Such developed specific surface area of nanotubes is of great interest especially for some electrochemical applications, hence, the activated nanotubes have been used for electrode materials in supercapacitors. Indeed, ability of the charge accumulation was extremely enhanced. In the case of $A/Co_xMg_{(1-x)}O$ sample, the capacitance increased almost 7 times from 15 (for non-activated nanotubes) to 90 F/g (after chemical activation), especially in alkaline medium (6M KOH). Ability for charge accumulation of this material in alkaline, acidic and organic medium at 2 and 10 mV/s scan rate of potential was investigated by voltammetry and galvanostatic charge/discharge in 6 M KOH. Capacitors could be easily charged/discharged at high current density with a characteristic rectangular shape. Galvanostatic charging up to 1 V gives also a correct shape of the curve without any resistive character. In 1 M H_2SO_4 the values reached 85 F/g for $A/Co_xMg_{(1-x)}O$ and 95 F/g for A/CoSi700 activated material. In organic electrolytic solution (1.4 M $TEABF_4$ in acetonitrile) the values of capacitance reached 65 F/g for both types of activated nanotubes (Figure 9).

For the first time the activation of MWNTs has been successfully realized with a significant increase of specific surface area up to 1050 m^2/g. Activation method by KOH seems to be well adapted for development of surface area of carbon nanotubes as well as for opening their tips. Activated material preserves nanotubular shape confirmed by TEM observation and the nitrogen adsorption analysis. Creation of defects on the walls of tubes supplied the development of microporosity responsible for an excellent capacitance behaviour. The values of capacitance increased after the activation process especially for $A/Co_xMg_{(1-x)}O$ nanotubes from 15 F/g to 90 F/g in alkaline medium (6M KOH). In organic electrolyte the capacitance of both activated material reached 65 F/g with a box-like shape characteristics.

5. Conlusions

The mesoporous character of carbon nanotubes plays a dominant role in their electrochemical properties. Compared to conventional carbon materials, carbon nanotubes have a higher rate of electron transfer. Their entangled network and central canal are at the origin of pseudocapacitive effects, allowing an easy access of the ions to the electrode/electrolyte interface.

Relatively high values for lithium insertion degree up to x = 2.1 (in Li_xC_6) but with a significant hysteresis confirm that lithium insertion follows faradaic redox reactions but with a capacitive character (so called pseudoeffects). The lack of voltage plateau during lithium deinsertion, i.e. the continuous change of lithium/carbon interaction energy, is related with the evacuation of Li_n^+ clusters from mesopores. The mesopores are also responsible for an important irreversible capacity due to the easy access of solvated ions to the active surface where they are decomposed. Large hysteresis and irreversible capacity are two major drawbacks which preclude carbon nanotubes from any practical application in lithium-ion batteries.

Nanotubular carbon materials seem to be the most attractive for capacitor applications especially after modification such as electrodeposition of a thin layer of conducting polypyrrole (PPy). Values of specific capacitance are markedly enhanced due to the contribution of pseudofaradaic properties of PPy. The open entangled network of the nanocomposite favours the formation of a three-dimensional electrical double layer in the bulk.

The activation of MWNTs has been successfully realized with a significant increase of specific surface area up to 1050 m^2/g. Activation method by KOH seems to be well adapted for development of surface area of carbon nanotubes as well as for opening their tips. Activated material preserves nanotubular shape confirmed by TEM observation and the nitrogen adsorption analysis. Creation of defects on the walls of tubes supplied the development of microporosity responsible for an excellent capacitance behaviour.

Such significant modification of surface microporosity for essentially mesoporous nanotubular material is of great interest for many electrochemical applications (supercapacitor electrodes, supports for catalysts, gas storage).

6. Acknowledgements

This work was supported by the NATO Science for Peace Programme (SfP 973849).

7. References

1. Ago, H., and Yamabe, T. (1999) Frontiers of carbon nanotubes and beyond, in K. Tanaka, T. Yamabe and K. Fukui (eds.), *The Science and Technology of Carbon Nanotubes,* Elsevier, pp. 164-183.
2. Tans, S.J., Devoret, M.H., Dai, H., Thess, A., Smalley, R.E., Geerligs, L.J., and Dekker, C. (1997) Individual single-wall carbon nanotubes as quantum wires. *Nature,* **386**, 474-477.
3. McEuen, P.L. (1998) Carbon-based electronics. *Nature,* **393**, 15-17.
4. Soneda, Y., Szostak, K., Delpeux, S., Bonnamy, S., and Béguin, F. (2001) High yield of multiwalled carbon nanotubes from the decomposition of acetylene on Co/MgO catalyst. *Proc. Carbon'01*, Lexington, KY, American Carbon Society, pp.30-31.
5. Frackowiak, E., Gautier, S., Gaucher, H., Bonnamy, S., and Béguin F. (1999) Electrochemical storage of lithium in multiwalled carbon nanotubes. *Carbon,* **37**, 61-69.
6. Leroux, F., Méténier, K., Gautier, S., Frackowiak, E., Bonnamy, S., and Béguin, F. (1999) Electrochemical insertion of lithium in catalytic multi-walled carbon nanotubes. *J. Power Sourc.* **81-82**, 317-322.
7. Wu, G.T., Wang, C.S., Zhang, X.B., Yang, H.S., Qi, Z.F., He, P.M., and Li, W.Z. (1999) Structure and lithium insertion properties of carbon nanotubes. *J. Electrochem. Soc.* **146**, 1696-1701.
8. Béguin, F., Metenier, K., Pellenq, R., Bonnamy, S., and Frackowiak, E. (2000) Lithium insertion in carbon nanotubes. *Mol. Cryst. Liq. Cryst.* **340**, 547-552.
9. Gao, B., Bower, C., Lorentzen, J.D., Fleming, L., Kleinhammes, A., Tang, X.P., McNeil, L.E., Wu, Y., and Zhou, O. (2000) Enhanced saturation lithium composition in ball-milled single-walled carbon nanotubes. *Chem. Phys. Lett.* **327**, 69-75.
10. Claye, A.S., Fischer, J.E., Huffman, C.B., Rinzler, A.G., and Smalley, R.E. (2000) Solid-state electrochemistry of the single wall carbon nanotube system. *J. Electrochem. Soc.* **147**, 2845-2852.
11. Che, G., Lakshmi, B.B., Fisher, E.R., and Martin, C.R. (1998) Carbon nanotubule membranes for electrochemical energy storage and production, *Nature,* **393**, 346-349.
12. Che, G., Lakshmi, B.B., Martin, C.R., and Fisher, E.R. (1999) Metal-nanocluster-filled carbon nanotubes: catalytic properties and possible applications in electrochemical energy storage and production, *Langmuir* **15**, 750-758.
13. Nützenadel, C., Züttel, A., Chartouni, D., and Schlapbach, L. (1999) Electrochemical storage of hydrogen in nanotube materials. *Electrochem. Solid-State Lett.,* **2**, 30-32.
14. Nützenadel, C., Züttel, A., and Schlapbach, L. (1999) Electrochemical storage of hydrogen in carbon single wall nanotubes. *Electronic properties of novel materials-science and technology of molecular nanostructures.* Kuzmany H et al editors. American Institute of Physics, Melville, NY (USA) pp. 462-465.
15. Qin, X., Gao, X.P., Liu, H., Yuan, H.T., Yan, D.Y., Gong, W.L., and Song, D.Y. (2000) Electrochemical hydrogen storage of multiwalled carbon nanotubes. *Electrochem. Solid-State Lett.* **3**, 532-535.
16. Rejalakshmi, N., Dhathathreyan, K.S., Gowindraj, A., and Statishkumar, B.C. (2000) Electrochemical investigation of single-walled carbon nanotubes for hydrogen storage. *Electrochim. Acta* **45**, 4511-4515.
17. Tibbetts, G.G., Meisner, G.P, and Olk, C.H. (2001) Hydrogen storage capacity of carbon nanotubes, filaments, and vapor-grown carbon fibers. *Carbon* **39**, 2291-2301.
18. Hernadi, K., Fonseca, A., Nagy, J.B., Bernaerts, D., Fudala, A., and Lucas, A.A. (1996) Catalytic synthesis of carbon nanotubes using zeolite support. *Zeolite* **17**, 416-423.
19. Hernadi, K., Fonseca, A., Piedigrosso, P., Delvaux, M., Nagy, J.B., Bernaerts, D., and Riga, J. (1997) Carbon nanotubes production over Co/silica catalysts. *Catalysis Lett* **48**, 229-238.
20. Hamwi, A., Alvergnat, H., Bonnamy, B., and Béguin, F. (1997) Fluorination of carbon nanotubes. *Carbon* **35**, 723-728.
21. Colomer, J.F., Piedigrosso, P., Willems, I., Journet, C., Bernier, P., Van Tendeloo, G., Fonseca, A., and B'Nagy, J. (1998) Purification of catalytically produced multi-wall nanotubes. *J. Chem. Soc. Faraday Trans.* **94**, 3753-3758.
22. Kyotani, T., Tsai, L., and Tomita, A. (1996) Preparation of ultra fine carbon in nanochannels of an anodic aluminum oxide film. *Chem Mater* **8**, 2109-2113.
23. Conway, B.E. (1999) *Electrochemical supercapacitors – scientific fundamentals and technological applications*, New York, Kluwer Academic/ Plenum.
24. Niu,C., Sichel, E.K., Hoch, R., Moy, D., and Tennet, H. (1997) High power electrochemical capacitors based on carbon nanotube electrodes. *Appl. Phys. Lett.* **70**, 1480-1482.
25. Frackowiak, E., Méténier, K., Bertagna, V., and Béguin, F. (2000) Supercapacitor electrodes from multiwalled carbon nanotubes. *Appl. Phys. Lett.* **77**, 2421-2423.
26. Frackowiak, E., and Béguin, F. (2001) Carbon materials for the electrochemical storage of energy in capacitors. *Carbon,* **39**, 937-950.
27. Frackowiak, E., Jurewicz, K., Delpeux, S., and Béguin, F. (2001) Nanotubular materials for supercapacitors. *J. Power Sourc.* **97-98**, 822-825.

28. Jurewicz, K., Delpeux, S., Bertagna, V., Béguin, F., and Frackowiak, E. (2001) Supercapacitors from nanotubes/polypyrrole composites. *Chem. Phys. Lett.* **347**, 36-40.
29. Frackowiak, E., Jurewicz, K., Szostak, K., Delpeux, S., and Béguin F. (2002) Nanotubular materials as electrodes for supercapacitors *Fuel Process. Tech.* **77**, 213-219.

AUTHOR INDEX

SUBJECT INDEX